AN ENCYCLOPEDIA

of

CANDY AND ICE-CREAM MAKING

by

SIMON I. LEON

Candymaker

Consultant to the Confectionery Industries

edited by

S. Joseph Leon

1959

CHEMICAL PUBLISHING CO., INC.

212 Fifth Avenue, New York, N. Y.

CHEMICAL PUBLISHING CO., INC.
New York N. Y.

Printed in the United States of America

PREFACE

Who is the master confectioner?

He is the man who is an expert in his trade, who knows his business from A to Z, and can teach others to become proficient in this field.

The author of this book began as a candymaker in early youth and has devoted his entire life to candymaking. He has written this book so that others may benefit from fifty-two years of experience and hard work.

Many good candy books have been published, but as a rule they do not cover the retail and smaller wholesale manufacturing field completely. For example: A certain formula, though it may be the best, calls for steam or vacuum cooking. This will not suit the candymaker who uses open-fire cooking. Here is only one example in a hundred where a formula that is suitable for the large candy factory will not give good results in the medium shop.

The large candy manufacturer has his highly paid executive to take care of all his problems, but the smaller manufacturer has his hands full with making and selling his products. He is only too likely to neglect the candymaking end. That is why this book should be of particular interest to the retail and medium-wholesale candymaker. The man who must know his business from A to Z should find this work a treasury of candymaking.

The author is not an historian, but reviewing his career, he finds that he has lived through revolutionary changes in confectionery manufacturing. In Lublin, Poland, in the year 1906, he made his first batch of hard candy. There were six small candy shops in town making mostly hard candy and fruit-filled, wrapped caramels. Two of them also made a few molded solid chocolate bars. All the work was done by hand. Coke or coal was used for fuel in cooking the batches. There were no thermometers. The cold-water test was the standard method for cooking batches and many of them were either burned or cooked too low. Sticky hard candy was the problem of the day. There were no batch warmers for spinning hard candy. In order to keep the batch warm, one had to warm up the pouring slab.

Candy was made in Eastern Europe in this primitive fashion. There were boom seasons and off seasons, but surprisingly enough, candy was popular the year round.

In 1910, the author reached metropolitan Warsaw. He found hundreds of small candy shops scattered throughout the city and its suburbs. Thousands of pounds of candy were produced daily in the same primitive way. Most of the hard candy, solid and filled goods, were hand-wrapped by young girls. But even then, there were the first pioneers—two or three more or less mechanized candy factories complete with smoke stack and factory whistle. There were two modern, medium-sized chocolate shops—Warshawianka and Viktor—which produced a quality product that competed successfully with Swiss chocolate. A number of shops specialized in caramels cupped and packed in wooden fiber boxes. About three or four chocolate shops made chocolate coatings from the bean on and sold them to small candy shops throughout Poland. But on the whole, improvements in ingredients, equipment, and production methods were few.

To move from this milieu into the United States was like exchanging a covered wagon for a transatlantic plane. In New York City, before the First World War, the author found modern six-story plants which employed more than 300 workmen. All machinery was automatic, and most of the basic raw material shot the chutes from the top floor directly into the cooking kettles. The large-scale production methods were impressive, but the quality of the product left much to be desired. Many large plants in America produced for the five and dime chain and variety stores. Not only hard candy, but also crystallized creams, jellies, gums, and fudge retailed for 20 cents a pound, and fancy-package chocolates went for 30 cents. This was possible because the cost of raw materials and ingredients was low, labor was cheap, and working hours were long. The speed-up reigned. Sanitary conditions were only too likely to be neglected. Since large orders went to the lowest bidder, prices were figured on cost of labor and material. The way to eliminate the competitor was to beat him at price rather than quality. So a quality product was out of the question. Packaging was cheap and unimaginative; most candies were sold in bulk.

This was the wholesale candy-manufacturing business in America—or a large part of it—before the First World War. Over on the retail manufacturing end, conditions differed. Several large companies with chain stores and many independent confectionery shops offered a quality product at a higher price and were fairly successful.

Recalling the situation as it was in 1914, the confectioner who has lived through these past forty years can only be impressed by the considerable gains the industry has made. New labor-saving equipment, new ingredients, improved packaging, improved sanitary conditions, and higher wages have resulted in an improved product. The lesson of these years is the importance of offering a quality product at a fair price. For the most part, the firms which were able to do this survived the depression. *An Encyclopedia of Candy and Ice-Cream Making* comes out of a conviction that quality and quality again is the answer. It is meant to help the reader keep a clean shop and produce candy that will hold up.

Candy is a sweet food, mild in digestion, pleasant and palatable in taste. The proficient candymaker must first understand what he is making and become familiar with all the raw ingredients that go into his product. He must also be something of an artist, since blending color and flavor is nothing less than an art.

Before he can hope to make good candy, the confectioner must master the technical details. What is candy made from? What makes it plastic and keeps it commercial? What makes it attractive to the eye? What chemical changes does the batch undergo? How to avoid spoilage? Success depends on finding answers to these and other questions.

In this volume, every kind of candy, ice cream, and several allied confections gets specific and detailed treatment. The book contains more than 600 proven formulas.

INTRODUCTION

During the past fifty years, the confectionery industry has accelerated quantity production. From cooking on a coal stove to continuous process cooking, from the hand-operated candy drop roller to the continuous seamless automatic rope-spinning and cutting machine is a long step forward. However, this new labor-saving machinery has not eliminated the demand for the competent, all-around candymaker. On the contrary: Because of a shortage of willing replacements, the all-around confectioner is at a premium in America today.

Before World War I, the young helper usually became a full-fledged candymaker in five or six years. During this period of apprenticeship, he learned how to cook, color, flavor, stripe, and spin hard candy, how to make nut brittles, and how to prepare caramels and fudge. He was not only a good cream man but also made nougat, paste, jellies, and gumdrops. He learned all about chocolate and was usually able to run an efficient chocolate department. In the years that followed, plant expansion, technological improvements, the great economic depression, and World War II sent these all-around men into smaller retail manufacturing shops, where they have become a vanishing breed. Because their demand for all-around candymakers is limited, the larger plants have not felt this shortage, but confectioners who operate smaller and medium-sized establishments are finding it difficult to locate versatile experts. In view of this shortage and the difficulty in recruiting competent replacements, it is essential that the small and medium-sized operators learn every phase of quality production.

The aspiring confectioner must first learn the raw materials that go into candy. Sugar is the most vital ingredient. There are two kinds of sugar made in America: beet sugar, produced in the western states, and cane sugar, refined in all parts of the country. For confectionery purposes, the standard granulated type of cane sugar is preferred. Unlike beet sugar, it does not foam when boiling and has a pure, snow-white color. When powdered sugar is called for, the fine 6X type is preferred. Powdered sugar is mixed in centers which go into filled hard candies. A partial ingredient in most filled hard-candy centers, it is one of the main ingredients in chocolate straws and other crunchy, filled hard candies. Powdered sugar is also used in making Easter-egg icings and for dusting when making after-dinner and plaited mints.

Corn syrup is the second basic ingredient. It is produced by numerous corn refineries in all parts of America. Although these refineries recommend several brands of corn syrup for confectionery use, the standard 42-degree corn syrup is the most dependable brand and the only one that can be used in making all kinds of candy without creating complications. A high-conversion corn syrup, "enzyme converted," is particularly useful in making jellies and gumdrops.

Invert sugar is a sugar solution brought to high conversion by adding a doctoring agent, such as citric or tartaric acid, to sugar and water. With few exceptions, the formulas in this encyclopedia do not call for invert sugar as an ingredient, since the writer believes that it is safer and more economical to add the doctoring agent directly to the batch and thus insure the inversion of each specific batch, i.e., keep it from graining or crystallizing.

Various milks—evaporated, condensed, skimmed, and powdered—are used in candymaking. Evaporated milk gives the best results in toffee, caramel, and fudge manufacture; it is also preferable for ice cream and sherbets. Eighteen percent and 20 percent sweet creams are used in candymaking; 20 and 40 percent sweet creams go into ice cream. Pure salted dairy butter is used in making quality caramels, fudge, and butter creams. There are two types of fats that the confectioner should be familiar with. The 96-degree refined coconut-butter fat is used for caramels and all chewing candies and the 76-degree refined coconut-butter fat is used for filled centers and salted nuts.

For all centers and soft candies, pure extract of vanilla and pure fruit flavors or oils give superior results. For hard candies, however, a good imitation flavor may prove better. Flavor manufacturers will furnish confectioners with data on the strength and behavior of their products. Acids are used to support fruit flavors; that is, they bring out a more pronounced flavor. Tartaric acid gives better results when used in making quality hard candies, but for creams and soft candies, citric acid, which is much cheaper, can also be used effectively. Only certified pure-food colors should be used for coloring all kinds of candies. The paste form of color, already mixed, is preferred to the powder. It is used as is for hard candies, but is usually diluted with a little water when mixed into soft-candy batches. The quantities of colors to be used must be left to the confectioner's discretion. He must mobilize whatever artistic skill he has and learn in time to achieve both deep and delicate coloring.

After he has familiarized himself with the basic ingredients, the confectioner must learn the basic procedures.

Cooking is the first method he must master. With the exception of certain types of marshmallows, all candy must be cooked. Despite recent experimentation with cold methods of making candy, there is as yet no satisfactory alternative.

Proper cooking methods are particularly important in making durable hard candy. Cooking may be considered the main project in the hard-candy department.

Open-fire cooking will produce the highest-quality hard candy. This method involves no machinery—just a kettle and stove. The confectioner should use enough water so that all the sugar will be thoroughly dissolved. He should stir the batch constantly until it starts to boil. He should then cover it up for a few minutes and, while the kettle is hot, he should wash down the sides of the batch with a scrubbing brush to prevent graining. The batch should be cooked at 320 to 340°F. If this temperature is exceeded, the batch will be burned. The extremely high temperature gives candy its crispness. After hard candy is cooked, it loses about 5 percent in weight in contrast to creams, jellies, and other soft candies which gain approximately 10 percent in weight after cooking because some of the water remains in the finished product. Hard candy which gets the open-fire-type cook retains only about 1 percent moisture. It is crisp and, when properly flavored, becomes a memorable eating experience.

A small batch cannot be cooked in a large pan nor can a large batch be cooked in a small pan. In the first case, the batch will scorch and in the second case, the batch will overflow when it starts to boil. Therefore, in open-fire cooking particularly, the right-sized copper pan must be selected. When the sugar, water, and corn syrup are placed in the pan, they should fill no more and no less than three quarters of the pan. The pan should sit about halfway in the inner part of the gas stove. If these precautions are followed, the sides of the pan will not be scorched or burned.

Hard candy may also be cooked by the vacuum methods. There are three kinds of vacuum cooking: open-fire vacuum cooking, steam vacuum cooking, and continuous steam vacuum cooking. As continuous steam vacuum cooking is more suitable for large-scale production and does not concern the smaller or medium-sized confectionery establishment, this encyclopedia will confine itself to the first two methods.

When vacuum cooking is used for hard candy, it is usually cooked about 20° lower than is the case in open-fire cooking. The resulting product is clearer and drier, but has a dull finish. Hard candy cooked by vacuum is more plastic, but it does not have the bite or the crisp quality that the open-fire type cook gives. However, vacuum cooking is faster and, therefore, more popular with the larger factories.

Under vacuum, the temperature needed to extract moisture from a sugar solution is not as high as that needed to produce this result under normal atmospheric pressure. In open-fire cooking without vacuum, the confectioner must cook a hard-candy batch at 300 to 340°F, depending on the amount of corn syrup it contains, in order to evaporate the water in the batch. In open-fire vacuum cooking, the batch is cooked at some 20° lower temperature, i.e., at 280 to 320°F, depending again on the amount of corn syrup it contains. In this method, the batch is first cooked by open fire to approximately 280°F and then finished by vacuum without additional cooking. A dome-shaped vacuum unit equipped with cooking coils is used. The vacuum cooker is clamped to the top of the open-fire cooking pan.

The procedure is as follows: Sugar, water, and corn syrup are cooked together to approximately 280°F. The dome-shaped vacuum cooker is attached to the kettle without delay and the vacuum is started and applied for 7 to 10 minutes. Residual heat in the batch causes continuous evaporation of water from the batch without the application of additional heat. The cooling water of the vacuum pump condenses the resulting vapor, which is removed by the same force. The resulting batch is lighter in color and keeps drier than open-fire-cooked hard candy.

More corn syrup is used in this method than in straight open-fire cooking, because when hard candy is cooked by open fire all the way, the batch is heated to approximately 315°F. This heat converts part of the cooked batch, i.e., numerous sugar particles become scorched and will not crystallize again. The batch has undergone inversion and is secure against graining. But in open-fire vacuum cooking, because the high temperature is not necessary, inversion does not occur. More corn syrup must, therefore, be added to keep the batch from graining.

Steam vacuum cooking is done with a three-in-one unit, the steam vacuum batch cooker. This unit includes an open-type dissolving pan, a vacuum cooker, and a vacuum pump. The procedure is as follows: The sugar, water, and corn syrup are dissolved in a clear solution in the dissolving pan. The solution is then pumped into the vacuum pan. The vacuum pump sets to work and the cooking is completed under vacuum, which differentiates this method from the previous one. The batch is ultimately cooked at 280 to 290°F and then drawn off at the bottom of the vacuum pan into a receiving pan. It is now ready for the slab. The candymaker must be certain that the vacuum remains steady at about 28 inches and does not fall off during the transfer process. If he does this, the cooking will be finished perfectly.

Several basic formulas for hard candy mixtures follow. These formulas will enable the candymaker to act independently and still be accurate. All individual hard candy formulas in this encyclopedia are based on the following.

OPEN-FIRE COOKING: FORMULA A

Sugar	25 lb
Corn Syrup	5 lb
Water	5 qt

The batch is cooked at 320°F.

OPEN-FIRE COOKING: FORMULA B

Sugar	30 lb
Water	6 qt
Cream of Tartar	3 level teaspoonfuls

Regular mixtures are cooked at 330°F.
Barley-sugar mixtures are cooked at 340°F.

OPEN-FIRE VACUUM COOKING: FORMULA A

Sugar	28 lb
Corn Syrup	12 lb
Water	6 qt

The batch is cooked at 280°F.

OPEN-FIRE VACUUM COOKING: FORMULA B

Sugar	40 lb
Water	7½ qt
Cream of Tartar	4 level teaspoonfuls

The batch is cooked at 300°F.

STEAM VACUUM COOKING: FORMULA A

Sugar	200 lb
Corn Syrup	100 lb
Water	40 qt

The batch is cooked at 290°F.

Vacuum cooking is usually associated with hard candy, but in the past few years, vacuum pans have been developed for cooking cream fondant batches, caramels, taffies, and toffies. Only factories geared for large-scale production and marketing low-priced products will be interested in these newly developed methods. The formulas in this encyclopedia call for open-fire cooking of these items.

Candies containing cooking starch as a basic ingredient cannot be cooked on the open fire, but must be cooked by steam. Open-fire cooking will scorch the starch and make a clear white batch impossible. Gumdrops, certain jellies, and basic batches for marshmallows are in this class.

Steam for cooking is produced in three ways. The larger confectioner may maintain his own engine room in the basement and employ a licensed engineer who operates a boiler that sends steam to all kettles equipped for steam cooking. Some large factory buildings supply steam to tenants who require it and charge for it on a rental basis. The smaller confectioner may be equipped with one or two steam-fitted kettles and a gas engine. The engine is fired until it produces 90 to 100 pounds of steam. This will supply the kettles with enough live steam to cook any sized batch by fast steam cooking.

Steam-fitted kettles have a double jacket. A water pipe brings water to the kettle. An inlet valve fills the space between the two jackets with water; an outlet valve lets the water out when there is an overflow of steam; a steam valve brings the steam to the water between the two jackets. The live steam causes water to boil. The constantly boiling water between the two jackets produces the heat that keeps the batch inside the kettle cooking until it reaches the desired temperature. When the steam does not enter the kettle fast enough, the outlet valve is opened for several minutes to let the dead steam escape. This permits the live steam to enter faster and makes possible speedier and more efficient cooking.

If the gas engine is in good condition, it will produce the required 90 to 100 pounds of steam 30 to 40 minutes after it is first lit. The gas burners under the engine may be turned off about 10 minutes before the last batch is finished cooking. This makes the workroom cooler and more comfortable in hot weather.

Batches which contain cooking starch as a basic ingredient must be stirred while they are cooking to prevent scorching. The best plan is to equip the kettles with automatic stirring paddles. When steam cooking caramels, fudge, and toffies, large amounts of evaporated milk and cream go into the batch. In this case, it is particularly necessary to equip the kettles with proper stirring paddles to prevent scorching. However, since scorching is less probable in steam cooking than in open-fire cooking, it will usually suffice to stir the batch with a long mixing stick.

Soft candies containing cooking starch as a basic ingredient must be steam cooked; all other soft candies can be cooked by steam or open fire. Steam cooking is labor saving. It makes possible larger batches and speedier production.

Most of the caramels, toffees, and fudge listed in this encyclopedia are made with milk or cream. Whether cooked by steam or open fire, the batch should always be started with enough milk or cream to stir up the sugar well until it dissolves. The balance of the milk or cream should be added to the cooking batch in a slow stream to keep it from curdling. The batch should be finished with a slow cook to prevent scorching.

Lecithin is usually used in making candies than contain butterfat. Caramels, toffies, fudge, and butterscotch are in this class. Lecithin helps the other ingredients assimilate the butterfat. Unless this happens, the butterfat will separate; it will come out on top of the batch and eventually run off as waste. However, caramels and toffies also contain hard fat or coconut butter. Because most hard fats already contain lecithin in their composition, it is usually not necessary to use it in making these particular items. Only when the hard fat in the caramels and toffees does not contain lecithin should it be added to the batch.

Preparing cream centers for chocolate coating is another basic operation. Unless the centers are well prepared, they will be unfit for consumption. The formulas in this encyclopedia call for two types of cream centers for chocolate coating: hand-rolled centers and centers cast in starch.

To make centers cast in starch, the cream of the basic formula is remelted to semiliquid consistency and then flavored and molded in starch impressions. When casting the centers, the starch should be clean and dry. The starch room should be cooled to about 60°F. Whether the centers are hand- or machine-coated, they should be dry and free from lumpy starch when placed on the boards for coating. Otherwise, the chocolate will not cover them well. If the centers are greasy and their moisture content is too high, the chocolates will be streaky or they may ferment.

To make hand-rolled centers, the basic formula varies partially from the running cream formula, i.e., the formula for centers cast in starch. This basic formula, which is the author's own, is fully described in the encyclopedia. Hand-rolled cream centers have several advantages over centers cast in starch. Their production is faster and easier. The cream need not be remelted; after the batch is creamed and sweated down for 10 to 15 minutes and then kneaded, the cream is cut and rolled directly from the batch either by hand or by a hand-roll machine. This method not only makes the remelting process unnecessary, but it also eliminates casting the centers in starch and cleaning them from starch when they have firmly set. Above all, the quality of chocolates containing hand-rolled centers is superior. The one disadvantage of hand-rolled cream centers is that they cannot be made in fancy shapes, such as are possible with starch-cast centers.

Several precautions should be taken when preparing hand-rolled creams for chocolate coating. When the cream is ready for the cutting and rolling process, it should be of a fair, firm consistency—i.e., neither too hard nor too soft. When the centers are cut and rolled either by hand or by machine, they are still slightly warm. They should not be chocolate-coated until they are completely cooled down. This takes 2 to 3 hours. Centers which are left to be coated the next day should be covered with sheets of wax paper so that they will retain some of their moisture.

One essential precaution: No cream centers, whether hand-rolled or cast in starch, should lie around uncoated longer than 28 hours. If they are made Monday, they must be finished Tuesday. Otherwise their quality is jeopardized.

Selecting the right coating for the various centers calls for thought. The confectioner should use the best-quality chocolate coatings and make certain that they contain lecithin. For lecithin is also essential in chocolate production. It acts as an emulsifier and assures the confectioner that, when the coating is remelted, it will be of the proper consistency for coating the centers. Butter creams, mints, and other full sweet centers go well with a semi-sweet or bittersweet chocolate. Chewy centers, fruit-flavored creams, jellies, pastes, and marshmallows taste best with a dark, sweet coating or with milk chocolate. Fruits, nuts, and cordials should only be coated in the highest quality vanilla sweet or milk chocolate.

Experienced chocolate coaters should do the hand coating. If machine coating is adopted, the best type of enrober should be used—one that has a long enough tunnel and enough cold air to cool the chocolates properly while they travel from the enrober to the packing room. The enrober should also have an attached tempering unit. For without such a unit, there is an element of risk. Some operators think they can temper chocolate by feeling it. This test works well with hand-dipped chocolates, but it is too risky for machine-coated chocolates. When centers are hand dipped, they stay for some time on the dipping trays. If they do not set immediately, they will set eventually. But in enrobing, even with the longest tunnel available, the chocolates travel no more than 5 minutes from the tunnel to the packing room. They either set in this period or become scrap. Tempering by an automatic and adjustable gage will do the required job, whereas tempering by hand may prove costly guesswork. Only an experienced operator should run the enrobing machine. The temperature of the chocolate coating and packing rooms should always be kept near 60°F.

Chocolates should never be packed in boxes before they have firmly set and cooled. This process usually takes 2 hours after they are coated. Chocolates containing hand-rolled cream centers or cream centers cast in starch should not be placed in a cold room immediately after they are packed. They should stay in a room of normal temperature for at least 2 days in order to allow the invertase, which is added to the cream, to react and soften the cream centers. After 2 days, they can go in a cold storage room.

Certain marshmallow formulas in this encyclopedia call for a hot room; these cannot be made unless there is a well-equipped hot room in the factory. Smaller concerns find it almost impossible to maintain hot rooms, since this means they must operate their own engine room. Most retail manufacturers and a large percentage of medium-sized wholesale manufacturers buy their marshmallows from specialty houses. However, there is a possibility that future research will develop formulas for producing marshmallows without a hot room.

Confectionery machinery demands careful attention. Machines should be thoroughly cleaned and lubricated at least once a week. Bearings that are exposed to heat should be lubricated with heat-resisting grease. Once a week, drop rollers and dies should be removed for thorough cleaning. They should be freely greased with slab oil to prevent the candy from sticking to them. Wrapping machines should be thoroughly cleaned after each day's work. Particular care and attention should be given this machine. The confectioner will observe that the spindles, axles, and bearings collect a tarlike substance. This consists of candy dust, moisture, and lubricating grease. The following cleaning procedure should be adopted: The machine should be sprayed with boiling soap water about once a week. The solution should contain one large handful of soft-soap powder to a pail of hot water. After the machine is sprayed, it should be dried and the grease nipples thoroughly lubricated.

Pans, kettles, tools, measuring equipment, and weighing equipment should be washed, scoured, and rinsed at the end of each workday. All formulas must be followed with scrupulous accuracy. It is, therefore, essential that the weighing and measuring equipment is checked for accuracy at least once a week.

An effort has been made in this introduction to give an over-all picture of basic operations. Perhaps it will also give some insight into the more common problems that come up in making candy. However, the confectionery industry, like every other, has not yet solved all its problems. In the course of various operations, the reader may face enigmatic situations for which there are as yet no solutions. To mention but a few: How can hard candy be made so that it will always keep dry, crisp, and fresh in all climates? How can nut brittle be made so that it will always stay fresh, crisp, and crunchy? How can after-dinner mints be made so that they will not harden or stick together? The author believes that the answer to these and related problems lies in developing good stabilizers that will act as homogenizers and keep all the ingredients in the finished product from separating. Separation of ingredients causes most candy troubles. The ingredients will not always hold together. That is why hard candies grain and stick together. That is why taffies, toffies, caramels, and nougats end by graining, melting, or sticking to wrappers. That is also why cream centers harden, crystallize, and ferment.

Invertase is used as a softening agent in cream centers. When one of these centers is opened up after a short period of time, the confectioner or the consumer will find that part of the center is really running liquid and a large amount of the cream around the sides of the center is semicrystalline. It now appears that the center is divided into three different textures: liquid, semisoft, and semicrystalline. A good stabilizer would homogenize the center to only one texture: semisoft. It would prolong the shelf life of chocolates and prevent fermentation.

Another urgent need is for homogenized chocolate coating that will always stay in good condition. Such a coating would not turn grey, nor would the cocoa butter in it separate from the other ingredients. In extreme heat, the cocoa butter separates and causes what is commonly called the fat bloom. It melts, leaks out from the chocolate, and often causes an entire layer of chocolates to form one sticky mass. When the cocoa butter leaks out, the centers dry out and the chocolates become inedible. There is some evidence that fat bloom can be eliminated by a powerful stabilizer. The author has experimented with two different kinds of coating. One coating contained lecithin and the other did not. If identical centers are covered with the two different coatings and kept apart under identical conditions, the results vary. At the first stroke of heat the sample without lecithin becomes scrap, but the sample that contains the lecithin comes through in better, though by no means perfect, shape. The experiment indicates that lecithin, a relatively weak stabilizer and emulsifier, does make some difference and does have some effect in preventing ingredients from separating and chocolates from sticking together. This suggests that a stabilizer more powerful than lecithin would solve this major problem.

A final word about the formulas in this encyclopedia. In formulating recipes, several factors must be kept in mind. These include the desired appearance of the finished product, its durability and shelf life, its probable selling price, the problems that will be encountered in shipping it, and, above all, its eating quality. When these factors have been considered and all the ingredients formulated, still another question remains: How large should the batch be?

For several reasons, it is not practical to formulate all recipes so that the finished products will be of uniform weight. For example, some hard candy flavors are more popular and sell faster than others. Therefore, these should be made in larger batches. After they are finished some hard candies are cut into large pieces and some into miniature pieces. When it is spun out, the life of a hard-candy batch is short. It will stay soft and pliable for 20 to 40 minutes and then harden into scrap. Hard candy which will be cut into miniature pieces should be made in smaller batches than hard candy which will be cut into large pieces, because cutting into miniature pieces takes longer. The time element also figures in making creams, jellies, and gum drops. Creams which are molded in larger shapes are made in larger batches, but creams which are molded in smaller shapes can only be made in small batches, because it takes longer to cast them in molded starch. If the batch is too large and is not finished in time, the melted cream hardens into scrap; if the confectioner tries to soften it by applying more heat, the result will be tough and hard cream centers. Fruit jellies cooked on the open fire are made in small batches, but jellies and gum drops containing cooking starch are steam-cooked in large batches. The double-jacketed kettles used in steam cooking usually come in larger sizes. Also, steam cooking is relatively expensive, and the confectioner wants to compensate for his investment in machinery and maintenance by cooking in large batches and cutting labor costs. For example, the largest cream-fondant batch that can be cooked on the open fire would weigh about 100 pounds and take about 45 minutes to cook. In contrast, a proper-sized, steam-equipped kettle could cook a 300-pound cream-fondant batch in approximately 40 minutes.

In the formulas contained in this encyclopedia, the size of the batch given varies with the usual demand for the candy and the usual difficulties encountered in producing it. In general, formulas for the more popular items call for larger batches and formulas for the more difficult products call for smaller batches. Effective formulas are like tailor-made clothes. They come in various sizes and are made to order. Whether or not formulas for "fruit tablets" and "barley-sugar cuts" yield the same poundage is relatively unimportant. What matters is that the formulas yield products which retain their quality for a reasonable length of time.

Where the confections belong to the same group, are produced with about the same amount of difficulty, and have about the same popularity, the quantities in the formulas are roughly the same. The recipes have been formulated with the smaller and medium-sized factory in mind. Larger confectioners, who wish to adjust these formulas to their own needs, should take the quantities listed in them as a guide and increase them proportionately. For example, if a formula calls for a 50-pound batch of butter creams and a reader wants to steam cook a 300-pound batch, he should increase every ingredient sixfold. Conversion from smaller to larger batches seldom raises chemical problems, but usually raises equipment problems. In this case, the problem is the cream beater. A quality fondant can only be made with a round, open-type cream beater. The largest cream beater of this type is 6 feet in diameter; it can handle a maximum batch of 100 pounds. Therefore, the confectioner who wants to steam cook a 300-pound batch must use three cream beaters. After the batch is cooked, it should be poured in equal amounts into all three cream beaters. The procedure described in the original formula can then be followed.

Most of these formulas call for open-fire cooking, although other methods are sometimes specified. If a reader is working on a product that can be cooked either by open fire or by steam and wants to convert an open-fire formula to a steam formula, he should first adjust the quantities to give him the larger batch and then adjust his equipment. For example, a small batch of hard or chewing candy may be pulled on the hook by hand, but a pulling machine is needed to pull a large batch. A small batch of jellies of the pectin or agar-agar type that are molded in starch when they are finished would fill about twenty starch boards, but a larger batch might require forty to sixty starch boards. A small batch of jellies molded in starch may be cast manually by funnel, but a larger batch would require a depositing machine, and a still larger batch would require the giant Mogul. Adjustment from open-fire to steam methods always involves larger quantities and additional equipment.

Larger manufacturers who are interested in quality as well as large-scale production may find it profitable to adjust retail-type candy to wholesale production methods.

HOW TO USE THIS ENCYCLOPEDIA

THE ENCYCLOPEDIA OF CANDY MAKING is an alphabetical listing of confections, ingredients, and processes designed to meet the needs of the smaller and medium-sized manufacturer who wants to know what goes into a particular confection and how to handle the ingredients effectively. All entries are in one alphabetical list. The terminology found in this encyclopedia is commonly used in shops and factories throughout America.

Most of the entries in this encyclopedia consist of formulas followed by explicit directions. Standard abbreviations are used in both.

There are general discussions of the major types of confections (e.g., BRITTLES); these are supplemented by entries which deal with the individual numbers that belong to the major types (e.g., BLACK-WALNUT BRITTLE, BRAZIL-NUT BRITTLE). When a confection belongs to a distinct group, it will be found under the group name (e.g., BARLEY-SUGAR CUTS) instead of the flavor name (e.g., clove). However, if the reader looks under the flavor first, a cross reference (CLOVE BARLEY-SUGAR CUTS: *See* BARLEY-SUGAR CUTS, CLOVE) will guide him to the proper entry. Where the confection does not belong to a distinct group, it will be found under the flavor name (e.g., BUTTERSCOTCH TOFFEE).

If the reader is looking for information on chocolate-covered confections, he should consult the various entries under CHOCOLATE-COVERED. However, if he looks under the name of the center first, a cross reference (BRAZIL NUTS, CHOCOLATE-COVERED: *See*

CHOCOLATE-COVERED BRAZIL NUTS) will guide him to the proper entry. If the reader wants information on salted nuts, he should look under SALTED. However, if he looks under the name of the nut first, a cross reference (e.g., BRAZIL NUTS, SALTED: *See* SALTED BRAZIL NUTS) will guide him to the proper entry. The abundant cross references are listed to prevent confusion and guide the reader quickly to the information he wants.

Special ingredients are usually described in detail under the particular confection in which they are used. For example, bob syrup, which is used in making cast creams, is discussed in the entry which describes the manufacture of cast creams (i.e., BOB SYRUP: *See* CAST CREAMS). Special processes are explained either under their own names (e.g., CRYSTALLIZING) or under a particular confection in whose manufacture they are used (e.g., BUILD-UP COATINGS: *See* CHOCOLATE ALMONDS, PAN-COATED).

Items, referred to in the formulas or directions, which have been described in more detail elsewhere in this Encyclopedia are printed in capital letters with or without (*See*) after them.

Illustrations of the machines commonly used in candymaking are inserted according to their basic name in alphabetical order. The courtesy line after the description of a machine will indicate the name and address of its manufacturer.

AN ENCYCLOPEDIA OF CANDY AND ICE-CREAM MAKING

A

AFTER-DINNER MINTS

After-dinner mints are so named because they are usually served when dinner is over. They usually sell well during the winter months, but their sale increases considerably when summer comes.

To keep the after-dinner mint color a clear white, the batches should be cooked rapidly. The faster a batch cooks the clearer and whiter its color will be. Cooking converts part of the sugar by breaking up some crystals and inverting them into syrup. The faster a batch cooks the fewer crystals will be broken up. In addition to a clearer, whiter color, rapid cooking gives the batch a modified graining consistency that hastens the creaming of after-dinner mints, which is the final stage in their production. Graining takes place before the after-dinner mints are creamed and are ready for packing.

Graining and creaming go hand in hand. Before any batch can be creamed, it must first undergo limited graining. Therefore, it is important not to prolong the cooking of the batch.

While it is necessary to break a certain amount of crystals to keep the batch from getting too grainy, breaking up too many crystals will deprive the batch of all graining consistency. This is harmful to both hard candies and creamed candies. In hard candies, it causes sweating and sticking; in creamed candies (which include mints), it sometimes stops a batch of fondant from creaming altogether and it can also prevent a batch of after-dinner mints from creaming. That makes the batch scrap.

Although the average candymaker is not a chemist and cannot measure the chemical processes involved in graining and creaming, he can use his skill and experience and develop his own independent system for cooking the batch of after-dinner mints. He must set his own time for cooking a batch. He must add just enough water and "doctor" to the batch, cool the batch rapidly after it is poured on the cooling slab, and bring it to the right degree of hardness to be ready for the pulling machine.

During all these operations, the candymaker must have two objectives in mind: (1) The batch must not grain *before* it is finished and ready for the setting process, and (2) it must grain *after* setting overnight and be ready for breaking apart the following morning.

The manner in which the mint flavor is added to the batch of after-dinner mints is also important. Although the following formulas state that the mint flavor is stirred into the batch while it is being folded on the slab, this direction should not be observed by smaller confectioners who do not have water-cooled slabs or have slabs with low water pressure which prolongs the time it takes to cool the batch before it goes to the pulling machine. These candymakers should keep in mind that excessive stirring of the batch on the slab may cause graining. For them the following two alternative ways of adding flavor are recommended: (1) The flavor can be added gradually during the pulling process. It can be held ready in the measuring glass, and when the batch begins to pull on the machine, the flavor may be added gradually in a fine stream until all of it is in the batch. (2) Again the flavor can be mixed in while the batch is being pulled, but in a different way. Before the batch goes to the pulling machine, a small deep hole can be formed in it by hand. The flavor can be poured into this opening and enough powdered sugar sprinkled on top of the flavor to absorb all of it. The opening can then be closed by folding a part of the batch over the top of the hole. The batch can then be placed on the pulling machine in the usual manner.

Some precautionary measures: Pulling should not be overdone. From 5 to 7 minutes is the specified time. When a batch is pulled on the pulling machine two changes occur. Pulling makes the batch fluffy; it changes color from gray to white and becomes lighter in weight. This is the first change. The second change is called graining. While pulling, the batch is being agitated and is, therefore, coming nearer the graining point. The effect of pulling on after-dinner mints is somewhat like that of cream beating on creams and fondants. Because creams and fondants are closely "doctored" and cooked at lower temperatures, the agitation of the cream-beating machine causes the batch to cream to a lump. After-dinner mints, however, are of a higher cook and more strongly "doctored"; therefore, the agitation may not grain the batch immediately, but it can and will cause the batch to grain prematurely. That is why a 7-minute pull is the limit.

Immediately after creaming, the after-dinner mints should be packed in airtight cans. When they are first creamed and broken apart, they are not as soft as they should be. A good after-dinner mint melts in the mouth, but will not ripen until it is covered in cans for a day or two. If the jars or the cans are left open or are not airtight, the mints will dry out, lose their flavor, and become inedible.

The after-dinner mint department in the average candy factory is usually housed jointly with the hard-candy section. The same cooking stoves, cold-water cooling slabs, and pulling machine are used. Space in the hot room is usually borrowed from the gum and marshmallow departments. The mint-cutting machine is the only piece of equipment that is used exclusively for making after-dinner mints. This is a two-in-one machine. It consists of two sets of changeable, brass, cutting rollers which are connected with a moving belt. The first roller takes in the batch, sizes it, and carves it into six lengths, i.e., vertically. The carved candy then travels to the second roller, which cuts it the other way, i.e., in the width, horizontally. The cut candy then travels on the belt to an end table, where strips measuring the length of the starchboards are chopped off with a long knife and placed on the starchboards. The candy is then ready for the warming room.

There is quite a demand for miniature after-dinner mints, which usually sell for a higher price. All after-dinner mint formulas can be used in making miniature mints. They are made in the manner specified in the formulas, but are cut smaller. For this purpose, a special set of two rollers with smaller carving space is used. The cutting machine can be quickly reconverted. The original rollers are taken out of the frame and the miniature cutting rollers are inserted in their place, fastened, and bolted.

The mint-cutting machine has a ten-year life span; it should be checked periodically. The confectioner should see that the cutting knives slice the candy evenly and thoroughly so that it breaks apart in good shape and does not become scrap. In plants where after-dinner mints are only one of the regular products and are not made in large quantities, breaking apart is accomplished by hand and creates no production problem. However, in factories that specialize in after-dinner mints, breaking apart by hand is a costly procedure. For these plants, mint-breaking machines have been designed. These machines are not all built alike, but all serve the same purpose and some do a good job.

Crystallized after-dinner mints are more attractive than the usual variety. When a good crystallizing job is done, the mints will be sparkling bright. Any after-dinner mint formula may be consulted in making crystallized mints. The only necessary change is that the batch should be cooked 2 to 4° higher. The overnight crystal is preferred. The crystal syrup should be cooked in the afternoon and applied when still lukewarm. This will give the mints a more sparkling finish.

It is important to pack the finished product in airtight tins in order to keep it fresh, soft, and creamy. Large tins are used for bulk, and fancy, 1-pound lithographed tins are used for packaged goods.

For the various after-dinner mint formulas, see the individual names, such as CREAM MINTS, BUTTER MELLOW MINTS, etc. *See also* JELLIED MINTS.

ALL-SUGAR CREAM: *See* CREAMS.

ALMOND BRITTLE

The ENGLISH WALNUT BRITTLE (*see*) formula is used, substituting 8 pounds of raw almonds for English walnuts.

ALMOND COATING: *See* BURNT ALMONDS.

ALMOND CREAM CUTS

Sugar	40 lb
Water	7 qt
Cream of Tartar	3 teaspoonfuls
Corn Syrup (melted)	5 lb

Almonds (raw, cut up fine)	5 lb
Invertase	1 teaspoonful
Vanilla Flavor	3 teaspoonfuls

The sugar, water, and cream of tartar are heated in a clean copper pan and stirred until dissolved. When the batch begins to boil, it is covered and allowed to steam for a few minutes. The sides are washed down and a thermometer is placed in the batch, which is cooked to 254°F. Now, the melted corn syrup is stirred in rapidly and the batch is poured out on a dampened, cooled cream beater and allowed to cool for about 8 minutes. The chopped almonds, invertase, and flavor are added. Then the beater is set in motion and run until the batch creams to a lump. It is then allowed to sweat back for 20 minutes, transferred to the rolling table, and dusted with flour. The batch is kneaded; rolled out; and cut into small triangles. When cool and dry, these triangles are placed in crystal pans which are only half filled, and crystallized with a cool crystal, using the overnight process (*see* CRYSTALLIZING). The cuts are then drained, dried, and packed as usual. This batch will yield approximately 60 pounds of candy.

ALMOND CRISP PATTIES

Sugar	6 lb
Corn Syrup	4 lb
Water	1 qt
Almonds	10 lb
Salt	2 teaspoonfuls

The sugar, corn syrup, salt, and water are heated in a copper pan and stirred until dissolved. When the batch begins to boil, it is covered and allowed to steam for a few minutes. The sides are washed down and a thermometer is placed in the batch, which is cooked to 240°F. The almonds are added gradually and with constant stirring. The batch is cooked until the almonds are roasted to a medium shade of brown. It is then turned out on a greased warmed slab and the edges are folded together until the batch is partly cool. The batch is then kept folded in a warm spot and a few small pieces are taken from the batch. These pieces are placed in round fritter rings and flattened out with a hand scraper to fill the rings. The process is continued, working rapidly, until the entire batch is used up. After the patties are cool, they are wrapped in moisture-proof cellophane. This batch will yield about 19 pounds of candy.

ALMOND DAINTIES

Center

Corn Syrup	8 lb
Sugar	32 lb
Water	5 qt
Ground Almonds	4 lb
Salt	1 tablespoonful
Oil of Bitter Almonds	3 teaspoonfuls

The sugar, corn syrup, water, and salt are heated in a large copper pan and stirred until dissolved. When the batch begins to boil, it is covered and allowed to steam for a few minutes. The sides are washed down and a thermometer is placed in the batch, which is cooked to 300°F. The ground almonds are stirred in gradually and the batch is stirred and cooked for 7 more minutes, or until the almonds are roasted in the batch. The flavor is then stirred in, the batch is poured out on a cold greased slab, and the edges are folded together. This folding process is continued until the batch is cool enough. It is then taken over to the spinning table in front of the batch warmer, passed through a peach-stone roller machine, and allowed to cool on the table. The almond candies are broken apart and put through a sieve; then they are ready for the revolving pan.

PANNING

Powdered Gum Arabic (dissolved in 3 qt of water)	3 lb
Sugar	20 lb
Corn Syrup	20 lb
Water	3 qt
Colors (yellow, green, red)	1 teaspoonful of each
Liquid Flavors (lemon, bitter almond, cinnamon)	3 teaspoonfuls of each
Vanilla Powder	¼ teaspoonful

The gum-arabic solution is heated slowly and strained through a fine sieve. The sugar and water are placed in a copper pan on the stove, stirred until dissolved, and brought to a boil. The batch is then removed to a stand near the revolving pan; the gum-arabic solution and the corn syrup are stirred in and mixed thoroughly. The syrup is then divided into four parts as follows: One remains white and is flavored with vanilla powder; the second is colored yellow and flavored lemon; the third is colored green and flavored bitter almond; and the fourth portion is colored red and flavored cinnamon.

About one fourth of the almond candies is placed in the revolving pan, which is set in motion. One fourth of the white syrup is applied to the candies and mixed thoroughly. They are then covered with enough powdered sugar to absorb the added syrup and allowed to run until they are dry. This process is continued until all the white syrup is used up. The almonds are then removed and spread out on clean trays. The same process is followed with the other three colors and the remainder of the almond candies is handled and finished in the same way. They are then ready for the final polishing process.

POLISHING

The covered almond dainties are placed in the standard polishing pan (*see* REVOLVING PAN), the surface of which is covered and hardened with the usual hard wax. The pan is set in motion and some small pieces of hard wax are added (*see* PANNED GOODS). The dainties are polished while the pan is running in steady motion for 1 to 1½ hours. When the candies appear glossy and well-polished, they are removed and spread out on clean trays lined with wax paper, covered until dry, and finally packed in boxes. The batch will yield approximately 85 pounds of candy.

ALMOND NUT SQUARES: *See* NUT SQUARES, ALMOND.

ALMOND PASTE

Sugar	18 lb
Water	4 qt
Blanched Almonds (chopped fine)	25 lb
Oil of Bitter Almonds	1½ oz

The sugar and water are heated in a large copper pan and stirred until dissolved (212°F). The chopped almonds are added and the heat is reduced to a low flame. The batch is stirred constantly and is cooked until the mass comes loose and does not adhere to the pan. The pan is then moved near the marble slab, and the flavor is stirred in. The batch is turned out on the clean marble slab, spread out, and allowed to cool. It is then kneaded until smooth and placed in clean, covered cans to be used as desired. The batch will yield approximately 42 pounds of candy.

ALMOND PASTE BONBONS: *See* BONBONS, ALMOND-PASTE.

ALMOND POLISHING: *See* JORDAN ALMONDS and ALMOND DAINTIES.

ALMONDS, CHOCOLATE-COVERED: *See* CHOCOLATE-COVERED ALMONDS.

ALMONDS, SALTED: *See* SALTED ALMONDS.

ANGORA PASTE

Maple Sugar	6 lb
Corn Syrup	2 lb
Water	2 qt
Arrowroot	4 oz
Citric Acid	1 teaspoonful

The sugar, corn syrup, and water are placed in a copper pan on the stove over slow heat and stirred and cooked to 240°F. The arrowroot is mixed with water into a smooth paste and is added to the cooking batch, together with the citric acid. The batch is stirred constantly and cooked over low heat to 229°F or until it conforms with the usual jelly test (*see* ORANGE JELLY). It is then removed from the stove and poured out into shallow pans which are dusted with a liberal amount of powdered sugar. The batch is allowed to set until it is hard and is then sliced and cut into small squares. Each piece is rolled in a 50–50 mixture of powdered sugar and magnesia powder. The paste is placed on clean trays until it is dry and then packed in boxes as usual. This batch will yield approximately 10 pounds of candy.

ANISE GUMDROPS: *See* GUMDROPS, ANISE.

ANISE SALT-WATER TAFFY: *See* SALT-WATER TAFFY, ANISE.

ANISEED DROPS

The same process is followed as for MINT CLEAR DROPS (*see*) but for flavor, oil of aniseed is substituted. A small lump is taken from the batch and pulled to white on the hook. This is then used to stripe the main batch with fine stripes all around. The batch is kept in flat shape while it is passed through the drop machine and finished as usual. The batch will give about 19 pounds of candy.

APPLE PASTE, RUSSIAN-TYPE

Batch 1	
Apples (quality, ripe)	10 lb
Water	4 qt
Sugar	10 lb
Batch 2	
Agar-Agar	6 oz
Water	12 qt
Sugar	25 lb
Corn Syrup	15 lb
Citric Acid (dissolved in water)	5 oz
Apple Flavor	4 teaspoonfuls

The apples and 1 gallon of water are placed in a copper pan, covered, and allowed to boil until steaming. They are then removed from the stove, crushed with a paddle, and put through a meat-chopping machine. The pulp is placed back into the pan and the sugar added. The batch is stirred constantly, cooked to a firm paste or about 222°F, and then set aside.

Then the second batch is started by heating the agar-agar and the water in a large copper pan and stirring until dissolved. When boiling for a few minutes, the sugar is stirred in gradually. The stirring is continued while the corn syrup is added and the batch is cooked to 224°F. It is then removed from the stove and poured into the pan on top of the apple paste. The flavor is added, mixed thoroughly, and the batch is cooled down to about 100°F. The dissolved citric acid is then added and mixed well. The batch is poured out on a slab lined with brown packing paper between iron bars and spread out evenly to about ½ inch thickness. It is allowed to set overnight, then sliced and cut into small oblongs, and rolled in a mixture of powdered sugar and cornstarch. After thorough drying, the paste is packed in boxes lined with wax paper. The batch will yield about 80 pounds of candy.

APRICOT JELLY

Apricot Pulp	10 lb
Sugar	9 lb
Agar-Agar	4 oz
Water	8 qt
Sugar	12 lb
Corn Syrup	2 lb
Yellow Color	1 teaspoonful
Apricot Flavor	2 teaspoonfuls

The apricot pulp is drained through a wire sieve and then heated in a copper pan. Then the sugar is stirred in. The batch is cooked to a firm jelly or about 222°F, with constant stirring, and then set aside.

The agar-agar and water are heated in a copper pan and stirred constantly. After boiling a few minutes, the sugar is stirred in gradually. When it starts boiling again, the corn syrup is stirred in, and the batch is cooked to 223°F. It is then removed to a stand. The apricot jelly is stirred in, and the color and flavor are added and mixed thoroughly. The jellies are then cast in starch and run into fancy shapes. After setting overnight, they are sifted free of starch, dampened, and rolled in fine granulated sugar. When dry, they are placed in crystal pans and crystallized as usual (*see* CRYSTALLIZING). This batch will yield approximately 45 pounds of candy.

APRICOT MELLOWS

The APRICOT JELLY (*see*) batch is processed in the usual manner. Twenty pounds of small-size marshmallow drops are dipped into the jelly and then rolled in fine granulated sugar. The dipped mellows are spread out on boards which are surfaced with a 1-inch spread of fine granulated sugar. They are allowed to dry overnight and are then packed in the usual manner. The batch yields approximately 75 pounds of candy.

APRICOT PECAN GOODIES

GUM-ARABIC MARSHMALLOWS (*see*)	10 lb
Apricot Pulp	10 lb
Sugar	12 lb
Pecan Pieces	20 lb
Apricot Flavor	3 teaspoonfuls

The marshmallows are sieved and brushed free of starch. They are placed on boards ready for dipping. The apricot pulp and sugar are heated in a copper pan and stirred until dissolved. The sides are washed down and the batch is stirred constantly and cooked to 234°F, or until it forms a medium ball when tested in cold water (*see* COLD-WATER TEST). The batch is then moved near the slab, where the pecan pieces are spread out. The marshmallows are dipped rapidly in the apricot batch with a fork and rolled in the pecan pieces. They are shaped round and placed on boards lined with heavy wax paper. They are allowed to cool until dry and are then packed in boxes. The batch will yield approximately 50 pounds of candy.

APRICOTS, CHOCOLATE-COVERED: *See* CHOCOLATE-COVERED APRICOTS.

APRICOT SHERBET

Fresh Milk	20 qt
Sugar	20 lb
100-Grade Citrus Pectin	4 oz
Lemon Juice	¾ qt
Yellow Color	½ teaspoonful
Pure Apricot Flavor	½ oz
Apricot Pulp	6 qt

The pectin is mixed dry with 5 pounds of the sugar and placed in a copper pan on the stove over low heat. Four quarts of milk are added rapidly and the mixture is stirred and heated to the boiling point. It is then poured into the mixing pan. The balance of the milk and sugar, the juice, color, and flavor are placed into the same pan and the mix is stirred until all of the sugar is dissolved. The mix is then cooled rapidly and strained into the freezer. The freezing is started and when the sherbet is about halfway frozen, the apricot pulp is added and the freezing concluded (*see* VANILLA ICE CREAM NO. 1). The sherbet is stored and ripened. The batch will yield approximately 38 quarts of sherbet.

B

BANQUET MINT WAFERS

Sugar	40 lb
Water	10 qt
Silver Sheet Gelatin	2 oz
(soaked in 1½ cups of hot water)	
Mint Flavor	4 teaspoonfuls
Liquid Colors	2 teaspoonfuls

The sugar and water are placed in a clean copper pan on the stove and stirred until dissolved. When the batch begins to boil, it is covered and allowed to steam for a few minutes. The sides are washed down and a thermometer is placed in the batch, which is cooked to 246°F. The batch is then poured out on a dampened cooled cream beater and sprinkled with a little cold water on top. After cooling the batch to about 100°F, the soaked sheet gelatin is placed on top and the machine is started and kept in motion until the batch turns to a creamy lump. It is then allowed to sweat back for about 30 minutes. The cream fondant is placed in a steam-bath copper pan and stirred gradually until melted. Four teaspoonfuls of oil of peppermint are stirred in gradually. The melted cream is kept warm. Some is poured into a warm funnel and dropped on rubber mats or heavy wax paper spread out on the table. When cooled off on the top side, the mints are turned over and allowed to cool on the table. They are then packed in flat pans or glass dishes (one layer only). The batch will yield approximately 44 pounds of candy.

Note: These mint wafers may be made in assorted colors of white, yellow, green, and pink. They are usually colored in delicate pastel shades.

The melting of the cream requires great care and experience.

It must be melted just thin enough to cast and heated to the proper temperature (90 to 100°F) so that the wafers are hard enough when they set and do not turn too spotty when they are cooled and packed.

BARLEY-SUGAR CUTS, CINNAMON

The formula for BARLEY-SUGAR CUTS, LEMON is followed, but the citric acid is omitted. Red color and oil of cinnamon are stirred in while the edges are being folded. The batch is finished as usual. It will yield approximately 23½ pounds of candy.

BARLEY-SUGAR CUTS, CLOVE

The formula for BARLEY-SUGAR CUTS, LEMON is followed with the following exceptions: The main batch remains clear white. Oil of clove is used for flavoring, and two small portions are taken off the main batch. One portion is colored a deep red and the other is pulled white. The two pieces are shaped into two round sticks and put together and stretched out, doubled up four times and cut in four parts. The main batch is shaped square and stripes are placed on all four sides. The batch is cut and completed as usual. It will yield approximately 23½ pounds of candy.

BARLEY-SUGAR CUTS, LEMON

Sugar	25 lb
Water	3½ qt
Cream of Tartar	2½ teaspoonfuls
Oil of Lemon	3 teaspoonfuls
Powdered Citric Acid	1 oz

The sugar, water, and cream of tartar are placed in a copper pan on the stove and stirred until dissolved. When the batch begins to boil, it is covered and allowed to steam for a few minutes. The sides are washed down and a thermometer is placed in the batch, which is cooked to 338°F. The batch is poured out on a greased slab and the edges are folded together while stirring in the oil of lemon and the citric acid. This folding and kneading is continued until the batch is cool enough, when it is taken over to the spinning table in front of the batch warmer. A small piece is taken off and pulled by hand to a perfect white. The main batch is formed in a flat shape, and the white lump is then drawn in tiny little stripes and put all around the main batch on both sides. It is then spun into narrow strips and passed through a chip-cutting machine (*see* CONTINUOUS HARD-CANDY CUTTER). After cooling on the table, the candy is packed in tins. This batch will yield approximately 23½ pounds.

BARLEY-SUGAR CUTS, LIME

The formula for BARLEY-SUGAR CUTS, LEMON is followed. However, oil of lime is used for flavoring and enough green coloring is stirred into the batch to make a delicate green shade. The batch will yield approximately 23½ pounds.

BARLEY-SUGAR CUTS, PEPPERMINT

The formula for BARLEY-SUGAR CUTS, LEMON is followed. Oil of peppermint is stirred in for flavoring. A small red stripe is prepared. The batch is pulled on the hook to white. The red stripe is then arranged in tiny stripes and put all around the main batch. The batch is then finished in the usual manner. It will yield approximately 23½ pounds of candy.

BARLEY-SUGAR CUTS, WINTERGREEN

The formula for BARLEY-SUGAR CUTS, LEMON is followed, except that the batch is flavored with oil of wintergreen. A small portion is taken off and colored red for the stripe. The main batch is pulled on the hook to white, kneaded, and formed square. The red is divided into two wide and four thin stripes. Two sides of the batch are striped with wide single stripes and two sides with the thin stripes. The batch is then handled and finished as usual. It will yield approximately 23½ pounds.

The six barley-sugar cuts are mixed together and packed in tins.

BERNESE CHOCOLATES

Sugar	35 lb
Cream of Tartar	2 teaspoonfuls
Water	6 qt

Melted Corn Syrup	4 lb
Condensed Milk	5 lb
Vanilla Flavor	4 teaspoonfuls
Invertase	2 teaspoonfuls

The sugar, cream of tartar, and water are placed in a copper pan on the stove and stirred until dissolved. When the batch begins to boil, it is covered up and allowed to steam for a few minutes. The sides are washed down and a thermometer is placed in the batch, which is cooked to 260°F. Then the melted corn syrup and condensed milk are added gradually and the heat is reduced to 242°F. The batch is then poured out on a cooled, dampened cream beater and the top is sprinkled with a few drops of cold water. After the batch is cooled down rapidly to about 105°F., the flavor and invertase are added. The beater is set in motion and the creaming process is continued until the batch forms a lump. It is then kneaded and passed through the cut-roll machine, which cuts it into small balls. When cool enough, they are dipped in fondant-processed dark sweet chocolate. After setting, they are packed in the usual way. The batch will yield approximately 48 pounds of candy.

BERNESE-CHOCOLATE BELMONTS

The formula for BERNESE CHOCOLATES is followed up to the creaming process. Four pounds of chopped black walnuts are then kneaded into the cream. The batch is then cut and rolled into small balls. When cool, these balls are coated in Swiss milk chocolate, which is processed as follows: Roasted, ground almonds are mixed with the chocolate while it is being cooled. After the centers are dipped, they are allowed to set and are then packed in the usual manner. The batch will yield approximately 52 pounds of candy.

BITTER-ALMOND OIL: *See* MARZIPAN CANDIES.

BITTER-SWEET TOPPING FOR SUNDAES

Fondant-Processed Sweet Chocolate	2 lb
Bitter Chocolate	1 lb
Liquid Honey	5 lb
Simple Syrup	½ qt

Both kinds of chocolate are melted in a double boiler. The honey is poured in a small copper pan, which is placed on the stove over low heat. The chocolate is stirred gradually into the honey, the simple syrup is added, and all of the ingredients are mixed thoroughly. The mixture is then poured into a clean stone crock, and it is ready for use.

The batch will yield approximately 4 quarts of bitter-sweet topping.

BLACK-WALNUT BRITTLE

The ENGLISH WALNUT BRITTLE formula is duplicated, using black walnuts instead of the English walnuts. The batch will yield approximately 20 pounds of candy.

BLACK-WALNUT BUTTER CREAMS

The formula for BUTTER CREAMS: VANILLA is followed. While the batch is in the creaming process, 4 pounds of chopped black walnuts are added. The batch is then finished in the usual manner.

Note: Butter creams are to be chocolate coated by hand. For approximate yield of chocolates, *see* HAND ROLLED CREAM CENTERS: VANILLA NUT.

BLACK-WALNUT SQUARES

Sugar	8 lb
Corn Syrup	12 lb
Molasses	2 qt
Coconut Butter	1 lb
Water	2 qt
Salt	3 teaspoonfuls
Hand-Picked Black Walnuts	22 lb

All of the listed ingredients, with the exception of the black walnuts, are placed in a large copper pan on the stove and stirred until dissolved. The sides of the pan are washed down and the batch is stirred and cooked to 252°F. The black walnuts are then stirred in gradually and mixed thoroughly. The batch is then turned out on a greased slab lined with iron bars and is spread out evenly to about ½ inch thickness. It is cut with an adjustable cutter into 1½ inch squares. These are allowed to cool until hard and they are then cupped or wrapped as usual.

This formula should yield approximately 45 pounds of squares.

BLACK-WALNUT OPERA CHOCOLATES: *See* OPERA CHOCOLATES, BLACK-WALNUT.

BOARDWALK FUDGE, CHOCOLATE

Sugar	16 lb
Invert Sugar	4 lb
Light Fresh Cream	4 qt
Salt	1 tablespoonful
Evaporated Milk	4 lb
Dairy Butter	2 lb
Cream Fondant	16 lb
Bitter Chocolate (cut up)	4 lb
Vanilla Flavor	2 teaspoonfuls

The sugar, invert sugar, cream, and salt are placed in a large copper pan on the stove and stirred until dissolved. When the batch begins to boil, the sides are washed down and stirring is continued. The evaporated milk is added gradually and the butter is stirred in toward the end of cooking. The batch is cooked until it forms a soft ball when tested in cold water. It is then moved to a stand near the slab and the chocolate is added rapidly. When this is melted, the fondant and flavor are stirred in rapidly and mixed thoroughly. The batch is then poured into greased pans; each pan is filled to the top. The fudge is allowed to set overnight and is then ready for use.

This batch should yield 52 pounds.

BOARDWALK FUDGE, CHOCOLATE NUT

The formula for BOARDWALK FUDGE, CHOCOLATE is followed. Three pounds of chopped Brazil nuts are stirred into the batch before pouring. When the pans are filled they are topped with selected Brazil nuts. The batch is finished in the usual way and should yield 48 pounds of candy.

BOARDWALK FUDGE, MAPLE WALNUT

Sugar	12 lb
Maple Sugar	4 lb
Invert Sugar	4 lb
Light Fresh 18% Cream	4 qt
Salt	1 tablespoonful
Evaporated Milk	4 lb
Dairy Butter	2 lb
Cream Fondant	16 lb
Maple Flavor	4 teaspoonfuls
English Walnuts	4 lb

The sugar, maple sugar, invert sugar, cream, and salt are placed in a large copper pan on the stove and stirred until dissolved. When the batch begins to boil, the sides are washed down. Stirring is continued while the evaporated milk and butter are added. The batch is cooked until it forms a medium ball when tested in cold water. It is then moved close to the slab and the cream fondant and flavor are added, stirred, and mixed thoroughly. One half of the walnuts is chopped and stirred into the batch, which is then poured into greased pans. Each pan is filled to the top, and walnut halves are used for topping. The fudge is allowed to set overnight and may then be used as desired.

This batch should yield 48 pounds of candy.

BOARDWALK FUDGE, PISTACHIO

The formula for BOARDWALK FUDGE, VANILLA is followed. When the batch is ready for pouring, it is colored a bright green and flavored with pistachio. One and one half pounds of chopped pistachio nuts are added and mixed thoroughly. The batch is then poured out into greased pans and topped with chopped pistachio nuts. The fudge is allowed to set overnight and is then ready to use as desired.

This formula should yield 45½ pounds of candy.

BOARDWALK FUDGE, VANILLA

Sugar	16 lb
Invert Sugar	4 lb

Salt	1 tablespoonful
Light Fresh Cream	4 qt
Evaporated Milk	4 lb
Dairy Butter	2 lb
Cream Fondant	16 lb
Vanilla Flavor	4 teaspoonfuls

The sugar, invert sugar, salt, and cream are placed in a large clean copper pan on the stove and stirred until dissolved. When the batch begins to boil, the sides are washed down and the evaporated milk is stirred in rapidly. Stirring is continued while the butter is added. The batch is cooked until it forms a soft ball when tested in cold water. It is then moved to a stand near the slab and the cream fondant is added rapidly and stirred while the vanilla flavor is added. Stirring is continued until all ingredients are well mixed. The batch is then poured out on greased pans and allowed to set overnight. It is now ready for use as desired.

This batch should yield 44 pounds of candy.

BOARDWALK FUDGE, VANILLA NUT

The formula for BOARDWALK FUDGE, VANILLA is followed. Three pounds of pecan pieces are added to the batch before it is poured out on the pans. Each pan is filled to the top and topped with some pecan halves. The fudge is then allowed to set overnight and is ready for sale.

This formula should yield 47 pounds of candy.

BOB SYRUP: *See* CAST CREAMS.

BOMBAY BARS

Sugar	18 lb
Corn Syrup	12 lb
Coconut Butter	2½ lb
Fresh 18% Cream	8 qt
Evaporated Milk	12 lb
Salt	2 tablespoonfuls
Vanilla Flavor	4 teaspoonfuls
Macaroon Coconut	2 lb
Roasted Cashews	3 lb
Melted Bitter Chocolate	2 lb
MARSHMALLOW WHIP (*see*)	2 lb

The sugar, corn syrup, coconut butter, salt, and cream are placed in a large copper pan on the stove and stirred until dissolved. When the batch begins to boil, the sides are washed down. Stirring is continued and the evaporated milk is added slowly. The batch is stirred continuously and cooked until it forms a medium hard ball when tested in cold water. The batch is moved near the slab and the flavor is stirred in. One third of the batch is poured out on the greased slab lined with high iron bars and spread out to approximately 1/4 inch thickness. The balance of the main caramel batch is then divided into two parts. One part is poured out into a smaller pan and the coconut is added to this and mixed thoroughly. This portion is then poured out on top of the batch on the slab and spread out evenly to cover the entire batch. The marshmallow whip is added to the final third of the original batch and mixed through thoroughly. The melted chocolate and nuts are also added and mixed through this portion, which is then turned out on top of the two-layer batch on the slab and is rolled out to cover the entire top. The batch is allowed to cool overnight; then it is sliced and cut into bars of the desired size. The bars may be coated with chocolate or wrapped in moisture-proof cellophane.

This formula should yield approximately 60 pounds of bars.

BONBON CREAM

Sugar	40 lb
Silver Sheet Gelatin	1 oz
(dissolved in one cup of warm water)	
Water	8 qt

The sugar and water are placed in a clean copper pan on the stove and stirred until dissolved. When the batch begins to boil, it is covered and allowed to steam for a few minutes. The sides are then washed down and a thermometer is placed in the batch, which is cooked to 246°F. The batch is poured on a cooled, dampened cream beater and allowed to cool to about 100°F. The dissolved gelatin is added, and the beater is then set in motion and allowed to run until the batch creams to a lump. The cream is placed in a clean metal container and kept covered until it is used for dipping bonbons.

This batch will yield approximately 45 pounds of cream.

BONBONS, ALMOND PASTE FOR

Sugar	20 lb
Water	4 qt
Cream of Tartar	1½ teaspoonfuls
Melted Corn syrup	4 lb
Bleached, Ground Almonds	5 lb
Oil of Bitter Almonds	3 teaspoonfuls
Invertase	1 teaspoonful
Glycerin	1 teaspoonful to every 5 lb of bonbon cream

The sugar, water, and cream of tartar are placed in a copper pan on the stove and stirred until dissolved. When the batch begins to boil, it is covered and allowed to steam for a few minutes. The sides are then washed down and a thermometer is placed in the batch, which is cooked to 246°F. The melted corn syrup is stirred in rapidly, the batch is poured out on a cooled, dampened cream beater, and allowed to cool for 7 minutes. The nuts, flavor, and invertase are added and the beater is set in motion and kept running until the batch creams to a lump. The batch is then allowed to sweat back for fifteen minutes; then it is transferred to the working table where it is kneaded, rolled, and cut into small triangles, which are ready to use as centers for bonbons.

The batch will yield about 35 pounds of centers.

BONBONS, ALMOND-PASTE, GREEN

The process for BONBONS, WALNUT-PASTE CENTER is followed, but the cream is colored pastel green and flavored with oil of bitter almonds. The small triangular centers are then dipped and touched off with a zig-zag design. The bonbons are completed as usual.

BONBONS, CHOCOLATE COCONUT

MOLASSES COCONUT CENTER BONBONS are dipped. The cream is colored and flavored with cocoa powder until the bonbons are of a bright chocolate color. They are then finished in the usual manner. Seventeen pounds of centers will yield about 25 pounds of bonbons after dipping. *See* COMMERCIAL CREAM COATING.

BONBONS, COCONUT-CENTER, MAPLE

MOLASSES COCONUT BONBON CENTERS (*See*) are dipped into the bonbon coating. The cream is colored and flavored with mapleine. The bonbons are then completed in the usual manner. *See* COMMERCIAL CREAM COATING.

Seventeen pounds of centers will yield about 25 pounds of bonbons after they are dipped.

BONBONS, FILBERT-PASTE CENTER, ORANGE

The same process is followed as for BONBONS, WALNUT-PASTE CENTER. Orange emulsion is added to the cream while stirring in the glycerin. Some small designs can be made with the dipping fork after dipping each bonbon.

BONBONS, FILBERT-PASTE CENTER, PINK

The process for BONBONS, WALNUT-PASTE CENTER is followed. The cream is colored a pastel pink and flavored with cherry. Some small pieces of glacéed cherries are cut up and used as toppers. The bonbons are then finished and packed as usual.

BONBONS, PECAN-PASTE CENTER, MAPLE

The process for BONBONS, WALNUT-PASTE CENTER is followed. The cream is flavored and colored with mapeline. The dipped bonbons are topped off with small pecan halves.

BONBONS, PECAN-PASTE CENTER, VIOLET

The process for BONBONS, WALNUT-PASTE CENTER is followed, but the cream is colored a bright violet and flavored with oil of violet while the glycerin is stirred in. The pecan paste oblong centers are dipped and touched up with a small fancy design. The bonbons are then finished as usual.

BONBONS, WALNUT-PASTE CENTER, WHITE

About 10 pounds of BONBON CREAM are placed in a clean double boiler and melted over very low heat. Then 2 teaspoonfuls of glycerin are stirred in gradually. The cream is stirred to a thin consistency. (If it is too thick, it may be reduced with a little water.) The melting pan is moved to the working table. When the cream is lukewarm, it is ready for dipping. The walnut-paste centers are than dipped with a dipping fork into the white cream and deposited on dipping trays lined with wax paper. The bonbons are allowed to cool, and they are then packed in boxes.

For about 10 pounds of centers 10 pounds of dipping cream should be sufficient.

BONBONS, WALNUT-PASTE CENTER, YELLOW

The same process is followed as for the BONBONS, WALNUT-PASTE CENTER, WHITE, but the cream is colored a delicate yellow and flavored lemon. Walnut-paste creams are used as centers. Some glacéed pineapple is cut into very small pieces and used to top off the bonbons, which are then dipped and finished as usual.

About 10 pounds of centers should require 10 pounds of dipping cream.

BOSTON BEANS

Centers

Sugar	30 lb
Corn Syrup	10 lb
Water	4 qt
Salt	1 tablespoonful
Butter	1 lb
Ground Roasted Peanuts	4 lb
Vanilla Flavor	3 teaspoonfuls

The sugar, corn syrup, water, and salt are placed in a copper pan on the stove and stirred until dissolved. When the batch begins to boil, it is covered and allowed to steam for a few minutes. The sides are washed down and a thermometer is placed in the batch, which is cooked to 300°F. The butter and peanuts are stirred in gradually and allowed to cook 5 minutes longer or until the peanuts are well roasted; then the flavor is stirred in. The batch is poured out on a cold greased slab, and the edges are folded together. The batch is folded and cooled rapidly until it is firm enough to handle. It is then taken over to the spinning table in front of the batch warmer and passed through a round, small cherry roller and allowed to cool on the table. The candies are broken apart and put through a sieve; they are then ready for the panning process.

The same panning process is followed as described under ALMOND DAINTIES, but the gummed syrup is colored a pale red and flavored vanilla. When the beans are sugar covered and dry, they are ready for polishing.

The same polishing process is used as described under ALMOND DAINTIES. The candies are handled and finished as usual.

This batch should yield approximately 100 pounds.

BRAZIL NUTS, CHOCOLATE-COVERED: *See* CHOCOLATE-COVERED BRAZIL NUTS.

BRAZIL NUTS, SALTED: *See* SALTED BRAZIL NUTS.

BRAZIL-NUT BRITTLE

The formula for ENGLISH WALNUT BRITTLES is used, substituting chopped Brazil nuts for English walnuts.

The batch will yield approximately 20 pounds.

BRAZIL-NUT SQUARES: *See* NUT SQUARES, BRAZIL.

BRITTLES

Nut brittles are popular the year round. However, they sell best in the fall and winter months, particularly in October and November for Halloween and the Thanksgiving holidays. Most brittles are sold over the counter in bulk, but some peanut brittle is sold in one-pound packages. When made in bulk, all brittles should be packed in tins after they have cooled. When made for package goods, the brittles should be packed in parchment-board boxes lined with waxed paper and wrapped in moistureproof cellophane.

Success in making these goods depends on a rapid turnover. Large stocks cannot be carried on the shelves. If brittles are well-made and sold when they are fresh, the repeat business should be satisfactory.

Nut brittles and bars are related to the hard-candy family. Therefore, the rules and regulations that prevail in making hard candy also hold here. It is best to use the confectioners' type crystal "A" sugar. Cream of tartar is not used as a "doctor" in making brittles. Pure, refined, crystal-white corn syrup is the standard "doctor." Formulas must be followed carefully. Guess work is dangerous. Correct measurement of water is important. Insufficient water makes for grainy batches and excessive quantities of water will produce sticky brittles. In all formulas that call for molasses, the "open kettle" type must be used because it cooks lightly without scorching and has a delicious flavor. The right cooking finish is also a must. Cooking the batches too low will make them sticky and the nuts will be half raw rather than brittle. Whereas cooking the batches too high will give the nuts a burnt taste and the brittle will liquefy.

Where the cold-water test is called for, the candy that is being tested must crunch up in the mouth and not stick to the teeth. For cooking brittles, larger kettles are needed; when a formula calls for a 30-pound batch it is advisable to use a kettle large enough for an ordinary 50-pound batch. This is particularly true where bicarbonate of soda is mixed into the batch because its addition causes the batch to rise to more than double its ordinary size. For the same reason, a batch containing bicarbonate of soda must be poured on the greased slab very rapidly; otherwise, the kettle will overflow and the confectioner's hands may be burned. Stretching the batch very thin makes for better eating. If the candy is packed in tins as soon as it cools off, it will not get sticky. All wastes and scrap should be used up from one batch to the other. The best way to use up the scrap and avoid scorching is to add it to the batch when cooking is almost finished. Then stir the batch and finish cooking it in accordance with the usual procedure.

For the various brittle formulas, *see* the individual names, such as PEANUT BRITTLE, COCONUT BRITTLE, etc. *See* also NUT SQUARES.

BUILD-UP COATING: *See* CHOCOLATE ALMONDS, PAN-COATED.

BURNT ALMONDS

Water	1 qt
Powdered Gum Arabic (dissolved in 3 qt of water)	3 lb
Sugar	4 lb
Light Roasted Medium Almonds	50 lb
Granulated Sugar	80 lb
Corn Syrup	10 lb
Cocoa Powder	2 lb
Vanilla Flavor	1½ oz
Scarlet Red Color	1 tablespoonful

The gum arabic, sugar, and water are dissolved over a small flame and strained through a fine sieve.

The roasted almonds are placed in a warm revolving pan and are sprinkled lightly with the gum arabic solution. This process is continued until 1 quart of the solution is used up and the almonds are wet and fairly well covered with the solution. The temperature of the revolving pan is kept at about 110°F. Then the gummed almonds are sprinkled with 5 pounds of extrafine granulated sugar until they are completely covered with it. Then they are allowed to roll until dry. Now they are ready for the coating operations.

The syrup for the first operation is prepared as follows: 10 pounds of sugar, 1 pound of corn syrup, and 3 pints of water are placed in a copper pan on the stove and stirred until dissolved. The sides are washed down and the batch is cooked to 224°F. One teaspoonful of scarlet red color is stirred in. This warm syrup is then applied to the dry gummed almonds in a slow in-and-out motion, a little at a time. The almonds are allowed to roll dry from one charge to the other. The process is continued until all the syrup is used up and the almonds are rolled dry. They are then ready for the following second step.

Ten pounds of sugar, 2 pounds of corn syrup, and 1½ quarts of water are placed in a copper pan on the stove, stirred, and cooked to 226°F. Then 8 ounces of cocoa powder and 1 teaspoonful of red color are stirred in and mixed through well. The syrup is applied to the almonds exactly as in the first operation. The pan is kept warm and the almonds are allowed to roll dry between each charge of syrup. The almonds are now ready for the third step.

Now 15 pounds of sugar, 2 pounds of corn syrup, and 2 quarts of water are placed in a copper pan on the stove, stirred, and cooked to 236°F. Then 1 teaspoonful of red color and 8 ounces of cocoa powder are stirred in thoroughly. This mixture is again applied to the almonds, following the procedure of the first operation. The almonds are always rolled dry before successive charges of the syrup are applied.

The syrup for the fourth step is prepared as follows: 20 pounds of sugar, 2 pounds of corn syrup, and 3 quarts of water are placed in a copper pan on the stove, stirred, and cooked to 242°F. Then 2 teaspoonfuls of red color and 1 pound of cocoa powder are stirred in and mixed through well together with one half of the vanilla flavor. The syrup is applied to the almonds as before, about 1½ pints at a time, and very slowly. The almonds are allowed to run until they are dry between two applications of syrup.

For the last operation, 25 pounds of sugar, 3 pounds of corn syrup, and 3½ quarts of water are placed in a copper pan on the stove, stirred, and cooked to 240°F. Then 3 teaspoonfuls of red color and the balance of the vanilla flavor are stirred in. The syrup is again applied to the almonds in the final coating with extra care. One pint of syrup is applied at a time, and the almonds are allowed to roll dry after each application in order to obtain a fine coating. When the last syrup is all used up and the burnt almonds are dry, they are removed, sieved carefully, and placed on clean trays. The revolving pan is then brushed clean and the almonds are returned to the pan, which is set in motion. The almonds are sprinkled with some gum arabic solution, just enough to coat their rough surface thinly. The solution is added, a little at a time, until the almonds are properly coated. Prolonged rolling of the almonds is avoided. The pan is then stopped and the almonds are removed and placed on clean screen-bottom trays. They are allowed to dry overnight, and are then packed in boxes.

This batch should yield approximately 200 pounds of candy.

BURNT SUGAR: *See* SAVING WASTE.

BUTTER ALMOND TOFFEE

Sugar	17 lb
Corn Syrup	3 lb
Butter	2 lb
Water	2 qt
Salt	2 teaspoonfuls
Bicarbonate of Soda	2 teaspoonfuls
Roasted Chopped Almonds	4 lb

The sugar, corn syrup, water, and salt are placed in a copper pan on the stove and stirred until dissolved. When the batch begins to boil, it is covered and allowed to steam for a few minutes. The sides are washed down and a thermometer is placed in the batch, which is cooked to 300°F. The butter is stirred in gradually and the batch is stirred continuously while the roasted almonds are added. The batch is cooked until it forms a hard crisp layer when tested in cold water (*see* COLD-WATER TEST FOR CRISP HARD CANDY). The bicarbonate of soda is mixed in and the batch is then poured out on a warm greased slab and spread out evenly to a very thin sheet. It is then marked and cut rapidly with an adjustable cutter into squares of about 1½ inches. The squares are loosened underneath with a large knife and are broken apart and cooled. They are then coated in a dark vanilla sweet chocolate and sprinkled on top with ground roasted almonds. After setting, the toffee is packed in boxes. The batch will yield approximately 45 pounds of toffee.

BUTTER CHIPS

Sugar	20 lb
Water	3½ qt
Cream of Tartar	2 teaspoonfuls
Butter	1 lb
Salt	1 tablespoonful
Powdered Nutmeg	1 teaspoonful

The sugar, water and cream of tartar are placed in a copper pan on the stove and stirred until dissolved. When the batch begins to boil it is covered and allowed to steam for a few minutes. The sides are washed down and a thermometer is placed into the batch, which is cooked to 330°F. The butter, salt, and nutmeg are then stirred in and the batch is poured on a greased slab. The edges are folded together and kneaded until cool enough. The batch is then pulled on the hook until light, when it is taken over to the spinning table in front of the batch warmer and formed into a flat sheet. It is then passed through a chip-cutting machine and allowed to cool on the table. Finally it is packed in tins.

The batch will yield approximately 19½ pounds. It may be coated with milk chocolate.

BUTTER CREAM EGGS, HAND-ROLLED

Sugar	35 lb
Water	8 qt
Cream of Tartar	2 teaspoonfuls
Melted Corn Syrup	5 lb
Butter	3½ lb
Salt	2 teaspoonfuls
Vanilla Flavor	3 teaspoonfuls
Melted Bitter Chocolate	1 lb
Invertase	2 teaspoonfuls

The sugar, water, and cream of tartar are placed in a clean copper pan on the stove and stirred until dissolved. When the batch begins to boil, it is covered and allowed to steam for a few minutes. The sides are washed down and a thermometer is placed in the batch, which is cooked to 256°F. The melted corn syrup is stirred in rapidly, and the batch is poured out on a clean, cooled, dampened cream beater. The batch is cooled for about 10 minutes, and the flavor, salt, and invertase are added. The beater is then set in motion, and the butter (cut up in small pieces) is added. Creaming is continued until the batch forms a lump. About 5 pounds of the batch is set aside and melted chocolate is mixed through this portion. The remaining white cream is kneaded soft and the chocolate portion is placed in the center and covered with the white portion. The cream is then passed through the cut-roll machine for smaller eggs or cut and rolled by hand for larger ones. After the eggs have cooled on boards, they are coated in pure bitter-sweet chocolate. The finished eggs are allowed to set and cool on the dipping trays and they are then packed in boxes.

The batch will yield approximately 70 pounds of eggs.

BUTTER CREAMS, CHOCOLATE

Sugar	45 lb
Cream of Tartar	3 teaspoonfuls
Water	10 qt
Salt	3 teaspoonfuls
Butter	4 lb
Marshmallow Whip	2 lb
Vanilla Flavor	2 teaspoonfuls
Invertase	2 teaspoonfuls
Melted Bitter Chocolate	4 lb

The sugar, water, and cream of tartar are placed in a copper pan on the stove and stirred until dissolved. When the batch begins to boil, it is covered and allowed to steam for a few minutes. The sides are washed down and a thermometer is placed in the batch, which is cooked to 243°F. The batch is then poured out on a clean, cooled, dampened cream beater and cooled rapidly for about 6 minutes or until it registers 110°F. The salt, flavor, butter, and marshmallow whip are then added and the machine is set in motion. The invertase and chocolate are added while the batch is being creamed. When thoroughly creamed, the batch is kneaded smooth and put through the cut-roll machine. The creams are allowed to cool on the trays and they are then coated with a high-quality bitter-sweet chocolate. When the finished butter creams are cool enough, they are packed in boxes.

The batch will yield approximately 60 pounds.

BUTTER CREAMS, COCONUT

The formula for BUTTER CREAMS, VANILLA is followed all the way through. While the batch is being creamed on the beater, 3 pounds of macaroon coconut are added. The batch is then finished in the usual manner.

Note: Butter creams are to be chocolate coated by hand only.

BUTTER CREAMS, CUSTARD

The formula for BUTTER CREAMS, VANILLA is followed all the way through. While the batch is being creamed, 2 pounds of chopped pecans and 2 pounds of chopped seedless raisins are added. The batch is finished in the usual manner.

Note: Butter creams are to be chocolate coated manually.

BUTTER CREAMS, GLAZED

Sugar	40 lb
Corn Syrup	20 lb
Evaporated Milk	24 lb
Cooking Starch (dissolved in 2 qt of water)	1½ lb
Salt	1 tablespoonful
RUNNING CREAM (*see*)	25 lb
Vanilla Flavor	4 teaspoonfuls
Butter Flavor	2 teaspoonfuls

The sugar, corn syrup, salt, and milk are placed in a large copper pan on the stove and stirred until dissolved. The soaked cooking starch is added to the cooking batch while it is being stirred constantly. The batch is cooked to 244°F and removed from the heat. The cream and flavor are stirred in and mixed thoroughly. The batch is then cast in clean dry starch; small molds are used for the impressions. These molds should be suited to different seasons of the year, e.g., corn for Halloween, turkeys for Thanksgiving, etc. The creams are allowed to remain in the starch boards until they are firm. They are then sieved, brushed free of starch, and prepared for glazing. Approximately 25 pounds of the butter creams are placed in a revolving pan, and about ½ quart of confectioner's glaze is slowly poured over the creams and gently mixed through by hand while the pan is revolving slowly. When the candies are thoroughly covered with glaze, they are spread out on wax-paper-lined trays to dry. They are packed in the usual manner. This formula should yield approximately 100 pounds.

Note: These candies can be made in a variety of colors and flavors.

BUTTER CREAMS, VANILLA

Sugar	50 lb
Water	10 qt
Cream of Tartar	3 teaspoonfuls

Salt	3 teaspoonfuls
Butter	5 lb
Vanilla Flavor	4 teaspoonfuls
Invertase	2 teaspoonfuls

The sugar, water, and cream of tartar are placed in a large copper pan on the stove and stirred until dissolved. When the batch begins to boil, it is covered and allowed to steam for a few minutes. The sides are washed down and a thermometer is placed in the batch, which is cooked to 245°F. The batch is then poured out on a clean, cool, damp cream beater and cooled rapidly for about 6 minutes or until it registers 100°F. The salt, butter, flavor, and invertase are added, and the beater is set in motion and kept running until the batch creams. It is then kneaded smooth and put through the cut-roll machine. The small butter cream balls are placed on clean trays lined with wax paper and allowed to cool. They are then coated with high-quality bitter-sweet chocolate and allowed to cool on the trays. When cool enough, the butter creams are packed in boxes.

This formula should yield approximately 62 pounds of cream centers.

Note: Butter creams are to be chocolate coated by hand only.

BUTTER DELIGHTS

Sugar	22 lb
Corn Syrup	8 lb
Water	5 qt
Coconut Butter	8 oz
Salt	1 tablespoonful
Dairy Butter	1 lb
Vanillin Powder	¼ teaspoonful

The sugar, salt, corn syrup, coconut butter and water are placed in a clean copper pan on the stove and stirred until dissolved. When the batch begins to boil, it is covered and allowed to steam for a few minutes. The sides are washed down and a thermometer is placed in the batch which is cooked to 280°F. The butter is stirred in and the batch is cooked for another minute. It is then poured out on a cooled greased slab lined with iron bars. The edges are folded together while the vanillin is stirred into the batch. This folding process is continued until the batch is cool enough to handle. It is then pulled on the hook or pulling machine until it turns light and fluffy. The batch is taken to the spinning table in front of the batch warmer and formed into a round shape. The batch is spun out into long thick sticks of about ½ inch diameter. The candy is allowed to cool on the table and then cut into pieces of about 4 inch length. When cold the sticks are packed in tin cans.

The batch will yield approximately 30 pounds of candy.

BUTTER MELLOW MINTS

Sugar	20 lb
Water	3 qt
Cream of Tartar	2 teaspoonfuls
Salt	2 teaspoonfuls
Butter	1 lb
Oil of Peppermint	2 teaspoonfuls
Glycerin	½ oz

The sugar, water, cream of tartar, and salt are placed in a copper pan on the stove and stirred until dissolved. When the batch begins to boil, it is covered and allowed to steam for a few minutes. The sides are washed down and a thermometer is placed in the batch, which is cooked to 274°F. The butter is stirred and melted in rapidly and the batch is then poured out on a cold greased slab. The edges are folded together while the oil of peppermint is added and mixed through. The batch is cooled rapidly, and when cold enough, it is taken over to the pulling machine. The glycerin is added to the batch during the pulling process, and the pulling is continued for 7 minutes. The batch is then taken to the spinning table, formed flat, and passed through the mint-cutting machine. It is then placed on boards lined with wax paper and dusted with powdered sugar. The boards are placed in a warm room overnight. When creamed, the mints are broken apart and packed in airtight tins.

This batch should yield approximately 20 pounds of candy.

BUTTER TOFFEE: *See* MAPLE WALNUT BUTTER TOFFEE.

BUTTER TOFFEE, CHOCOLATE-COVERED: *See* CHOCOLATE-COVERED TOFFEE.

BUTTERSCOTCH, CHOCOLATE-COVERED: *See* CHOCOLATE-COVERED BUTTERSCOTCH.

BUTTERSCOTCH DAINTIES, CHOCOLATE

The formula for BUTTERSCOTCH DAINTIES, VANILLA is followed, but 1½ pounds of melted chocolate is stirred into the cooked batch before it is poured on the greased slab. It is then finished as usual.

The batch will yield approximately 30 pounds.

BUTTERSCOTCH DAINTIES, VANILLA

Sugar	20 lb
Water	3 qt
Cream of Tartar	2 level teaspoonfuls
Salt	2 teaspoonfuls
Butter	1½ lb
COCONUT CREAM (*see*)	10 lb
Vanilla Flavor	2 teaspoonfuls

The sugar, water, cream of tartar, and salt are placed in a large copper pan on the stove and stirred until dissolved. When the batch begins to boil, it is covered and allowed to steam for a few minutes. The sides are washed down and a thermometer is placed in the batch, which is cooked to 330°F. The butter is stirred in rapidly and the batch is cooked for another minute. It is then poured out on a greased slab, and the vanilla flavor is stirred in while the edges are folded together. The coconut cream is heated in a pan over low heat until almost hot.

When the batch on the slab is firm enough to handle, it is transferred to the spinning table in front of the batch warmer and formed into a flat batch. The warm cream center is placed in the center of the flat batch. It is then wrapped over and the edges are closed up. The batch is kept in the spinning shape and is then spun and cut in a cutting machine to the desired size. When cool, the dainties are packed in tins.

The batch will yield approximately 28 pounds of candy.

BUTTERSCOTCH SAUCE

Lecithin	¼ oz
Sugar	7 lb
Corn Syrup	3 lb
Fresh 18% Cream	2 qt
Salt	1 teaspoonful
150 Grade Citrus Pectin	2 oz
Butter	2 lb
Vanilla Flavor	1 teaspoonful

The pectin is mixed dry with 1 pound of the sugar. One pint of cream is placed in a small pan on the stove over low heat. The pectin sugar is added and stirred until dissolved. The balance of the sugar and cream and the salt are added. The batch is again stirred until the sugar is dissolved. The corn syrup and the butter are stirred in gradually, and the batch is cooked to 218°F. It is then removed to a cool spot and the flavor is stirred in. The sauce is cooled rapidly, poured into a clean stone crock, and covered. It is then ready to use as desired.

The batch will yield approximately 8 quarts of butterscotch sauce.

BUTTERSCOTCH SQUARES

Sugar	17 lb
Corn Syrup	8 lb
Water	3 qt
Butter	1 lb
Salt	1 tablespoonful
Vanilla Flavor	1 oz

The sugar, corn syrup, and water are placed in a copper pan on the stove and stirred until dissolved. When the batch begins to boil, it is covered and allowed to steam for a few minutes. The sides are washed down and a thermometer is placed into the batch, which is cooked to 290°F. The heat is shut off and the butter and salt are stirred in while the batch is brought to a boil again. The vanilla flavor is then stirred in and the batch is poured on a greased slab between iron bars. It is spread out evenly to ½ inch thickness and then loosened underneath and cut with a roller knife to 1 inch squares. It is then allowed to cool and finally packed in boxes lined with wax paper.

The batch will yield approximately 22 pounds.

BUTTERSCOTCH SUCKERS

Crystal A Sugar	20 lb
Brown Sugar	10 lb
Corn Syrup	10 lb
Water	4 qt
Salt	3 teaspoonfuls
Butter	2 lb
Vanilla Flavor	4 teaspoonfuls

The sugars, corn syrup, water, and salt are placed in a copper pan on the stove and stirred until dissolved. When the batch begins to boil, it is covered up and allowed to steam for a few minutes. The sides are washed down and a thermometer is placed in the batch, which is cooked to 290°F. The butter is then added and the batch is stirred and cooked for another 2 minutes. It is then poured out on a greased slab lined with iron bars, and the edges are folded together while the flavor is stirred in. The folding process is continued until the batch is cool enough to handle. It is then transferred to the spinning table in front of the batch warmer and formed into shape. The batch is then passed through an automatic sucker machine and the candies are allowed to cool on the table. They are then wrapped and packed as usual. This formula should yield approximately 45 pounds of suckers.

Note: The sucker machine inserts the sticks as it forms and cuts the suckers.

BUTTERSCOTCH TOFFEE

Sugar	12 lb
Corn Syrup	8 lb
Coconut Butter	8 oz
Water	2 qt
Salt	1 teaspoonful
Butter	1 lb
Evaporated Milk	2 lb
Vanilla Flavor	2 teaspoonfuls

The sugar, corn syrup, coconut butter, water, and salt are placed in a clean copper pan on the stove and stirred until dissolved. When the batch begins to boil, it is covered and allowed to steam for a few minutes. The sides are washed down and a thermometer is placed in the batch, which is cooked to 256°F. The thermometer is removed; the butter and milk are added gradually, and the batch stirred continuously under a slow fire until the batch forms a medium ball when tested in cold water. The batch is then poured on a greased slab between iron bars and the flavor is stirred in while the edges are folded together. The water is turned on under the slab to cool the batch rapidly. When it is cool enough, the batch is cut and wrapped by hand or machine.

This batch will yield approximately 25 pounds of candy.

BUTTERSCOTCH WAFERS, CHOCOLATE

Sugar	6½ lb
Corn Syrup	2½ lb
Salt	1 teaspoonful
Water	1 qt
Butter	12 oz
Melted Bitter Chocolate	10 oz
Vanilla Flavor	1 teaspoonful

The sugar, corn syrup, water, and salt are placed in a copper pan on the stove and stirred until dissolved. When the batch begins to boil, it is covered and allowed to steam for a few minutes. The sides are washed down and a thermometer is placed in the batch, which is cooked to 290°F. The butter, chocolate, and flavor are stirred in rapidly and mixed through well. The batch is then run into wafers through a funnel. The wafers are placed on a greased slab and when cool, they are loosened and packed in tins or jars.

This formula should yield approximately 10 pounds.

BUTTERSCOTCH WAFERS, CHOCOLATE NUT

The formula for BUTTERSCOTCH WAFERS, CHOCOLATE is followed, but 8 ounces of chopped pecans are stirred into the batch with the butter and chcolate. The batch is finished in the usual manner, and should yield approximately 10½ pounds of wafers.

BUTTERSCOTCH WAFERS, CINNAMON

The formula for BUTTERSCOTCH WAFERS, VANILLA is followed, but while stirring in the butter, 1 teaspoonful of red color and 1 teaspoonful of cinnamon oil are added. The batch is finished in the usual manner, and should yield approximately 9 pounds of wafers.

BUTTERSCOTCH WAFERS, VANILLA

Sugar	6 lb
Corn Syrup	2 lb
Water	1 qt
Salt	1 teaspoonful
Butter	8 oz
Vanilla Flavor	1½ teaspoonfuls

The sugar, corn syrup, water, and salt are placed in a small copper pan on the stove and stirred until dissolved. When the batch begins to boil, it is covered and allowed to steam for a few minutes. The sides are washed down and a thermometer is placed in the batch, which is cooked to 290°F. The butter and flavor are stirred in rapidly. A portion of the batch is then poured into a funnel and dropped out in wafers on a greased slab. This is continued until the entire batch has been used. The wafers are cooled, and after they are loosened with a large knife, they are packed in tins or jars.

This formula should yield approximately 9 pounds.

BUTTERSCOTCH WAFERS, VANILLA NUT

The formula for BUTTERSCOTCH WAFERS, VANILLA is followed, but 8 ounces of chopped walnuts are mixed through the batch when the butter is stirred in. The batch is then finished in the usual manner, and should yield approximately 9½ pounds of wafers.

C

CALIFORNIA FRUIT KISSES

Center

Sugar	3 lb
Corn Syrup	2 lb
Chopped Soft Figs	1 lb
Chopped Seedless Raisins	1 lb
Chopped Soft Dates	2 lb
Water	½ qt

The sugar, corn syrup, and water are placed into a copper pan on the stove and stirred until dissolved. When the batch begins to boil the sides are washed down and a thermometer is placed in the batch which is cooked to 236°F. The chopped fruit is then stirred in and mixed thoroughly. This center is kept warm until ready to use.

Jacket

Sugar	8 lb
Corn Syrup	10 lb
Molasses	1 qt
Water	1 qt
Coconut Butter	8 oz
Salt	2 teaspoonfuls
Vanilla Flavor	1 teaspoonful
Oil of Lemon	1 teaspoonful

All of the ingredients, excluding the flavors, are placed in a copper pan on the stove and stirred until dissolved. When the batch begins to boil, the sides are washed down. Stirring and cooking are continued and the batch is cooked until it forms a medium ball when tested in cold water. The batch is poured out on a greased cold slab between iron bars, and the edges are folded together while the flavors are stirred in. The batch is cooled until hard enough and then transferred to the pulling machine and pulled for 5 minutes. It is then placed on the spinning table and formed into a flat batch. The warm center is placed in the middle, the batch is wrapped and the ends are closed up. The batch is formed round and spun into kisses. It is then cut and wrapped as usual.

This batch will yield approximately 30 pounds of candy.

CANDIED APPLES

Water	1 qt
Sugar	8 lb
Corn Syrup	2 lb
Red Color	1 tablespoonful
Vanilla	1 teaspoonful
Apples	50–60 large or 75–80 small

The sugar, corn syrup, and water are placed in a small copper pan on the stove and stirred until dissolved. When the batch begins to boil, it is covered and allowed to steam for a few minutes. The sides are washed down while the red color and a thermometer are placed in the batch, which is allowed to cook to 300°F. The kettle is then removed and placed in the center of a greased slab, the flavor having been stirred in before removing from the heat. Lollipop sticks are placed in the center of each apple and each one is dipped into the red batch until it is well-covered. The apple is then held over the kettle until drained, and finally placed on the greased slab. This process is continued until all apples are well covered. After cooling on the slab, the apples are packed in tins.

This batch will yield about 50 to 60 large apples or 75 to 80 small apples.

CANDY CLEAR TOYS, HAND-MADE

Sugar	80 lb
Corn Syrup	20 lb
Water	12 qt

The sugar, corn syrup, and water are placed in a large clean copper pan on the stove and stirred until dissolved. When the batch is about at the boiling point it is removed from the stove. All the metal clear-toy molds are greased and set up in even rows on the marble. The large batch of melted syrup is used as a base. In a clean small copper pot with a good pouring point, about 5 to 6 pounds of the melted syrup are placed on the stove and cooked to 290°F. This hot candy is poured into the clear-toy molds on marble and allowed to cool until hardened. The molds are carefully opened and the candy toys are taken out and placed on waxed-paper-lined boards. After cooling, they are packed in tins. This process is continued until all the syrup from the base batch is used up.

For red toys, the small pouring batch is colored bright red while cooking on the stove. It is handled and finished as usual.

The batch will yield approximately 95 pounds.

CANDY CLEAR TOYS, MACHINE-MADE

Sugar	42 lb
Corn Syrup	8 lb
Water	6 qt
Flavor	1 oz
Color	1 teaspoonful

The sugar, water, and corn syrup are placed in a copper pan on the stove and stirred until dissolved. When the batch begins to boil, it is covered and allowed to steam for a few minutes. The sides are washed down, and a thermometer is placed in the batch, which is cooked to 310°F. The batch is then poured out on a greased slab, and the edges are folded together while the oil of lemon is stirred in. The folding and kneading are continued until the batch is cool enough; it is then transferred to the clear-toy machine in front of the batch warmer. It is passed through the clear-toy roller machine, allowed to cool on the table, and then packed into tins.

These toys are usually made in three colors: white, green, and red. The green is a light shade and flavored with oil of lime. The red is a bright shade and made with cherry or raspberry flavor.

The batch will yield approximately 47½ pounds.

CANDY CURLS, GREEN PISTACHIO

The CANDY CURLS, VANILLA formula is followed, but the batch is colored light green and made with pistachio flavor. It is finished in the usual manner.

The batch will yield about 14 pounds.

CANDY CURLS, PINK

The formula for CANDY CURLS, VANILLA is used, but the batch is colored bright pink and flavored with oil of rose. It is completed in the usual manner and will yield approximately 14 pounds.

CANDY CURLS, VANILLA

Sugar	15 lb
Water	2 qt
Cream of Tartar	1½ teaspoonfuls
Vanilla Flavor	2 teaspoonfuls

The sugar, water, and cream of tartar are placed in a copper pan on the stove and stirred until dissolved. When the batch begins to boil, it is covered and allowed to steam for a few minutes. The sides are washed down and a thermometer is placed in the batch, which is cooked to 330°F. The batch is then poured out on a greased slab, and the edges are folded together while stirring in the vanilla flavor. The kneading and folding are continued until the batch is cool enough; then the batch is transferred to the hook and pulled partly while the air is twisted out, until it turns a glossy white. It is then taken over on the spinning table in front of the batch warmer, where a candy-curler machine is attached to the table. The batch is kneaded and shaped round, and then spun in fine strings; the strings are wrapped around the curler. The curls are continuously removed from the machine and allowed to cool on the table in front of an electric fan. They are packed into tins or glass jars.

The batch will give about 14 pounds.

CANDY CURLS, VIOLET

The CANDY CURLS, VANILLA formula is followed, but the batch is colored a bright purple and made with violet flavor. It is completed in the usual manner and will yield approximately 14 pounds.

CARAMEL APPLES

Sugar	5 lb
Corn Syrup	7 lb
Salt	1 teaspoonful
Coconut Butter	4 oz
Fresh Light Cream	2 qt
Evaporated Milk	3 lb
Vanilla Flavor	1 teaspoonful
Medium-Size Firm Apples	75

The apples are placed on skewer sticks on clean boards. The sugar, corn syrup, salt, coconut butter, and cream are placed in a clean copper pan on the stove and stirred until dissolved. When the batch begins to boil, the evaporated milk is stirred in. The batch is cooked to a soft ball when tested in cold water and removed to a stand near the greased slab. The flavor is then stirred in. The apples are dipped in the caramel on the greased slab and allowed to cool on the slab. They are then packed on trays lined with wax paper.

CARAMEL DIPPING: *See* LOG-CABIN PECAN ROLL and LOG NUT ROLL.

CARAMEL LOLLIPOPS: *See* LOLLIPOPS, CARAMEL.

CARAMEL PASTE

Sugar	20 lb
Evaporated Milk	10 lb
Potato Flour	5 lb
Corn Syrup	40 lb
Condensed Milk	20 lb

The sugar and evaporated milk are placed in a large copper pan on the stove, stirred, and cooked to a soft ball. About 5 pounds of this caramel are taken out into a small pot and mixed with the potato flour to a thick paste. This paste is stirred back into the batch, and then the corn syrup and condensed milk are stirred in gradually. The batch is then heated (but not cooked) on a very low flame. It is then poured out into large tubs and allowed to set overnight. It is now ready for use.

This formula should yield approximately 100 pounds of paste.

CARAMEL PECAN ICE CREAM

Fresh 20% Cream	16 qt
Fresh 40% Cream	2 qt
Condensed Milk	2 qt
100 Grade Citrus Pectin	4 oz
Sugar	12 lb
Caramel Color	12 oz
Vanilla Flavor	1 oz
Butter	1 lb
Chopped Pecans	2½ lb
Salt	1 tablespoonful

The pecans and butter are placed in a small pan on the stove over low heat and stirred until the nuts are toasted to a medium-brown color. They are then salted, placed in a flat pan, and set aside to cool. The pectin is mixed dry with 2 pounds of the sugar and placed in a double boiler on the stove with 2 quarts of the light cream. This mixture is stirred until it dissolves and begins to boil. It is then poured into a large ice-cream can. The balance of the sugar and the light cream, the heavy cream, and the condensed milk are added. This mixture is stirred with a long paddle until the sugar is dissolved. Now the color and the flavor are stirred in, and the mix is then strained into the freezer. Before the freezer is shifted to high speed, the toasted nuts are added. Otherwise the process for making VANILLA ICE CREAM NO. 1 is followed.

The batch will yield approximately 40 quarts.

CARAMELLOWS

Sugar	6 lb
Corn Syrup	5 lb
Coconut Butter	8 oz

Salt	1 teaspoonful
Fresh Light Cream	2 qt
Evaporated Milk	2 lb
CARAMEL PASTE (*see*)	1 lb
Vanilla Flavor	2 teaspoonfuls
Small DIPPING MARSHMALLOWS (*see*)	5 lb

The sugar, corn syrup, coconut butter, salt, and cream are placed in a copper pan on the stove and stirred until dissolved. When the batch begins to boil, the evaporated milk is stirred in gradually. The stirring is continued while the caramel paste is added. The batch is cooked to a soft ball when tested in cold water. The vanilla flavor is stirred in and the batch is removed to a spot near the greased slab. The marshmallows are brushed off clean and dipped with a dipping fork into the hot caramel so that the caramel covers the marshmallow completely. The dipped marshmallows are deposited on the greased slab. This process is continued until the caramel cools off in the batch, and it is too cold to dip. The dipped marshmallows are wrapped in moisture-proof cellophane, and the caramel left over is saved and used up in any batch of chocolate caramels.

This batch should yield approximately 22 pounds of candy.

CARAMELS: *See* CHEWING CANDIES and Individual CARAMEL Formulas, e.g., CARAMEL APPLES. *See also* CHOCOLATE ANNEX CARAMELS, CARO WHIRLS, ITALIAN CARAMELS, LONDON CARAMELS, PAN CARAMELS, and VARIETY CRAMELS.

CARO WHIRLS, CHOCOLATE

Sugar	11 lb
Corn Syrup	9 lb
Coconut Butter	12 oz
Salt	1 tablespoonful
Fresh Cream	3 qt
Evaporated Milk	10 lb
CARAMEL PASTE (*see*)	3 lb
Cut-Up Bitter Chocolate	2½ lb
Vanilla Flavor	2 teaspoonfuls

The sugar, corn syrup, salt, coconut butter, and cream are placed in a large copper pan on the stove and stirred until dissolved. When the batch begins to boil, the evaporated milk is stirred in, and the stirring is continued while the caramel paste is added. The batch is cooked to a medium ball when tested in cold water. The chocolate and flavor are mixed in. The batch is then poured out on a greased slab lined with iron bars and spread to about 1/4 inch thickness. It is then allowed to cool on the table while the center is being made.

Center

Sugar	8 lb
Corn Syrup	4 lb
Coconut Butter	8 oz
Salt	1 teaspoonful
Fresh Cream	2 qt

All of the ingredients are placed in a copper pan on the stove and stirred until dissolved. When the batch begins to boil, the sides are washed down and the batch is stirred and cooked to a firm ball or to 252°F. It is then poured out on a cold greased slab, and the edges are folded together while 1 teaspoonful of pink color and 2 teaspoonfuls of strawberry flavor are stirred in. The batch is then cooled rapidly, and when cool enough, it is taken over to the pulling machine and pulled for 5 minutes. It is then transferred to the slab, stretched out thin, and spread on top of the caramel to about 1 inch from the edges. The caramel is then folded up with the pink inside, formed round, spun out, and cut and wrapped as usual.

This batch should yield approximately 50 pounds of candy.

CARO WHIRLS, VANILLA

Sugar	12 lb
Corn Syrup	8 lb
Coconut Butter	1 lb
Salt	1 tablespoonful
Fresh 18% Light Cream	3 qt
Evaporated Milk	6 lb
Vanilla Flavor	3 teaspoonfuls
CARAMEL PASTE (*see*)	4 lb

The sugar, corn syrup, salt, coconut butter, and cream are placed in a large copper pan on the stove and stirred until dissolved. When the batch begins to boil, the evaporated milk is added gradually, and the stirring is continued while the caramel paste is added. The batch is cooked to a firm ball when tested in cold water. The vanilla flavor is then stirred in, and the batch is poured out on a greased slab lined with iron bars. It is spread out evenly over the entire slab to about 1/4 inch thickness. It is then allowed to cool while the center is being made.

Center

Sugar	8 lb
Corn Syrup	4 lb
Coconut Butter	8 oz
Salt	1 teaspoonful
Fresh Cream	2 qt

All of the ingredients are placed in a copper pan on the stove and stirred until dissolved. When the batch begins to boil, the sides are washed down, and the stirring is continued. The batch is cooked until it forms a firm ball when tested in cold water or to about 252°F. It is then poured out on a cold greased slab, and the edges are folded together while the batch is cooled rapidly. When cool enough, it is taken over to the pulling machine and pulled for about 5 minutes. It is then transferred to the slab, stretched out thinly and spread over the caramel to about 1 inch from the edges. The caramel is folded with the white inside, formed round, spun out, cut and wrapped by machine.

This batch should yield approximately 50 pounds.

CASHEW NUTS, SALTED: *See* SALTED CASHEW NUTS.

CASHEW-NUT BRITTLE

Sugar	6 lb
Corn Syrup	6 lb
Cashew Nuts	8 lb
Water	1 qt
Salt	2 tablespoonfuls

The sugar, corn syrup, water, and salt are placed in a copper pan on the stove and stirred until dissolved. When the batch begins to boil, it is covered up and allowed to steam for a few minutes. The sides are washed down, and a thermometer is placed in the batch, which is cooked to 260°F. The thermometer is removed and the flame lowered. The cashews are added gradually, and the stirring and cooking continued until the nuts are roasted and the candy breaks crisp when tested in cold water. The batch is poured out on a greased slab, the edges are folded together, and the batch is kept soft in the center of the slab. One half of the batch is rolled out thin with a rolling pin, turned on the opposite side, and stretched as thin as possible. The other half of the batch is processed in the same way. After cooling on the slab, the candy is packed into tins.

The batch will yield approximately 20 pounds of brittle.

CAST COCONUT CREAM EGGS: *See* COCONUT CREAM EGGS, CAST.

CAST LIQUID CHERRIES: *See* LIQUID CHERRIES, CAST.

CAST MAPLE CREAMS: *See* MAPLE CREAMS, CAST.

CAST VANILLA CREAMS: *See* VANILLA CREAMS, CAST.

CHERRIES, CHOCOLATE-COVERED: *See* CHOCOLATE-COVERED CHERRIES.

CHERRY COCONUT STARS: *See* COCONUT STARS, CHERRY.

CHERRY EGGS, HAND-ROLLED

The formula for PINEAPPLE EGGS is followed. The batch is colored red while cooling on the cream beater. After the batch has cooled for 7 minutes, the beater is set in motion and the batch is flavored with 3 pounds of glacéed cherry pieces, 3 teaspoonfuls of cherry flavor, and 2 teaspoonfuls of citric acid solution. The marshmallow whip and invertase are added and the batch is creamed to a lump. It is then kneaded soft and put through the cut roll machine to the desired size. When cool, the eggs are coated with dark vanilla sweet chocolate and marked with the letter "C" on top. When the finished eggs are cool and set, they are packed in boxes.

The batch will yield approximately 70 pounds of cherry eggs.

CHERRY GOODIES: *See* MARZIPAN CHERRY GOODIES.

CHERRY SEA-FOAM FUDGE: *See* SEA-FOAM FUDGE, CHERRY.

CHERRY SEA-FOAM KISSES: *See* SEA-FOAM KISSES, CHERRY.

CHERRY SHERBET

Fresh Milk	20 qt
Sugar	20 lb
100 Grade Citrus Pectin	4 oz
Lemon Juice	1½ qt
Red Color	1½ teaspoonfuls
Cherry Flavor	1 oz
Maraschino Cherries (chopped, with the juice)	6 qt

The pectin is mixed dry with 5 pounds of the sugar, and placed in a copper pan on the stove over low heat. Four quarts of milk are poured into the pan rapidly, and the mixture is stirred and heated to the boiling point. It is then poured into the mixing can. The balance of the milk and sugar, the juice, color, and flavor are added to the batch, and the mix is stirred until all the sugar is dissolved. The mix is then cooled rapidly and strained into the freezer as usual. The freezing process is begun and continued until the sherbet is frozen about half-way. The chopped cherries with the juice are added, and the freezing is completed. The storing and ripening of the sherbet are carried out in the usual manner (*see* VANILLA ICE CREAM NO. 1).

The batch will yield approximately 36 quarts.

CHERRY SOUR BALLS: *See* SOUR BALLS, CHERRY.

CHERRY STICK CANDY: *See* STICK CANDY, CHERRY.

CHERRY SYRUP

Ripe, Pitted Cherries	5 lb
Sugar	5 lb
Water	½ qt
SIMPLE SYRUP (*see*)	6 qt
FRUIT ACID SOLUTION (*see*)	2½ oz
Pure Wild-Cherry Flavor	4 teaspoonfuls
Liquid Red Color	1 teaspoonful

The cherries, sugar, and water are placed in a small copper pan on the stove over very low heat and heated to the boiling point. The cherries are then covered and allowed to cook for 5 minutes. They are then removed from the stove and cooled rapidly. The mixture is then placed in a large wooden mixing bowl and mashed to a pulp. The simple syrup is poured into a clean stone crock, and the cherry pulp is stirred in rapidly. The acid solution, flavor, and color are then stirred into the syrup and mixed thoroughly. The crock is covered, and the cherry syrup is ready for use.

The batch will yield approximately 10 quarts of cherry syrup.

CHERRY VANILLA ICE CREAM

Fresh 20% Cream	4 qt
Fresh 40% Cream	2 qt
Condensed Milk	2 qt
Sugar	8 lb
100 Grade Citrus Pectin	2 oz
Maraschino Cherries (with the juice)	4 qt
Vanilla Flavor	1 oz

The pectin is mixed dry with 2 pounds of the sugar and placed in a double boiler on the stove together with 1 quart of the light cream. This mixture is stirred until dissolved and heated to the boiling point. It is then poured into the mixing can, and the balance of the light cream and sugar, the condensed milk, and the heavy cream are added to the batch and mixed with a long paddle until the sugar is dissolved. The vanilla flavor is stirred in, and the mix is strained into the freezer. When the cream freezes about half-way, the cherries are mashed in the juice and added into the freezer. Otherwise the same freezing process is followed as for VANILLA ICE CREAM NO. 1 (*see*).

The batch will yield approximately 25 quarts.

CHEWING CANDIES

Chewing candies are usually made in bulk and sometimes serve as centers for chocolate dipping. It is very important that these candies have a lasting shelf life. They must stay chewy to be salable. When they become grainy they lose their flavor and good taste. Correct "doctoring" is, therefore, necessary. When too much corn syrup is used in a formula the candy is tough and not sweet enough. However, when not enough corn syrup is used the candy will grain in a short period of time. Cooking at the right temperature is the next important point. During hot weather, the cooking should be at higher temperatures; during cool weather, the cooking should be at medium temperatures and during cold weather, at low temperatures. Wrapping is another problem. All salt-water taffies, kisses, caramels, and nougats must be wrapped, and it is advisable to wrap them before they get sticky.

There are four kinds of chewing candies: toffies, taffies, nougats, and caramels. Although these candies are classed together, they have their own characteristics to some extent and should be studied individually.

Toffies are made of sugar, corn syrup, fresh cream, evaporated milk, coconut and dairy butter, and flavors. These are the main ingredients. The formulas for these goods are well balanced; it is necessary to prepare the batches carefully, check the ingredients, and get correct weights and measurements. When first starting to cook the batch, moderate heat is essential so that enough time is allowed for the batch to dissolve perfectly. When the batch starts to boil, the heat can be increased slightly. When the batch begins to cook thickly, the heat is decreased again and the cooking is finished with slow heat. Flavors are very important, and must be carefully chosen. When toffies are first tasted, the rich creamy texture is noted, but after the consumer chews a while he notes the flavor and flavor is the last thing he remembers. This lingering sensation of flavor is sometimes called the aftertaste.

With few exceptions, taffies do not contain butterfat. For their taste they rely mostly on molasses, honey, and flavors. Therefore, these main ingredients must be carefully selected. Many brands of so-called molasses do not have the true molasses flavor and scorch when the batch nears finishing. Molasses of this type will yield a golden color, but will not give the real, old-fashioned molasses flavor. When buying molasses, it is important to watch for the code. "Open-kettle" molasses is the most suitable type.

Nougats depend mostly on honey and egg white (albumen) for their body and quality. These two ingredients must be carefully selected. Whenever a formula is balanced or adjusted, the honey and albumen should not be cut down. Instead they should be increased, if necessary. Many imitations of honey are on the market, but these should not be used. Honey should be bought from a reputable firm and guaranteed to be pure honey. Synthetic albumen, which usually contains mixtures of soya whip and certain chemicals, should also be avoided. Flavors and nuts give nougats the right finishing touch and here again careful selection is essential.

Caramels derive their body mostly from butterfat. Great care should be taken in selecting the cream and evaporated milk that go into this product. Fresh 18 to 20% cream is preferable. The most suitable evaporated milk is the canned type. It is not advisable to carry large stocks of evaporated milk and run the risk of spoilage. The 84% type of hard fat or coconut butter is preferable. There is a pure vanilla extract and an imitation vanilla extract. The pure vanilla extract gives the best flavor, but it is not highly concentrated and thus a larger quantity must be used. This makes for soft, reduced batches. The imitation extract is highly concentrated and its flavoring strength is three times that of pure vanilla, but it does not give the fine, delicate vanilla flavor of the pure extract. Therefore, a happy medium—a pure vanilla extract strengthened with vanillin powder—about 2 ounces of powder to 1 gallon of extract—should be used, which will give a combination vanilla that has twice the strength of regular vanilla and much of its purity.

For the various chewing-candy formulas, *see* the *Individual Names,* such as BUTTERSCOTCH TOFFEE, FRENCH NOUGAT, MOLASSES COCONUT TAFFY, etc. *See also* COLONIAL NOUGAT.

CHOCOLATE ALMOND BARK

Dark Vanilla Sweet Chocolate	15 lb
High-Quality Milk Chocolate	10 lb
Freshly Roasted Almond Pieces	12 lb

The chocolate is melted in a steam-bath pan as usual. It is then cooled to about 84°F and further cooled in small portions on the dipping table. Some cooled almonds are then added to the chocolate and mixed thoroughly. This mixture is spread out on wax-paper-lined trays to about ½ inch thickness and allowed to cool in a room of 64°F. It is then cut and broken into small sheets, which are packed in boxes.

Note: This bark can also be made with a variety of other nuts, such as pecans, roasted cashews, and filberts.

The batch will yield 37 pounds of chocolates.

CHOCOLATE ALMOND CLUSTERS

Dark Vanilla Sweet Chocolate	22 lb
Roasted Almond Pieces	18 lb

The chocolate is melted in a steam-bath pan over low heat (about 120°F); it is stirred gradually and cooled to about 88°F. A small portion (about 2 pounds) of the melted chocolate is cooled by mixing it with a hand scraper in the center of the dipping marble or metal table. It is cooled until it becomes pliable. About an equal amount of the almond pieces is then mixed with the chocolate and formed rapidly with the right hand into small clusters. These are deposited on wax-paper-lined trays. This process is continued until all the chocolate and almonds are used up. The dipped clusters are allowed to cool to about 64°F. They are then packed in boxes.

The batch will yield approximately 40 pounds of chocolates.

CHOCOLATE ALMOND CRUNCH, SUNDAE TOPPING

Sugar	6 lb
Corn Syrup	3 lb
Fresh 18% Cream	2½ qt
Salt	1 teaspoonful
150 Grade Citrus Pectin	1 oz
Chopped Bitter Chocolate	1 lb
Chopped Roasted Almonds	3 lb
Vanilla Flavor	1 teaspoonful
Glycerin	4 teaspoonfuls

The pectin is mixed dry with 2 pounds of the sugar. One quart of cream is placed in a clean copper pan on the stove, and the pectin sugar is added. The batch is stirred until the sugar is dissolved. The remaining sugar and cream and the salt are added to the batch and again stirred until the sugar is completely dissolved. The corn syrup and chopped chocolate are added gradually, and the batch is stirred continuously and cooked to 217°F.

It is then moved to a cool spot, and the chopped roasted almonds are stirred in rapidly. The flavor and glycerin are added and mixed thoroughly. The sundae topping is then poured into clean glass jars and closed with air tight lids. It is then ready for use.

The batch will yield approximately 8 quarts of chocolate almond cruch sundae topping.

CHOCOLATE ALMONDS, PAN-COATED

For making chocoate panned goods the STANDARD smooth-surfaced REVOLVING PAN (*see*) is adjusted to run 24 to 30 revolutions per minute. It must be located in a room of 64°F.

For controlled results, it is necessary to feed cool dry air of about 65°F to the revolving pan. The cooled air may be supplied by a mechanical blower with an outlet of about 6 inches. The air is forced through the outlet and into the revolving pan. The air pipe which brings the cool air from the blower into the revolving pan should be about 3 inches in diameter. The air pipe must be so constructed that the stream of air may be diverted from the revolving pan. This diversion of air is periodically necessary when processing panned chocolate coating.

The almonds must be roasted dry. Two to three pounds of chocolate coating to each pound of almonds is considered a good average. The chocolate used may be dark sweet or it may be blended with milk chocolate.

To start the operation, the interior of the standard revolving pan is lined with a thin layer of chocolate coating. A fine brush is used. The cool air is then turned on to harden the chocolate.

About 30 pounds of roasted almonds are then placed into the revolving pan, which is tempered to about 70°F. The pan is set in motion. The chocolate to be applied is tempered to about 100°F and poured on the almonds in a fine slow stream out of an automatic chocolate sprayer. The hand is worked back and forth over the almonds so that they do not cling together. As soon as the chocolate is well distributed over the almonds, a stream of cool air is immediately diverted into the pan, but as soon as the chocolate begins to set, the air is diverted from the pan, and the almonds are allowed to run until they acquire a smooth coating.

When two charges of the chocolate tempered at 100°F are

applied and the points of the almonds are properly covered, additional charges of chocolate are added, about 1 quart at a time. The chocolate is allowed to set after each charge, and the cold air is kept blowing into the revolving pan after each charge of coating. The confectioner should avoid running the pan with the air turned off for too long a period, since the ensuing frictional heat would rub off the coating from the ends of the almonds.

When the almonds have enough chocolate to give them the desired size, a small amount of melted cocoa butter is applied to them. The melted cocoa butter is stirred through with the hand while the pan is in motion. As soon as the cocoa butter is set and the surface of the chocolate almonds is smooth, the almonds are removed, spread out on clean trays lined with wax paper, and allowed to set for about 1 hour. They are then ready for polishing.

For polishing panned chocolates a special polishing revolving pan is required. It should be equipped with fingers or indentations which cause the almonds to run when the pan is in operation. The polishing pan should be located in the same room and give about 32 revolutions per minute. It should also be equipped to receive cool dry air.

The polish is prepared as follows: 4 pounds of gum arabic and 2 pounds of granulated sugar are dissolved in 4 quarts of water, heated over low heat, and strained through a fine sieve. Five pounds of corn syrup are stirred into the gum solution. After removing the foam from the top, the solution is placed in glass jars. It is cooled to room temperature and is then ready for use.

The chocolate-coated almonds are placed in the polishing pan, which is set in motion. The polish solution is sprayed on the rolling almonds in very small amounts, about a cupful at a time, while the hand is run through the rolling almonds. The almonds are covered thinly with the solution. Cool air is then fed into the pan, and the almonds are kept rolling until they are dry. This process is repeated about three times. After each operation, air is applied until the almonds roll dry.

The polished almonds are carefully removed with a flat basin and spread out on clean trays lined with wax paper. They are then allowed to dry and are packed in boxes.

By following the directions described, a complete assortment of panned chocolate for a bridge or television package can be produced. When the price of chocolate is prohibitive for some manufacturers, it is more economical to apply a preliminary or build-up coating, i.e., a thin coating of cocoa butter and then a mixture of cocoa powder and powdered sugar. The following basic method can be used.

The centers are placed into the smooth STANDARD REVOLVING PAN (*see*) which is set in motion with the air turned off. A very thin coating of melted cocoa butter is applied to the centers, just enough to moisten them. After the centers are covered with the melted cocoa butter, a mixture consisting of 10 pounds of sifted powdered sugar and 10 pounds of sifted cocoa powder is applied on the wet centers, just enough to absorb the cocoa butter. Cool air is then blown into the revolving pan, and the centers are allowed to roll and form a firm coating. An additional charge of cocoa butter is then applied and covered with more of the powdered sugar-cocoa mixture. Again the cool air is blown into the pan, and the centers are allowed to run until a firm coating is developed.

After a sufficiently thick coating has been attained, successive charges of properly tempered chocolate coating may be applied to produce a smooth coating. Sufficient chocolate coating should be added to give the chocolates the desired size. The following assortment of centers is suggested: filberts, pecans, roasted jumbo peanuts, cashew buds, large seedless raisins, glacéed cherries, and glacéed pineapple cubes.

CHOCOLATE ANNEX CARAMELS

Sugar	11 lb
Corn Syrup	15 lb
Coconut Butter	1 lb
Salt	1 tablespoonful
Fresh 18% Light Cream	4 qt
Evaporated Milk	8 lb
CARAMEL PASTE (*see*)	5 lb
Bitter Chocolate (cut up)	4 lb
Vanilla Flavor	3 teaspoonfuls
DIPPING MARSHMALLOWS (*see*)	3 lb

The sugar, corn syrup, salt, coconut butter, and cream are placed in a large copper pan on the stove and stirred until dissolved. When the batch begins to boil, the evaporated milk is stirred in gradually. Stirring is continued while the batch is cooking. The caramel paste is added and the batch is cooked until it forms a medium ball when tested in cold water. The batch is then removed to a stand, and the chocolate and flavor are added and stirred until completely melted. After adding the marshmallows and mixing well, the batch is poured out on a greased slab between iron bars and spread out evenly with a palette knife. It is allowed to cool until it is hard enough to handle. Then it is passed through a sizing machine, cut into square caramels, and wrapped as usual.

This batch should yield approximately 48 pounds of candy.

CHOCOLATE BANQUET MINTS

Sugar	30 lb
Water	7 qt
Silver Sheet Gelatin	1 oz
(soaked in 1 cup of warm water)	
Oil of Peppermint	4 teaspoonfuls

The sugar and water are placed in a copper pan on the stove and stirred until dissolved. When the batch begins to boil, it is covered and allowed to steam for a few minutes. The sides are washed down and a thermometer is placed in the batch, which is cooked to 245°F. The batch is poured out on a cooled, dampened, cream beater and allowed to cool to 110°F. The soaked gelatin is added and the machine is set in motion and continued to operate until the batch creams to a lump. It is then allowed to sweat back for 15 minutes and remelted in a steam bath as usual (*see* CAST CREAM CENTERS). The mint flavor is stirred in gradually and the cream is then cast through a funnel on heavy wax paper on the table. The creams are allowed to cool and set, and they are then turned upside down and placed on trays. The mints are coated with bitter-sweet chocolate, allowed to set, and packed as usual.

This formula should yield approximately 35 pounds of mint wafers or 43 pounds of machine chocolate coated wafers, or 50 pounds of hand-dipped wafers.

CHOCOLATE BOARDWALK FUDGE: *See* BOARDWALK FUDGE, CHOCOLATE.

CHOCOLATE BOTTOMED COCONUT KISSES: *See* COCONUT KISSES, CHOCOLATE BOTTOMED.

CHOCOLATE BUTTER CREAMS: *See* BUTTER CREAMS, CHOCOLATE.

CHOCOLATE BUTTER CRUNCH CLUSTERS

Sugar	25 lb
Water	3 qt
Cream of Tartar	2 teaspoonfuls
Butter	1 lb
Ground Roasted Almonds	1 lb
Salt	2 teaspoonfuls

The sugar, water, and cream of tartar are placed in a large copper pan on the stove and stirred until dissolved. When the batch begins to boil, it is covered and allowed to steam for a few minutes. The sides are washed down and a thermometer is placed in the batch, which is cooked to 330°F. The thermometer is removed and the fire is reduced to low. The butter and almonds are stirred in and cooked another 4 minutes. The batch is then poured out on a greased slab lined with iron bars, and the edges are folded together. The batch is cooled rapidly and, when cool enough, it is pulled on the hook or machine until light and spongy. It is then transferred to the spinning table in front of the batch warmer, formed into a flat batch, and spun out into fine thin chips. These are allowed to cool on the table, and they are then broken and crushed with a rolling pin into fine pieces, which are coated in clusters with a high-quality milk chocolate. The clusters are allowed to set and are then packed as usual.

The batch will yield approximately 50 pounds of chocolates.

CHOCOLATE BUTTER SQUARES

The same process is followed as for BUTTERSCOTCH SQUARES, VANILLA. Two pounds of bitter chocolate, cut up fine, are added to the batch together with the butter for color and flavor. It is then finished in exactly the same manner as the vanilla formula.

The batch will yield approximately 22½ pounds.

CHOCOLATE BUTTERNUT TOASTIES

Center

Fresh Peanut Butter	5 lb
Melted Sweet Chocolate	2 lb
Finely Chopped Roasted Almonds	2 lb
Dutch Cocoa Powder	1 lb
Nutmeg Flavor	1 teaspoonful

All of the listed ingredients are placed in a small copper pan on the stove over low heat and stirred continuously until they are almost hot. This batch is then kept warm for use as a center.

Jacket

Sugar	20 lb
Water	3 qt
Cream of Tartar	2 teaspoonfuls
Oil of Rose	1 teaspoonful
Pink Color	As desired for a delicate pink

The sugar, water, and cream of tartar are placed in a copper pan on the stove and stirred until dissolved. When the batch begins to boil, it is covered and allowed to steam for a few minutes. The sides are washed down and a thermometer is placed in the batch, which is cooked to 336°F. The batch is then poured out on a greased slab, and the edges are folded together while the flavor is stirred in. The batch is colored a delicate pink and then folded with the iron bar until firm. It is then pulled on the hook for approximately 2 minutes and twisted until it becomes glossy. The batch is then transferred to the spinning table in front of the batch warmer, where it is kneaded smooth and formed flat. The warm center is placed in the middle of the flat batch, the edges are folded together and wrapped up, and the ends are closed. The batch is formed into shape and spun into medium-sized sticks with gloved hands. These are passed through a lump-cutting machine and allowed to cool on the air-conditioned table. The toasties are then packed in airtight cans.

This formula should yield 28 pounds of toasties.

CHOCOLATE BUTTERSCOTCH DAINTIES: *See* BUTTERSCOTCH DAINTIES, CHOCOLATE.

CHOCOLATE BUTTERSCOTCH WAFERS: *See* BUTTERSCOTCH WAFERS, CHOCOLATE.

CHOCOLATE CARMEL NOUGAT ROLL

Sugar	11 lb
Corn Syrup	13 lb
Salt	1 tablespoonful
Coconut Butter	1 lb
Fresh Cream	4 qt
Evaporated Milk	8 lb
Bitter Chocolate (cut up)	3 lb
Vanilla Flavor	2 teaspoonfuls
HONEY NOUGAT (*see*)	12 lb

The sugar, corn syrup, salt, coconut butter, and cream are placed in a copper pan on the stove and stirred until dissolved. When the batch begins to boil, the evaporated milk is stirred in. The stirring is continued, and the batch is cooked to a medium ball when tested in cold water. The chocolate and flavor are stirred in, and the batch is poured out on a greased slab and rolled out to about ⅛ inch thickness. The 12 pounds of honey nougat are rolled out quite thin and placed on top of the caramel sheet. The batch is folded up, shaped round, and rolled out in long sticks on the table. It is then cut into small rolls and packed in bonbon cups on flat caramel trays lined with wax paper.

This formula should yield approximately 46 pounds of candy.

CHOCOLATE CARAMEL SAUCE

Lecithin	¼ oz
Sugar	7 lb
Corn Syrup	5 lb
150 Grade Citrus Pectin	2 oz
Fresh 18% Cream	3 qt
Butter	2 lb
Chopped Bitter Chocolate	3 lb
Salt	1 teaspoonful
Vanilla Flavor	2 teaspoonfuls

Two pounds of the sugar are mixed dry with the pectin. One quart of cream is placed in a copper pan on the stove over medium heat. The pectin sugar is added, and the batch is stirred until the sugar is dissolved. The salt, corn syrup, lecithin, remaining sugar and cream, chopped chocolate, and butter are added to the batch which is stirred constantly and cooked to 218°F. It is then removed to a cool spot. The vanilla flavor is stirred in, and the batch is cooled rapidly. The sauce is then poured into a clean stone crock and covered. It is then ready for use.

The batch will yield approximately 10 quarts of chocolate caramel sauce.

CHOCOLATE CARO WHIRLS: *See* CARO WHIRLS, CHOCOLATE.

CHOCOLATE CHERRY CREAMS

Sugar	30 lb
Water	6 qt
Invert Sugar	5 lb
Chopped Glacéed Cherries	5 lb
Citric Acid	2 teaspoonfuls
Cherry Flavor	2 teaspoonfuls
MARSHMALLOW WHIP (*see*)	3 lb
Invertase	2 teaspoonfuls

The sugar, invert sugar, and water are placed in a copper pan on the stove and stirred until dissolved. When the batch begins to boil, it is covered and allowed to steam for a few minutes. The sides are washed down and a thermometer is placed in the batch, which is cooked to 246°F. The batch is poured out on a clean, cooled, dampened cream beater. The top is sprinkled with a few drops of cold water, and the batch is cooled down rapidly to 105°F. The marshmallow whip and invertase are added and the beater is set in motion. While creaming is continued, the cherries, citric acid, and flavor are added. The batch is creamed to a lump and allowed to sweat back for 15 minutes. It is then kneaded soft, put through a cut-roll machine, and formed into small balls. These are spread out on boards and, when dry, they are dipped in high-quality milk chocolate. The creams are allowed to set on the trays and then packed in the usual manner.

The batch will yield approximately 70 pounds of chocolates.

CHOCOLATE COATING: *See* CHOCOLATES.

CHOCOLATE COATING CARAMELS: *See* SAVING WASTE.

CHOCOLATE COCONUT BONBONS: *See* BONBONS, CHOCOLATE COCONUT.

CHOCOLATE COCONUT KISSES: *See* COCONUT KISSES, CHOCOLATE.

CHOCOLATE COCONUT PATTIES: *See* COCONUT PATTIES, CHOCOLATE.

CHOCOLATE COCONUT STARS: *See* COCONUT STARS, CHOCOLATE.

CHOCOLATE COLONIAL NOUGAT: *See* COLONIAL NOUGAT, CHOCOLATE.

CHOCOLATE COMMERCIAL FUDGE: *See* COMMERCIAL FUDGE, CHOCOLATE.

CHOCOLATE CORDIALS

Sugar	10 lb
Water	2 qt
Cream of Tartar	1 teaspoonful
Slow-Set Citrus Pectin	1 oz
Melted Bitter Chocolate	6 oz
Vanilla Flavor	½ teaspoonful
Invertase	½ teaspoonful

The citrus pectin is mixed dry with 3 pounds of sugar and then placed on the stove in a small copper pan which contains 1 qt of water. This mixture is stirred until dissolved. The balance of the sugar, water, and cream of tartar are added and the stirring is continued until these ingredients are completely dissolved. When the batch begins to boil, it is chilled with some cold water. The scum is removed and the sides are washed down. When the batch has reached 228°F, the chocolate is stirred in rapidly. The flavor and invertase are mixed in and the batch is cooled for 5 minutes. It is then cast in bottle-shaped impressions molded in clean, warm, dry starch which are placed in a warm room overnight. The cordials are then removed, brushed carefully, and coated carefully with dark, vanilla sweet chocolate. They are allowed to set and are then wrapped in colored aluminum foil and packed in boxes.

The batch will yield about 20 pounds of cordials.

CHOCOLATE CREAMS, HAND-ROLLED, OLD DUTCH TYPE

Sugar	45 lb
Light 18% Cream	9 qt
Salt	2 teaspoonfuls
Cream of Tartar	3 teaspoonfuls
Melted Corn Syrup	6 lb
Marshmallow Whip	2 lb
Melted Bitter Chocolate	5 lb
Vanilla Flavor	2 teaspoonfuls
Invertase	2 teaspoonfuls

The sugar, cream, and cream of tartar are placed in a large copper pan on the stove and stirred until dissolved. When the batch begins to boil, the sides are washed down. Stirring is continued and the batch is cooked to a soft ball or about 246°F. The melted corn syrup is then stirred in rapidly and the batch is poured out on a clean, cooled, dampened cream beater and cooled rapidly for about 7 minutes or to 110°F. The beater is set in motion and, while it is in operation, the marshmallow, chocolate, flavor, and invertase are added gradually. When the batch creams to a lump, it is allowed to sweat back for 15 minutes. It is then kneaded smooth and passed through the cut roll machine. The creams are allowed to cool on trays, and they are then coated with high-quality milk chocolate and packed as usual.

This batch should yield about 65 pounds of cream centers.

Note: For approximate yield of chocolates, *see* HAND ROLLED CREAM CENTERS, VANILLA NUT.

CHOCOLATE CREMO FUDGE: *See* CREMO FUDGE, CHOCOLATE.

CHOCOLATE CRYSTALLIZED MINTS: *See* CRYSTALLIZED MINTS, CHOCOLATE.

CHOCOLATE DE LUXE CARAMELS: *See* DE LUXE CARAMELS, CHOCOLATE.

CHOCOLATE DELIGHTS

Chopped PEANUT-BRITTLE (*see*) Crumbs	4 lb
Small Seedless Raisins	6 lb
High-Quality Milk Chocolate	10 lb
Toasted Coconut	2 lb

The toasted coconut, seedless raisins, and nut crumbs are thoroughly mixed in a large square tin. The milk chocolate is then processed in the same way as for clusters (*see* CHOCOLATE ALMOND CLUSTERS) and the mix is coated into small clusters. These are deposited on wax-paper-lined dipping trays. The candies are allowed to cool until set and are then packed in boxes.

This formula should yield approximately 22 pounds of candy.

CHOCOLATE DEPARTMENT: *See* SAVING WASTE.

CHOCOLATE ENROBER

This machine is of 34-inch size and covers all sorts of centers with chocolate. (Courtesy of the National Equipment Corporation, New York, New York.)

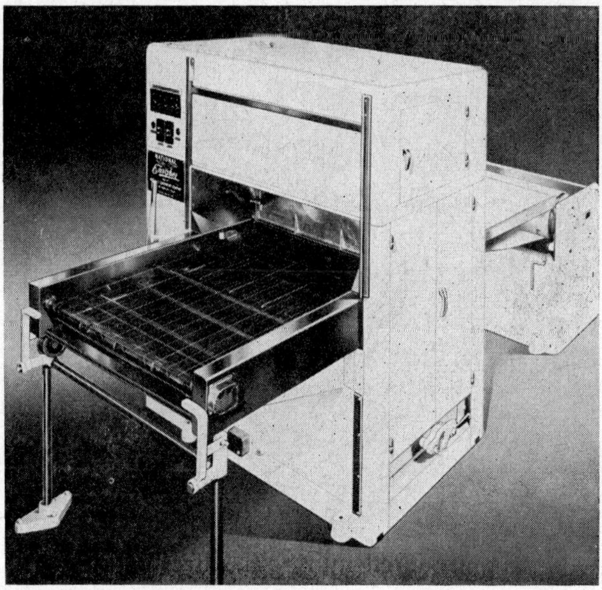

CHOCOLATE FRENCH CHEWING TAFFY: *See* FRENCH CHEWING TAFFY, CHOCOLATE.

CHOCOLATE FRENCH CRISP: *See* FRENCH CRISP, CHOCOLATE.

CHOCOLATE FRUIT MELLOWS: *See* FRUIT MELLOWS, CHOCOLATE.

CHOCOLATE FUDGE FOR COATING: *See* SAVING WASTE.

CHOCOLATE FUDGE SAUCE

Sugar	4 lb
Lecithin	1/4 oz
Corn Syrup	1 lb
Water	1/2 qt
Salt	1 teaspoonful
Fresh 18% Cream	1 1/2 qt
Chopped Bitter Chocolate	1 1/2 lb
Vanilla Flavor	1 teaspoonful
Glycerin	4 teaspoonfuls

The sugar, lecithin, corn syrup, water, and salt are placed in a copper pan on the stove and stirred until dissolved. The cream is stirred in gradually, and the batch is stirred and cooked to 224°F. The batch is removed to a cool spot and the chocolate is stirred in rapidly. When the chocolate is completely melted, the flavor and glycerin are stirred in. The sauce is cooled off, poured into a clean stone crock, and covered. It is then ready for use.

The batch will yield approximately 4 quarts of chocolate fudge sauce.

CHOCOLATE HOLLOW EASTER EGGS, RABBITS, ETC.

Dark Vanilla Sweet Chocolate	20 lb
Best-Grade Milk Chocolate	20 lb

The metal molds are heated enough to take the chill off. The chocolate is melted and cooled in the usual manner for chocolate coating. It is then poured into both halves of the molds. The molds are then shaken and turned upside down to empty the surplus chocolate and form a shell of chocolate in the mold. They are then moved to a cool room and allowed to remain there until very cold. The edges of the molds are bent slightly and the shell eggs are removed carefully. They are placed on clean trays and wrapped in tissue paper, ready for decorating or filling with chocolates. This process is continued until all the chocolate is used up. The same process is followed for solid Easter animals.

The batch will yield 40 pounds of hollow easter eggs, etc.

CHOCOLATE HONEY TABLETS

Sugar	20 lb
Corn Syrup	5 lb
Water	3 qt
Honey	2 lb
Bitter Chocolate	2 lb
Vanillin Powder	1 g

The sugar, water, and corn syrup are placed in a copper pan on the stove and stirred until dissolved. When the batch begins to boil, it is covered and allowed to steam for a few minutes. The sides are washed down and a thermometer is placed in the batch, which is cooked to 332°F. The heat is turned off. After the honey, chocolate, and flavor are stirred into the batch, the heat is turned on again, and the batch is stirred and cooked for another minute. Then it is poured out on a greased slab, and the edges are folded together, while the batch is cooling. The kneading is continued until the batch is cool enough; then it is taken over to the spinning table in front of the batch warmer. It is then passed through a tablet roller machine and allowed to cool on the table. The tablets are broken apart and put through a sieve and packed in tins.

This batch will yield approximately 27 pounds.

CHOCOLATE ICE CREAM NO. 1

Fresh 20% Cream	8 qt
Fresh 40% Cream	2 qt
Condensed Milk	4 qt
Chopped Bitter Chocolate	1½ lb
Sugar	12 lb
Pure Vanilla Flavor	1 oz

The chopped chocolate is melted in a double boiler on the stove. One quart of the light cream is stirred into the chocolate slowly and mixed to a paste. The creams, condensed milk, and sugar are placed in a steam-bath copper pan on the stove and stirred until dissolved. The chocolate paste is then added and the flavor is stirred in. The mix is then strained and cooled for about 1½ hours. It is then poured into the freezing machine and finished as usual. The same freezing process is used as for VANILLA ICE CREAM NO. 1 (*see*).

The batch will yield approximately 30 quarts.

CHOCOLATE ICE CREAM NO. 2

Fresh 20% Cream	16 qt
Fresh 40% Cream	2 qt
Condensed Milk	2 qt
Chopped Bitter Chocolate	1½ lb
100-Grade Citrus Pectin	3 oz
Pure Vanilla Flavor	1 oz
Sugar	12 lb

The pectin is mixed dry with 2 pounds of the sugar and placed in a double boiler on the stove with 2 quarts of the light cream. This mixture is stirred until it dissolves and begins to boil. The bitter chocolate is then stirred in slowly, mixed to a paste, and set aside. The light and heavy cream, condensed milk, and sugar are placed in a steam-bath copper pan on the stove and stirred until dissolved. The chocolate paste and flavor are stirred in, and the mix is strained and cooled as usual. It is then poured into the freezer, following the same process as under VANILLA ICE CREAM NO. 1 (*see*).

The batch will yield approximately 40 quarts.

CHOCOLATE ITALIAN CARAMELS: *See* ITALIAN CARAMELS, CHOCOLATE.

CHOCOLATE LADY CARAMELS: *See* LADY CARAMELS, CHOCOLATE.

CHOCOLATE LONDON CARAMELS: *See* LONDON CARAMELS, CHOCOLATE.

CHOCOLATE MARSHMALLOW PAN CARAMELS: *See* PAN CARAMELS, CHOCOLATE MARSHMALLOW.

CHOCOLATE MELTING AND MIXING MACHINE

This machine automatically melts and mixes chocolate for coating. (Courtesy of the Savage Brothers Co., Chicago, Illinois.)

CHOCOLATE MINT WAFER SANDWICHES

Sugar	40 lb
Water	9 qt
Silver Sheet Gelatin	1 oz
(soaked in two cups of warm water)	
Mint Flavor	5 teaspoonfuls
Coloring	Enough to yield pastel shade

The sugar and water are placed in a clean copper pan on the stove and stirred until dissolved. When the batch begins to boil, it is covered and allowed to steam for a few minutes. The sides are washed down and a thermometer is placed in the batch, which is cooked to 245°F. It is then poured out on a dampened, cool, clean cream beater and the top is sprinkled with a few drops of cold water. After the batch is cooled down rapidly to about 105°F, the soaked gelatin is added and the machine is set in motion. Creaming is continued until the batch forms a lump. Then it is allowed to sweat back for 15 minutes, placed in a steam-bath copper pan, and stirred continuously until remelted. The peppermint flavor is then stirred in rapidly and the cream is cast through a funnel on heavy wax paper which is spread out on the table. The cream is molded into medium-sized wafers, which are allowed to cool on the table. The wafers are then turned upside down and the chocolate is enrobed on them by hand. The chocolate is placed on the bottom of each wafer and covered up with another wafer, sandwich fashion. This process is continued until all the wafers have been used up. They are then allowed to set and packed in flat boxes.

Note: These may be made in pink, green or any other color desired. The cream of this formula is known as French cream.

The batch will yield approximately 55 pounds of chocolate mint wafer sandwiches.

CHOCOLATE NOUGAT PATTIES

Sugar	14 lb
Corn Syrup	6 lb
Water	3 qt
Coconut Butter	8 oz
Roman Punch Flavor	3 teaspoonfuls
MARSHMALLOW WHIP (*see*)	6 lb
Assorted Diced Fruits	5 lb

The sugar, corn syrup, water, and coconut butter are placed in a copper pan on the stove and stirred until dissolved. When the batch begins to boil, it is covered and allowed to steam for a few minutes. The sides are washed down, and a thermometer is placed in the batch, which is cooked to 254°F. The batch is removed from the stove and the marshmallow is stirred in gradually. The flavor is added while the diced fruits are stirred in. The batch is then dipped out on boards with a dampened ice-cream dipper to form patties, approximately 1 ounce each. The boards used for the patties are first lined with wax paper. After the nougat patties are allowed to set overnight, they are dipped in a high-quality milk chocolate. They are allowed to set on the dipping trays and are then wrapped and packed in boxes.

The batch will yield approximately 50 pounds of chocolate nougat patties.

CHOCOLATE NUT BOARDWALK FUDGE: *See* BOARDWALK FUDGE, CHOCOLATE NUT.

CHOCOLATE NUT BUTTERSCOTCH WAFERS: *See* BUTTERSCOTCH WAFERS, CHOCOLATE NUT.

CHOCOLATE NUT CREMO FUDGE: *See* CREMO FUDGE, CHOCOLATE NUT.

CHOCOLATE NUT DE LUXE CARAMELS: *See* DE LUXE CARAMELS, CHOCOLATE NUT.

CHOCOLATE NUT FUDGE

Sugar	20 lb
Corn Syrup	10 lb
Fresh 18% Light Cream	6 qt
Salt	1 tablespoonful
Vanilla Flavor	2 teaspoonfuls
Bitter Chocolate	4 lb
Caramel Paste	4 lb
Cream Fondant	18 lb
Black Walnuts	4 lb

The sugar, corn syrup, salt, and cream are placed in a large copper pan on the stove and stirred until dissolved. Stirring is continued and the batch is cooked until it forms a soft ball when tested in cold water. The batch is moved to a stand near the slab, and the caramel paste is stirred in while mixing in the bitter chocolate. The cream is then added and stirred until completely melted. Then the flavor is stirred in, the nuts are added and mixed thoroughly, and the batch is poured out on boards lined with wax paper. It is allowed to set overnight and then cut and packed as desired.

This batch should yield 65 pounds of candy.

CHOCOLATE NUT LONDON CARAMELS: *See* LONDON CARAMELS, CHOCOLATE NUT.

CHOCOLATE NUT SEA-FOAM FUDGE: *See* SEA-FOAM FUDGE, CHOCOLATE NUT.

CHOCOLATE NUT SEA-FOAM KISSES: *See* SEA-FOAM KISSES, CHOCOLATE NUT.

CHOCOLATE PAN CARAMELS: *See* PAN CARAMELS, CHOCOLATE.

CHOCOLATE PASTILLES, DUTCH TYPE, PANNED

Center

Sugar	40 lb
Cream of Tartar	3 teaspoonfuls
Salt	2 teaspoonfuls
Fresh 20% Light Cream	8 qt
Corn Syrup	10 lb
Melted Bitter Chocolate	4 lb
Vanilla Flavor	3 teaspoonfuls

The sugar, cream of tartar, salt, and the fresh cream are placed in a large copper pan on the stove and stirred until dissolved. When the batch begins to boil, the sides are washed down. Stirring is continued and the batch is cooked to 258°F or to a soft ball. The melted corn syrup is then stirred in and mixed through thoroughly. The batch is poured out on a dampened open-type cream beater and the water is turned on to keep the cream beater cool. The batch is allowed to cool about 7 to 8 minutes, and then the melted chocolate and the vanilla flavor are added. The beater is set in motion and kept running until the batch creams. It is then allowed to sweat back for about 1 hour, kneaded into a smooth batch, and put through a hand-roll machine in small pieces (about three hundred to the pound). These are allowed to set overnight. Then they are ready for panning.

For panning and polishing the directions for ALMOND DAINTIES (*see*) are followed. The candies are handled and finished as usual.

CHOCOLATE PEACH CREAMS

Sugar	30 lb
Cream of Tartar	2 teaspoonfuls
Water	7 qt
Melted Corn Syrup	4 lb
Peach Pulp	3 lb
Peach Flavor	3 teaspoonfuls
Citric Acid	2 teaspoonfuls
MARSHMALLOW WHIP (*see*)	2 lb
Invertase	2 teaspoonfuls

The sugar, cream of tartar, and water are placed in a copper pan on the stove and stirred until dissolved. When the batch begins to boil, it is covered up and allowed to steam for a few minutes. The sides are washed down and a thermometer is placed in the batch, which is cooked to 250°F. The melted corn syrup is stirred in rapidly, and the batch is poured out on a clean, cooled, dampened cream beater. The top is sprinkled with cold water and cooled rapidly to 110°F. The marshmallow whip and invertase are added and the beater is set in motion. The peach pulp, flavor, and citric acid are added while creaming is continued. The batch is creamed to a lump and allowed to sweat back for 15 minutes. It is then kneaded until soft, put through the cut-roll machine, and formed into small balls. After drying on boards, these balls are coated with dark vanilla sweet chocolate. They are then packed in the usual way.

The batch will yield approximately 65 pounds of chocolates.

CHOCOLATE PEANUT BUTTER KISSES

Sugar	8 lb
Corn Syrup	12 lb
Water	1 qt
Salt	1 teaspoonful
Coconut Butter	8 oz
Bitter Chocolate (cut fine)	2 lb
Vanilla Flavor	1 teaspoonful
Peanut Butter	6 lb

The sugar, corn syrup, water, salt, and coconut butter are placed in a copper pan on the stove and stirred until dissolved.

When the batch begins to boil, it is covered up and allowed to steam for a few minutes. The sides are washed down and a thermometer is placed in the batch, which is cooked to 254°F. The bitter chocolate and flavor are stirred in and, when completely melted, the batch is poured out on a cold, greased slab between iron bars. The edges are folded together and the batch is cooled rapidly. When cool enough, it is taken to the pulling machine and pulled for about 5 minutes. It is then transferred to the spinning table and formed out flat. The peanut butter is placed in the center, the batch is wrapped over, the ends are closed, and the batch is formed round. It is then spun out, cut into kisses, and wrapped as usual.

This batch will yield approximately 30 pounds of candy.

CHOCOLATE PEANUT CLUSTERS

Raw Spanish No. 1 peanuts are roasted in soft coconut oil to a light-brown color. They are spread out on heavy brown packing paper that absorbs the grease and left on the paper until they become cold. When they are ready for coating, the following method is considered effective. A small portion of chocolate (about 1 pound) is cooled off in the center of the coating marble. About 1 pound of the peanuts is mixed in to the cooled chocolate, which is kept in a pile. This pile is dipped out into small clusters, which are deposited on the dipping trays. This process is continued until all the chocolate and peanuts have been used up. The clusters are then cooled and packed.

Note: For 20 pounds of chocolates, use 10 pounds of peanuts and 10 pounds of coating.

CHOCOLATE PEANUT CRUNCH

Sugar	19 lb
Corn Syrup	5 lb
Salt	2 teaspoonfuls
Chopped Roasted Peanuts	5 lb
Bicarbonate of Soda	3 teaspoonfuls
Butter	1 lb
Water	3 qt

The sugar, corn syrup, salt, and water are placed in a copper pan on the stove and stirred until dissolved. When the batch begins to boil, it is covered and allowed to steam for a few minutes. The sides are washed down and a thermometer is placed in the batch, which is cooked to 290°F. The butter is stirred in and the peanuts are added while the batch is stirred continuously. It is cooked until it forms a hard crisp ball (*see* COLD-WATER TEST FOR CRISP HARD CANDY) when tested in cold water and is then poured out on a warm, greased slab and spread out to a thin sheet. This sheet is cut rapidly with an adjustable cutter into 1¾ inch squares. These are broken apart and allowed to cool on the boards. They are then dipped in dark vanilla sweet chocolate and the tops are sprinkled with a liberal amount of roasted ground peanuts. After setting, the crunch is packed in boxes.

The batch will yield approximately 55 pounds of chocolate peanut crunch.

CHOCOLATE PIGNOLIA CLUSTERS

Pignolia nuts which are grown mostly in Italy make an unusually fine candy. The nuts are usually roasted to a medium brown, cooled, and coated with a fine milk chocolate, following the almond clusters process (*see* CHOCOLATE ALMOND CLUSTERS).

Two pounds of chocolate will coat 1 pound of pignolia nuts.

CHOCOLATE RABBITS, CHICKS, ETC.

Sugar	10 lb
Water	2 qt
COCONUT CREAM (*see*)	20 lb

The molds used for these Easter animals are made of plaster of Paris in two halves. The molds are first cleansed thoroughly in warm water and then soaked in cold water. They are then lined up in an erect position, close to each other, with heavy weights on both ends, so that they remain close to each other.

The sugar and water are placed in a copper pan on the stove and stirred until dissolved. The sides are washed down and the batch is cooked to 240°F. It is then removed from the stove and the coconut cream is stirred in and mixed through thoroughly. The melted cream is placed in a funnel with a large opening and it is run into the soaked molds. Each mold is filled to the top. This casting process is continued until all the molds are filled. The edges are then trimmed with a small palette knife, and the molds are allowed to cool.

When the molds are opened, the rabbits are removed carefully and stood up on boards lined with wax paper. The molds are then again soaked in cold water to prepare for another operation and the same method is followed until all the cream is used up. A small knife is used to shave off the rough surfaces on each rabbit. They are then coated in pure bitter-sweet chocolate and allowed to set. They are then ready for decorating.

The batch will yield approximately 45 pounds of chocolate rabbits, etc.

CHOCOLATE RAISIN CLUSTERS

Fifteen pounds of small seedless raisins are soaked in medium-hot water for 10 minutes. They are drained in a sieve overnight. They are then dipped in a high-quality milk chocolate in small clusters, following the process used for CHOCOLATE ALMOND CLUSTER (*see*).

Fifteen pounds of raisins will use up 15 pounds of chocolate to make quality raisin clusters.

CHOCOLATE SALT-WATER TAFFY: *See* SALT-WATER TAFFY, CHOCOLATE.

CHOCOLATE SAUCE

Lecithin	¼ oz
Sugar	4½ lb
Corn Syrup	1½ lb
Water	½ qt
Salt	½ teaspoonful
Vanilla Flavor	½ teaspoonful
Chopped Bitter Chocolate	1 lb

The sugar, lecithin, corn syrup, water, and salt are placed in a small copper pan on the stove and stirred until dissolved. When the batch begins to boil, the sides are washed down and the batch is cooked to 224°F. It is then removed from the stove, and the chocolate and flavor are stirred in rapidly. The sauce is then cooled off and is ready to use.

The batch will yield approximately 4 quarts of chocolate sauce.

CHOCOLATE SCRAP: *See* SAVING WASTE.

CHOCOLATE SEA-FOAM FUDGE: *See* SEA-FOAM FUDGE, CHOCOLATE.

CHOCOLATE SNOW BALLS

Sugar	40 lb
Water	8 qt
Cream of Tartar	3 teaspoonfuls
MARSHMALLOW WHIP (*see*)	5 lb
Invertase	2 teaspoonfuls
Vanilla Flavor	4 teaspoonfuls
Macaroon Coconut	20 lb

The sugar, water, and cream of tartar are placed in a copper pan on the stove and stirred until dissolved. When the batch begins to boil, it is covered and allowed to steam for a few minutes. The sides are washed down and a thermometer is placed in the batch, which is cooked to 242°F. The batch is poured out on a clean, cooled, dampened cream beater and the top is sprinkled with a few drops of cold water. It is cooled rapidly to about 100°F and the marshmallow whip, invertase, and flavor are added. The beater is set in motion and allowed to run until the batch is creamed to a lump. Then the batch is allowed to sweat back for 20 minutes, placed in a steam-bath copper pan for remelting, and stirred over medium heat until it is melted sufficiently thin to run through a funnel. The batch is then cast into impressions of round shape with a pointed top. The balls are molded in clean, dry starch and allowed to set overnight. They are then removed, brushed free of starch, and dipped in a dark vanilla sweet chocolate. They are then rolled in fine macaroon coconut and allowed to set on the trays. The snow balls are packed in stock boxes.

The batch will yield approximately 85 pounds of chocolates.

CHOCOLATE STRAWS, LIME

The CHOCOLATE STRAWS, VANILLA (*see*) formula is duplicated, but the batch is colored light green and flavored with oil of lime. It is completed in the usual manner.

The batch will yield approximately 31 pounds.

CHOCOLATE STRAWS, STRAWBERRY

The CHOCOLATE STRAWS, VANILLA (*see*) formula is followed, but the batch is colored a light pink and flavored with strawberry flavor. It is completed as usual and will yield approximately 31 pounds.

CHOCOLATE STRAWS, VANILLA

Center

Bitter Liquor Chocolate	5 lb
6X Powdered Sieved Sugar	7 lb
Vanillin Powder	1 g

The chocolate and vanillin are placed in a clean double-jacket copper kettle over low heat and stirred until melted. The sieved powdered sugar is then stirred in and mixed to a stiff paste. The center is kept warm until needed.

Jacket

Sugar	20 lb
Water	3½ qt
Cream of Tartar	2 teaspoonfuls
Color	½ teaspoonful
Flavor	3 teaspoonfuls

The sugar, water, and cream of tartar are placed in a copper pan on the stove and stirred until dissolved. When the batch begins to boil, it is covered and allowed to steam for a few minutes. The sides are washed down and a thermometer is placed in the batch, which is cooked to 336°F. The batch is then poured out on a greased slab, and the edges are folded together while the vanilla flavor is stirred in. This folding and kneading process is continued until the batch is cool enough, when it is taken to the hook, pulled partly, and twisted to a glossy silvery color. It is then transferred to the spinning table in front of the batch warmer, where it is kneaded and formed into a flat batch large enough to hold the center. The warm chocolate center is then placed in the middle of the flat batch, and the edges are wrapped around and the ends closed up. The batch is shaped round and spun out in long strings of fine sticks. These are passed through a cutting machine and cut to about 4 inch lengths. After cooling on the table, the sticks are packed into tins.

The batch will yield approximately 31 pounds.

CHOCOLATE STRAWS, VIOLET

The CHOCOLATE STRAWS, VANILLA (*see*) formula is followed. The batch is colored bright purple or violet and flavored with oil of violet. It is completed and packed as usual.

The batch will yield approximately 31 pounds.

CHOCOLATE SYRUP NO. 1

Chopped Bitter Chocolate	3 lb
Lecithin	½ oz
Boiling Water	1½ qt
Fresh 18% Cream	3 qt
Sugar	10 lb
Salt	1 teaspoonful
Vanilla Flavor	4 teaspoonfuls

The chocolate is stirred into the hot water until it melts and turns into a paste. The fresh cream is mixed with the chocolate paste until the paste is dissolved. The sugar is then stirred in rapidly, and the lecithin and salt are added. The batch is stirred and cooked to 218°F. The batch is then removed from the stove, cooled, poured into a clean stone crock, and covered. The syrup is now ready for use.

The batch will yield approximately 10 quarts of chocolate syrup.

CHOCOLATE SYRUP NO. 2

Dutch Processed Cocoa Powder	1½ lb
Fresh 18% Cream	4 qt
Salt	1 teaspoonful
Sugar	10 lb
Vanilla Flavor	3 teaspoonfuls
Lecithin	¼ oz

One quart of the cream is heated, and the cocoa powder is mixed with this hot cream until it forms a paste. The sugar, lecithin, balance of the cream, and salt are placed in a copper pan on the stove. The cocoa paste is added, and the batch is stirred over low heat and cooked to 218°F. The batch is removed and cooled.

Then it is poured into a clean stone crock, covered, and used as desired.

The batch will yield approximately 9 quarts of chocolate syrup.

CHOCOLATE SYRUP NO. 3

Chopped Bitter Chocolate	2 lb
Hot Water	1½ qt
Simple Syrup	4 qt
Pure Vanilla Extract	4 teaspoonfuls
Glycerin	4 teaspoonfuls

The chocolate is stirred into the hot water until it melts and mixes to a thick paste. The simple syrup is poured into a clean copper pan on the stove, and the chocolate paste is added. The batch is stirred over low heat until the chocolate is dissolved. It is then removed from the stove and cooled. The flavor and glycerin are stirred in, and the syrup is poured into a clean stone crock and covered. It is then ready for use.

The batch will yield approximately 6 quarts of chocolate syrup.

CHOCOLATE TOFFEE, OLD DUTCH

Sugar	10 lb
Corn Syrup	10 lb
Coconut Butter	8 oz
Salt	1 tablespoonful
Fresh Cream	2 qt
Evaporated Milk	2 lb
Butter	1 lb
Bitter Chocolate (cut fine)	2½ lb
Vanilla Flavor	3 teaspoonfuls

The sugar, corn syrup, salt, coconut butter, and fresh cream are placed in a copper pan on the stove and stirred until dissolved. When the batch begins to boil, the evaporated milk is stirred in, and the stirring is continued while the butter is added. The batch is cooked until it forms a medium ball when tested in cold water (*see* COLD-WATER TEST) . The chocolate is then stirred in and, while it melts, the vanilla is mixed in. The batch is then poured out on a greased, cold slab and the edges are folded together. The batch is cooled rapidly and, when it is firm enough to handle, it is cut and wrapped as usual.

This batch will yield approximately 25 pounds of candy.

CHOCOLATE VARIETY FUDGE: *See* VARIETY FUDGE, CHOCOLATE.

CHOCOLATE WHIPPED-CREAM CENTERS

Sugar	35 lb
Water	7 qt
Cream of Tartar	2 teaspoonfuls
Melted Corn Syrup	5 lb
Marshmallow Whip	7 lb
Vanilla Flavor	4 teaspoonfuls
Invertase	3 teaspoonfuls

The sugar, water, and cream of tartar are placed in a copper pan on the stove and stirred until dissolved. When the batch begins to boil, it is covered and allowed to steam for a few minutes. The sides are washed down and a thermometer is placed in the batch, which is cooked to 248°F. The melted corn syrup is stirred in rapidly, the batch is poured out on a clean, cooled, dampened cream beater, and the top is sprinkled with a few drops of cold water. The batch is cooled down rapidly to 100°F, and the marshmallow whip, flavor, and invertase are added. The machine is set in motion and operation continued until the batch creams to a lump. The batch is allowed to sweat back for 15 minutes and is then transferred to the working table, which has been dusted with flour. The batch is kneaded until soft and is then put through the cut-roll machine. The centers are spread out on boards and, when cooled off and dry, they are coated with dark vanilla semisweet chocolate. The creams are allowed to set on the trays and they are then packed in the usual manner.

This formula should yield about 55 pounds of centers or 75 pounds of hand-dipped chocolates.

CHOCOLATE-COVERED ALMONDS

Almonds singly dipped are very expensive chocolates and should be handled with the greatest of care. The Valencia three-crown type is the preferred almond. The almonds must be roasted to a light-brown crisp and kept in airtight cans until used. Only the highest-grade dark sweet or milk chocolate should be used for dipping. The chocolate is cooled in small portions and adjusted to a rather thick coating with a few drops of water stirred into the coating. The almonds are then mixed into the chocolate, about a handful at a time, and picked out one at a time with the proper amount of chocolate around them. Each almond is deposited on the dipping tray and finished off with a string mark in the center.

It will take about 3 pounds of chocolate to coat 1 pound of almonds.

CHOCOLATE-COVERED APRICOTS

Fancy dry apricots are used and they are treated the same way as the CHOCOLATE COVERED FIGS (*see*). After they are drained and cooled, they are dipped in dark vanilla sweet chocolate.

Note: For 20 pounds of chocolates, use 10 pounds of apricots and 10 pounds of coating.

CHOCOLATE-COVERED BRAZIL NUTS

The same coating process is used as for CHOCOLATE-COVERED ALMONDS (*see*). The best grade of whole brazil nuts is preferred. Dark, sweet vanilla or milk chocolate coating may be used. A finishing mark of a long narrow ring over the entire length of the nut is the standard.

Two pounds of chocolate will coat 1 pound of Brazils.

CHOCOLATE-COVERED BUTTERSCOTCH

Sugar	12 lb
Corn Syrup	8 lb
Salt	2 teaspoonfuls
Coconut Butter	1 lb
Water	2 qt
Butter	1½ lb
Evaporated Milk	2 lb
Vanilla Flavor	3 teaspoonfuls

The sugar, corn syrup, coconut butter, salt, and water are placed in a copper pan on the stove and stirred until dissolved. When the batch begins to boil, it is covered and allowed to steam for a few minutes. The sides are washed down and a thermometer is placed in the batch, which is cooked to 260°F. The batch is then reduced with evaporated milk so that it will form a soft ball when tested in cold water (*see* COLD-WATER TEST). It is then poured out on a cold, greased slab between iron bars and spread out to ¼ inch thickness. When cool and hard, the batch is sliced and cut into small oblongs. These are coated with dark sweet vanilla chocolate and allowed to set on the dipping trays. The butterscotch is packed in stock boxes.

The batch will yield approximately 33 pounds of chocolates.

CHOCOLATE-COVERED BUTTER TOFFEE

Dairy Butter	3½ lb
Coconut Butter	2½ lb
Salt	2 teaspoonfuls
Sugar	25 lb
Corn Syrup	5 lb
Evaporated Milk	5 lb
Vanilla Flavor	3 teaspoonfuls
Oil of Lemon	¼ teaspoonful

The butter, salt, and coconut butter are placed in a copper pan on the stove over low heat and stirred until melted. The sugar and corn syrup are added and stirred until dissolved. When the batch begins to boil, it is covered and allowed to steam for a few minutes. The sides are washed down and a thermometer is placed in the batch, which is cooked to 290°F. The batch is then reduced with evaporated milk until it forms a soft ball when tested in cold water. The flavors are stirred in and mixed through well. The batch is poured out on a cold, greased slab between iron bars and spread out to about ¼ inch thickness. Then it is allowed to cool and, when hard enough, it is sliced and cut into small oblongs. These are coated with dark sweet vanilla chocolate. After they have been allowed to set on the trays, they are packed in the usual manner.

The batch will yield approximately 60 pounds of chocolates.

CHOCOLATE-COVERED CHERRIES

Whole glacéed cherries are preferred. They are first drained overnight and then coated heavily to prevent leaking in a high-quality milk chocolate. They are finished with a tight curl mark and otherwise completed in the usual manner.

It will take 2 pounds of chocolate to coat 1 pound of cherries.

CHOCOLATE-COVERED COCONUT CARAMELS

Sugar	12 lb
Corn Syrup	18 lb
Coconut Butter	1 lb
Water	2 qt
Macaroon Coconut	16 lb
Vanilla Flavor	4 teaspoonfuls

The sugar, corn syrup, coconut butter, and water are placed in a large copper pan on the stove and stirred until dissolved. When the batch begins to boil, the sides are washed down and the batch is cooked to 232°F. The flavor is added and the coconut is stirred in gradually in order to form a thick paste. The batch is then turned out on a greased slab between iron bars and spread out evenly to about ½ inch thickness. It is allowed to cool until hard; then it is sized, sliced, and cut into small squares. These are coated with a dark vanilla sweet chocolate. After setting, the caramels are packed in the usual manner.

The batch will yield approximately 75 pounds of chocolates.

CHOCOLATE-COVERED COCONUT-CREAM BARS NO. 1

Sugar	25 lb
Corn Syrup	5 lb
Water	4 qt
Macaroon Coconut	6 lb
FRENCH CREAM (*see*)	12 lb
Vanilla Flavor	4 teaspoonfuls
Invertase	1 teaspoonful

The sugar, corn syrup, and water are placed in a copper pan on the stove and stirred until dissolved. When the batch begins to boil, it is covered and allowed to steam for a few minutes. The sides are washed down and a thermometer is placed in the batch, which is cooked to 238°F. The batch is removed from the stove and the flavor is added. The french cream is added gradually and stirred until it is completely melted. The macaroon coconut and the invertase are added and mixed through thoroughly. The batch is poured out on a wax-paper-lined slab between iron bars and spread out evenly to about 3/4 inch thickness. It is then marked and cut into bars of about 3 1/2 inch length and 1 1/2 inch width. The bars are allowed to dry and then dipped in pure bitter-sweet chocolate. After they are set, the bars are wrapped and packed in the usual manner.

The batch will yield approximately 70 pounds of coconut-cream bars.

CHOCOLATE-COVERED COCONUT-CREAM BARS NO. 2

Sugar	16 lb
Corn Syrup	4 lb
Water	3 qt
Vanilla Flavor	4 teaspoonfuls
RUNNING CREAM (*see*)	6 lb
Macaroon Coconut	6 lb

The sugar, corn syrup, and water are placed in a large copper pan on the stove and stirred until dissolved. When the batch begins to boil, it is covered and allowed to steam for a few minutes. The sides are washed down and a thermometer is placed in the batch, which is cooked to 244°F. The batch is then moved near the slab and the cream stirred in until it has melted. The macaroon coconut is stirred in gradually and the flavor added. The batch is mixed thoroughly and then turned out on boards lined with heavy wax paper to approximately 1/2 inch thickness. After setting, the boards are turned over on the table, the paper is pulled off, and the coconut candy is cut into bars to the size desired. These are coated with bitter-sweet chocolate and wrapped and packed as usual.

This formula should yield approximately 45 pounds of chocolate-coated bars.

CHOCOLATE-COVERED COCONUT NOUGAT

Sugar	11 lb
Corn Syrup	14 lb
Water	2 qt
Gelatin (soaked in 2 cups of water)	2 oz

Coconut Butter	1 lb
MARSHMALLOW WHIP (*see*)	5 lb
Macaroon Coconut	5 lb
Vanilla Flavor	4 teaspoonfuls

The sugar, corn syrup, water, and coconut butter are placed in a copper pan on the stove and stirred until dissolved. When the batch begins to boil, it is covered and allowed to steam for a few minutes. The sides are washed down, and a thermometer is placed in the batch, which is cooked to 254°F. The gelatin is added and stirred constantly until completely dissolved. The marshmallow whip is then stirred in and beaten until fluffy. The batch should be adjusted to form a soft ball when tested in cold water (*see* COLD-WATER TEST). The coconut and flavor are stirred in, and the batch is poured out on a cooled, greased slab between iron bars and spread out evenly to approximately ¼ inch thickness. It is allowed to cool until hard; then it is sliced, cut into small oblongs, and coated with dark vanilla sweet chocolate. The nougats are packed in the usual manner.

The batch will yield approximately 55 pounds of chocolates.

CHOCOLATE-COVERED DATE DAINTIES

Soft Pitted Dates	10 lb
Cream Fondant (*see* FRENCH CREAM)	10 lb
Walnut Pieces	2 lb

The walnut pieces are kneaded into the fondant, and they are then rolled out and cut in small oblong pieces. The pitted dates are then filled with the cream oblongs and placed on boards lined with wax paper. When thoroughly dry, they are coated with high-quality milk chocolate and allowed to set. The dainties are then packed in boxes.

This formula should yield 35 pounds of chocolates.

CHOCOLATE-COVERED DATES

The best soft juicy dates are used. Each date is stuffed with a piece of walnut and is coated with high-quality milk chocolate, The standard zig-zag string is used.

Note: For 20 pounds of chocolates, use 10 pounds of dates, 2 pounds of nuts, and 8 pounds of chocolate.

CHOCOLATE-COVERED FIG PASTE

Soft Pulled Figs	12 lb
Sugar	5 lb
Water	½ qt
Corn Syrup	10 lb
Oil of Orange	2 teaspoonfuls
Sifted Powdered Sugar	5 lb

The figs are put through a meat-grinding machine. The sugar, water, and corn syrup are placed in a copper pan on the stove and stirred until dissolved. The figs are added gradually and stirred constantly over a slow fire. The batch is cooked until it forms a soft ball when tested in cold water (*see* COLD-WATER TEST). It is then removed from the stove. Then the flavor is added and the powdered sugar stirred in. The paste is then turned out on the slab, which has been dusted with some flour, and it is rolled out approximately to ¼ inch thickness. It is sliced and cut into small oblongs, placed on boards lined with wax paper, and allowed to dry for 3 hours. The paste is then coated with high-quality milk chocolate.

The batch will yield approximately 48 pounds of chocolates.

CHOCOLATE-COVERED FIGS

Oriental pressed or pulled figs are cut up in medium-size pieces and stewed for 5 minutes in a hot syrup of 3 pounds of sugar and 1 quart of water over low heat. The figs are then strained and allowed to drain until cool. They are coated with fine milk chocolate.

For 20 pounds of chocolates, use 10 pounds of figs and 10 pounds of coating.

CHOCOLATE-COVERED FILBERTS

The process for CHOCOLATE-COVERED ALMONDS (*see*) is used. Turkish Levant filberts are preferred, although satisfactory results may be obtained with the Oregon filberts also. The nuts are roasted to a medium brown and the skins are removed. They are usually dipped three together, forming a triangle, and finished with a small, tight curl mark over the top.

Two pounds of chocolate will coat 1 pounds of filberts.

CHOCOLATE-COVERED FRUIT-PUNCH CREAMS

Sugar	25 lb
Invert Sugar	2 lb
Fresh Light Cream	6 qt
Melted Corn Syrup	3 lb
Silver Sheet Gelatin	1 oz
(dissolved in 2 cups of warm water)	
Assorted Diced Fruits	4 lb
Roman Punch Flavor	4 teaspoonfuls

The sugar, invert sugar, and cream are placed in a large copper pan on the stove and stirred until dissolved. The batch is stirred continuously and cooked until it forms a medium soft ball when tested in cold water or until the thermometer registers 244°F. The melted corn syrup is then stirred in rapidly and the batch is poured out on a clean, cooled, dampened cream beater. The top is sprinkled with a few drops of cold water. The batch is cooled down rapidly to 105°F. The soaked gelatin is added and the beater is set in motion. The fruits and flavor are added while the batch is in the creaming process. After the batch has creamed to a lump, it is allowed to sweat back for 15 minutes. It is then kneaded soft, rolled out, and cut into small balls. These are allowed to cool on trays and are then coated with a high-quality milk chocolate. The creams are packed in the usual manner.

The batch will yield approximately 65 pounds of chocolates.

CHOCOLATE-COVERED HONEY COCONUT ROYALS

Coconut	10 lb
Honey	3 lb
Corn Syrup	10 lb
Sugar	5 lb
Water	1 qt
Vanilla Flavor	4 teaspoonfuls

The coconut and honey are placed in a medium open-type mixing machine which is started at low speed. The mixer is allowed to run while the batch is cooking.

The sugar, corn syrup, and water are placed in a copper pan on the stove and stirred until dissolved. The sides are washed down; the batch is cooked to 234°F. and then poured into the mixing machine with the coconut. The flavor is added and the mixer operated until the batch is mixed thoroughly. It is then turned out on a greased slab between iron bars, spread out evenly to approximately 1/4 inch thickness, and allowed to cool until firm enough to handle. It is then sliced, cut into small oblongs, and allowed to dry on the boards. The royals are then coated with a dark vanilla sweet chocolate. They are packed in the usual manner.

The batch will yield approximately 45 pounds of chocolates.

CHOCOLATE-COVERED HONEY NOUGAT

Sugar	12 lb
Corn Syrup	8 lb
Honey	5 lb
Coconut Butter	1 lb
Water	2 qt
MARSHMALLOW WHIP (*see*)	8 lb
Chopped Raw Almonds	5 lb
Vanilla Flavor	3 teaspoonfuls

The sugar, corn syrup, one half of the honey, the coconut butter, and water are placed in a copper pan on the stove and stirred until dissolved. When the batch begins to boil, it is covered and allowed to steam for a few minutes. The sides are washed down and a thermometer is placed in the batch, which is cooked to 258°F. The batch is removed from the stove and the marshmallow whip and flavor are stirred in gradually. The batch is then reduced with the balance of the honey so that it forms a soft ball (*see* COLD-WATER TEST), and the almonds are added and mixed in thoroughly. It is then poured out on a cooled, greased slab between iron bars, spread out evenly to approximately 1/4 inch thickness, and allowed to cool until hard. The batch is then sliced and cut into small oblongs, which are dipped in a dark vanilla sweet chocolate. After they are allowed to set, the nougats are packed in the usual manner.

The batch will yield approximately 60 pounds of chocolates.

CHOCOLATE-COVERED HONEYCOMB SPONGE

Sugar	10 lb
Corn Syrup	2½ lb
Water	1 qt
Molasses	½ qt
Salt	1 teaspoonful
Butter	4 oz
Vanilla Flavor	1 teaspoonful

The sugar, corn syrup, salt, and water are placed in a copper pan on the stove and stirred until dissolved. When the batch begins to boil, it is covered and allowed to steam for a few minutes. The sides are washed down and a thermometer is placed in the batch. The molasses is added gradually, and the batch is cooked to 286°F. The butter is stirred in rapidly, and the batch is poured out on a greased slab lined with iron bars. The flavor is stirred in as the edges are folded together. This folding is continued until the batch is cool enough to handle. Then it is pulled partly on the hook, transferred to the spinning table in front of the batch warmer, and formed into a flat batch. It is then wrapped around a lukewarm iron pipe 2 inches in diameter and 2 feet in length. One end of the batch is closed up and it is rolled around with the iron pipe inside it. The round mass is then moistened with a damp cloth on one side. The pipe is pulled out of the batch through the open end and some air is blown in through the open end of the batch. Then this open end is closed rapidly. The mass is then stretched out to about three times its original size and folded in half on the dampened side. Both ends of the batch are allowed to meet. The batch now consists of two elliptical strips side by side. Each strip is hollow inside. The folded batch is stretched out again to twice its size and doubled up, so that it now consists of two pairs of strips. Stretching has made the strips more rectangular than elliptical in shape. Stretching and folding are continued, folding sidewise and upward, until the batch consists of thirty-two strips, all hollow inside. The processed batch is kept in front of the batch warmer and spun out into long flat strips, which are passed through a chip-cutting machine and are then allowed to cool on the table. The chips are then broken apart and coated with dark, sweet chocolate. After they have set, they are packed in boxes.

Note: To make a good honeycomb sponge, it is necessary to warm up the front part of the spinning table, where the complicated stretching and folding are done. The original batch, which is usually kneaded and folded on the slab, is, in this instance, kneaded and folded on the front part of the spinning table. It requires expert skill to make this candy.

The batch will yield approximately 20 pounds.

CHOCOLATE-COVERED CREAM-FILLED LEMON STICKS

The CHOCOLATE-COVERED CREAM-FILLED MINT STICKS formula is followed, but the center is flavored lemon, with the addition of 2 teaspoonfuls of citric acid solution. The jacket batch is colored yellow and flavored lemon. The batch is processed and finished in the same manner as the mint and it will yield approximately 42 pounds.

CHOCOLATE-COVERED MAPLE COCONUT SQUARES

Maple Sugar	5 lb
Crystal "A" Sugar	10 lb
Corn Syrup	10 lb
Water	3 qt
Coconut Butter	12 oz
Vanilla Flavor	1 teaspoonful
Maple Flavor	3 teaspoonfuls
Macaroon Coconut	10 lb

The sugars, corn syrup, coconut butter, and water are placed in a copper pan on the stove and stirred until dissolved. When the batch begins to boil, it is covered and allowed to steam for a few minutes. The sides are washed down and a thermometer is placed in the batch, which is cooked to 234°F. The coconut is added and stirred in rapidly. The flavors are then added and mixed in thoroughly. The batch is turned out on a greased slab between iron bars and spread out to ½ inch thickness. After cooling on the slab until it sets, the batch is sliced and cut into small squares, which are allowed to dry on the boards. These are then coated with a high-quality milk chocolate and allowed to set on the trays. The coconut squares are packed in the usual manner.

The batch will yield approximately 55 pounds of chocolates.

CHOCOLATE-COVERED MARSHMALLOW BELMONTS

SNOW DRIFT MARSHMALLOWS (*see*) are used. One tablespoonful of freshly toasted macaroon coconut is mixed into each handful of the cooled tempered chocolate. The marshmallows are then dipped and finished roughly.

Note: The yield will be POUND FOR POUND. For example, 10 pounds of marshmallows and 10 pounds of chocolate will give 20 pounds of chocolate covered marshmallows.

CHOCOLATE-COVERED MARSHMALLOWS

SNOW DRIFT MARSHMALLOWS are excellent for chocolate coating, and dark vanilla sweet chocolate is proper to use for a perfect blend. The round ring mark for the top of each chocolate-covered marshmallow is the standard string. Before dipping, the marshmallows are brushed thoroughly free of starch. They are dipped in cooled tempered chocolate, allowed to set on trays, and packed in the usual manner.

Note: The yield is pound for pound. For example, 10 pounds of Snow Drift marshmallows and 10 pounds of chocolate will give 20 pounds of chocolate-covered marshmallows.

CHOCOLATE-COVERED CREAM-FILLED MINT STICKS

Center

Soft Coconut Butter	4 lb
Sifted Powdered Sugar	6 lb
Mint Flavor	2 teaspoonfuls

Jacket

Sugar	20 lb
Cream of Tartar	2 teaspoonfuls
Water	3½ qt
Mint Flavor	3 teaspoonfuls

The coconut butter is melted; the powdered sugar is then stirred in and the flavor added. Enough powdered sugar is used to make a thick paste. This center is kept fairly warm (almost hot) to use when desired.

To prepare the jacket, the sugar, water, and cream of tartar are placed in a copper pan on the stove and stirred until dissolved. When the batch begins to boil, it is covered and allowed to steam for a few minutes. The sides are washed down and a thermometer is placed in the batch, which is cooked to 324°F. The batch is then poured out on a cooled greased slab lined with iron bars, and the edges are folded together while the oil of mint is stirred in. The batch is cooled rapidly and then pulled partly by machine or by hand. It is then transferred to the spinning table in front of the batch warmer. One third of the batch is kept soft for the jacket and the larger part is formed flat. The warm center is placed in the center of the flat batch, and the long edges are dampened and folded over so that the center remains sealed. The batch is stretched out and folded over on the side so that both ends meet and stick together. The batch is stretched out and folded over again on top. This process is continued until the batch is folded over four times. The batch is then kept in square shape in front of the batch warmer until the small portion of the batch which has been set aside for the jacket is kneaded and formed in a flat shape large enough to take in the entire center part, wrapped around, and folded over. The batch is then spun out into fine round strips and cut into four-inch-long sticks on the continuous cutting machine. These are allowed to cool on the table and are then dipped in dark, sweet chocolate. They are rolled in chocolate sprinkles and allowed to set. The sticks are then packed in boxes.

The batch will yield approximately 42 pounds.

CHOCOLATE-COVERED MOLASSES COCONUT CARAMELS

Molasses	4 qt
Corn Syrup	12 lb
Sugar	8 lb
Coconut Butter	1 lb
Salt	2 teaspoonfuls
Evaporated Milk	5 lb
Macaroon Coconut	15 lb
Vanilla Flavor	2 teaspoonfuls

All of the ingredients except the coconut and the flavor are placed in a large copper pan on the stove and stirred until dissolved. When the batch begins to boil, the sides are washed down, and the batch is stirred and cooked to 234°F. or until it forms a soft ball when tested in cold water (*see* COLD-WATER TEST). The coconut and flavor are then stirred in and mixed thoroughly. The batch is turned out on a greased slab between iron bars and spread out evenly to approximately ½ inch thickness. It is allowed to cool until hard and then sized, sliced, and cut into small squares. These are coated with high-quality milk chocolate and allowed to set on dipping trays. The caramels are packed in stock boxes.

The batch will yield approximately 75 pounds of chocolates.

CHOCOLATE-COVERED MOLASSES COCONUT ROYALS

The formula for CHOCOLATE-COVERED HONEY COCONUT ROYALS (*see*) is followed, but 1½ quarts of molasses are mixed with the coconut instead of honey. Otherwise the batch is handled and finished in the same manner.

CHOCOLATE-COVERED ORANGE STICKS, CREAM-FILLED

The formula for CHOCOLATE-COVERED MINT STICKS, CREAM-FILLED is followed, but the center is flavored orange with an addition of 2 teaspoonfuls of citric acid solution. The jacket batch is colored and flavored orange. Otherwise the batch is completed in the usual manner and will yield approximately 42 pounds.

CHOCOLATE-COVERED NOUGATINES

Sugar	12 lb
Butter	3 lb
Salt	2 teaspoonfuls
Lightly Roasted Ground Almonds	5 lb

The sugar, butter, and salt are placed in a copper pan on the stove over low heat and stirred constantly until dissolved to a brown liquid. While stirring the batch, the almonds are added gradually. The batch is cooked until it becomes hard and crisp when tested in cold water (*see* COLD-WATER TEST). It is then poured out on a greased slab and the edges are folded together. The batch is kept warm and rolled out thin; then it is cut into 3/4 inch squares with an adjustable cutter. All this is done rapidly so that the batch will not spoil. The squares are then broken apart and allowed to cool on the table. They are then coated with a high-quality milk chocolate and packed in the usual manner.

The batch will yield approximately 27 pounds of chocolates.

CHOCOLATE-COVERED PASTE

LEMON AND ORANGE PASTE (*see*) are usually prepared for chocolate coating to complete the package of assorted chocolates. The pastes are brushed free of starch and dipped in dark vanilla sweet coating. They are usually coated roughly.

Note: For 20 pounds of chocolates, use 12 pounds of paste and 8 pounds of coating.

CHOCOLATE-COVERED PEANUT CHIPS

Sugar	12 lb
Corn Syrup	8 lb
Salt	2 teaspoonfuls
Water	2 qt
Virginia Roasted Chopped Peanuts	10 lb

The sugar, corn syrup, salt, and water are placed in a copper pan on the stove and stirred until dissolved. When the batch begins to boil, it is covered and allowed to steam for a few minutes. The sides are washed down, and a thermometer is placed in the batch, which is cooked to 280°F. While the batch is stirred constantly, the peanuts are added gradually and the batch is cooked until it becomes hard and crisp when tested in cold water. The batch is then poured out on a greased slab and the edges are folded together. It is kept warm while one half is rolled out with a rolling pin as thinly as possible and cut with an adjustable cutter into 3/4 inch squares. These squares are rapidly transferred to the table and the remaining portion of the batch is treated in the same way. The process must be rapid so that the candy will not spoil by becoming too hard. The squares are broken apart and, when cool, they are coated with a high-quality milk chocolate. They are then packed in the usual manner.

The batch will yield approximately 42 pounds of chocolates.

CHOCOLATE-COVERED PEANUT NOUGAT

Sugar	12 lb
Water	2 qt
Corn Syrup	13 lb
Coconut Butter	1 lb
Salt	1 teaspoonful
MARSHMALLOW WHIP (*see*)	5 lb
Freshly Roasted Spanish Peanuts	7 lb
Vanilla Flavor	4 teaspoonfuls

The sugar, corn syrup, water, coconut butter, and salt are placed in a copper pan on the stove and stirred until dissolved. When the batch begins to boil, it is covered and allowed to steam for a few minutes. The sides are washed down and a thermometer is placed in the batch, which is cooked to 254°F. The batch is then removed from the stove, and the marshmallow whip and flavor are stirred in gradually so that the batch will form a soft ball (*see* COLD-WATER TEST). A little water may be added if necessary to reduce the batch. The peanuts are stirred in and mixed through thoroughly. The batch is then poured out on a cooled, greased slab and spread out evenly to approximately ¼ inch thickness. It is allowed to cool until hard and then sliced, cut into small oblongs, and coated with a high-quality milk chocolate. The nougats are allowed to set and are then packed in stock boxes.

The batch will yield approximately 60 pounds of chocolates.

CHOCOLATE-COVERED PINEAPPLE

About ten pounds of glacéed pineapple slices are cut up; each slice is cut into about six pieces. These are coated with a high-quality dark vanilla sweet chocolate. The same finishing mark as for the CHOCOLATE-COVERED BRAZILS (*see*) is used. The chocolates are then cooled and completed in the usual manner. Ten pounds of pineapple will consume 10 pounds of chocolate.

CHOCOLATE-COVERED PLANTATIONS

Sugar	8 lb
Corn Syrup	12 lb
Molasses	2 qt

Coconut Butter	1 lb
Salt	2 teaspoonfuls
Water	1 qt
MARSHMALLOW WHIP (*see*)	8 lb
Evaporated Milk	2 lb
Vanilla Flavor	4 teaspoonfuls

The sugar, corn syrup, molasses, coconut butter, salt, and water are placed in a large copper pan on the stove and stirred until dissolved. When the batch begins to boil, the sides are washed down and the batch is cooked to 260°F. It is then removed from the stove and the marshmallow whip and flavor are stirred in gradually. The batch is reduced with the evaporated milk so that it forms a soft ball (*see* COLD-WATER TEST). It is then poured out on a greased slab between iron bars and spread out evenly to approximately 1/4 inch thickness. It is allowed to cool until hard; then it is sliced, cut into small oblongs, and coated with a high-quality milk chocolate. After the plantations are allowed to set on boards, they are packed in stock boxes.

The batch will yield approximately 50 pounds of chocolates.

CHOCOLATE-COVERED SPANISH APRICOT NOUGAT

Sugar	15 lb
Corn Syrup	5 lb
Water	2 qt
Apricot Pulp	10 lb
Chopped Almonds	4 lb
Macaroon Coconut	12 lb
Apricot Flavor	4 teaspoonfuls

The sugar, corn syrup, and water are placed in a copper pan on the stove and stirred until dissolved. When the batch begins to boil, it is covered and allowed to steam for a few minutes. The sides are washed down and a thermometer is placed in the batch, which is cooked to 280°F. The apricot pulp is added gradually, and the batch is stirred and cooked over low heat until it forms a firm medium ball when tested in cold water. It is then removed from the stove and the almonds, coconut, and flavor are added and mixed through thoroughly. The batch is then poured out on a greased slab between iron bars and spread out evenly to approximately ½ inch thickness. It is allowed to set overnight and is then sliced, cut into small oblongs, and allowed to dry. The nougats are then coated with a high-quality milk chocolate. When set, they are packed in boxes.

The batch will yield approximately 65 pounds of chocolates.

CHOCOLATE-COVERED TOFFEE CHIPS

Sugar	12 lb
Corn Syrup	8 lb
Salt	2 teaspoonfuls
Coconut Butter	1 lb
Fresh 20% Light Cream	6 qt
Chopped Raw Almonds	3 lb
Oil of Bitter Almonds	2 teaspoonfuls
Green Color	2 teaspoonfuls

The sugar, corn syrup, coconut butter, salt, and cream are placed in a clean copper pan on the stove and stirred until dissolved. When the batch begins to boil, the sides are washed down. The fire is reduced to medium, and the batch is stirred constantly and cooked until it forms a firm ball when tested in cold water (*see* COLD-WATER TEST). The nuts, flavor, and color are stirred in and mixed thoroughly. The batch is then poured out on a cold, greased slab between iron bars and spread out to approximately ⅛ inch thickness. When cool and hard, it is sliced and cut into small chips, which are coated with high-quality milk chocolate. The chips are allowed to set and are then packed in the usual manner.

The batch will yield approximately 42 pounds of chocolates.

CHOCOLATE-COVERED VANILLA CARAMELS

Sugar	8 lb
Corn Syrup	12 lb
Salt	3 teaspoonfuls
Coconut Butter	1½ lb
Evaporated Milk	15 lb
Vanilla Flavor	4 teaspoonfuls

The sugar, corn syrup, salt, coconut butter, and one half of the milk are placed in a copper pan on the stove over medium heat and stirred until dissolved. When the batch begins to boil, the fire is reduced to low. The batch is stirred constantly. Then the balance of the evaporated milk is stirred in gradually, and the batch is cooked until it forms a medium ball when tested in cold water. It is then removed from the stove and the flavor is stirred in. The batch is poured out on a cold greased slab between iron bars and spread out evenly to about 1/4 inch thickness. When cool, it is sliced and cut into small squares, which are coated with dark vanilla sweet chocolate and allowed to set on the dipping trays. They are packed in the usual manner.

The batch will yield approximately 48 pounds of chocolate-covered caramels.

CHOCOLATE-COVERED WALNUTS

The process for CHOCOLATE-COVERED ALMONDS (*see*) is followed. Chabert or Bordeaux walnut halves are preferred. The nuts are brushed off thoroughly before coating. They may be coated with dark or milk chocolate; a blend of 2 parts of dark and 1 part of milk chocolate is also very popular. The round ring mark is the standard for chocolate-covered walnuts. Three pounds of chocolate will coat 1 pound of walnuts.

CHOCOLATES

Assorted chocolates in fancy packages are very popular the year round. They outsell all fall and winter candies, particularly at Christmas time. Assorted chocolates are also the most profitable in the candy business. It is, therefore, in the interest of the average manufacturing confectioner to make this specialty of outstanding quality, which will attract repeat business.

Chocolate centers consist mostly of creams, chewy and hard masses, fruit jellies, and marmalades. Fruits and nuts usually top the assortment. In making and preparing these centers the confectioner should be quality minded. Whether they are starch cast or hand rolled, all creams should be made of fresh and pure ingredients under sanitary conditions. Only pure flavor should be used in the centers. It is best to use pure vanilla extract rather than cheaper imitations. In making cream centers, fresh fruit pulps supported by pure fruit flavors give best results. Hard and chewy centers should be made with fresh cream and butter and high-grade nuts flavored with pure oils and extracts. The finest glazed and preserved fruits should be coated with the best dark, sweet vanilla chocolate and the best pure milk chocolate. The finest first-grade nuts should be used. When coating roasted nuts, e.g., almonds or filberts, the roasting should not be overdone because a strong roast gives the nuts a burned taste. The chocolate coating must be carefully selected so that it blends with the center. For a sweet center, e.g., creams and chews, a semi-sweet chocolate is preferred. Fruits go well with milk chocolate. A quality mild, sweet chocolate (dark or milk) is preferred for nuts. Whether it is coated with dark sweet or milk chocolate, if the coating blends with the center and the assortment is arranged in a pleasing manner, the result will be a quality package.

Chocolate coating is a skill practiced mostly by female help. It takes endurance and patience, time, and long practice to learn how to do it properly. The chocolate-department executive should know all about chocolates and chocolate coating. He should be able to instruct the chocolate coaters. He should also be able to break in new chocolate dippers, know how and where to locate faults and shortcomings, and improve conditions in the chocolate-dipping department. For example, it is important that the chocolate is cooled to the right temperature. It takes almost an hour for chocolates to set cool enough for packing. If the centers are coated with warm chocolate, the chocolates become streaky and turn gray when they should be ready for packing. The result is, of course, scrap. It is also important that when cream centers are coated, the bottoms, as well as the entire center, are given a thorough covering to avoid leaky chocolates.

When, as today, skilled labor is at a premium and wages are high, most manufacturers of quality chocolates have adopted both machine and hand coating. Most modern enrobing machines are suitable for quality production. The enrober in the chocolate department should be used mostly for chewy and hard centers and cream centers cast in starch. Hand-rolled centers are best when coated by hand. Fruits and nuts should also be hand-coated, since most enrobing machines cannot put enough coating on them. The chocolate enrobing machine requires an expert operator. No other employees should attempt to operate this machine.

The chocolate is melted in a steam-bath melting pan or in a

regular chocolate melter. When first melted, it is heated to about 120°F. Before it is brought over to the coaters, it is cooled to about 86°F. The dipper then cools the chocolate to the right temperature before dipping the centers. Any experienced dipper knows when the chocolate is ready for dipping centers from the feel of it in her hand. An important point to remember: Whenever the centers are or the chocolate is too cold, the result will be streaky chocolates. Excessively warm centers or warm chocolate will yield gray chocolates. Therefore, the chocolate-coating room should never be below 60°F or above 66°F.

Leaky cream centers are caused mostly by "doctoring" the cream batch too heavily or not putting enough chocolate on the centers. The first mistake can be avoided only by following the formulas accurately. The second error can be avoided by employing only experienced coaters who know how to handle the centers.

Hot weather is the natural enemy of chocolates. It usually causes moldy, wormy chocolates and sunken bottoms. As yet there is no sure remedy for hot weather troubles. However, by taking certain precautions, these difficulties can be cut to a minimum. High-grade chocolate coating will prevent molding because it is better conched and milled for a longer period of time. This makes it more solid and, therefore, less susceptible to mold. Stock should be very low in the summer time. The best temperature for the stock room is 65° to 68°F. The room must be kept dark and dry to prevent sunken bottoms in cream centers. Excessive "doctoring" must be avoided because it makes the centers too moist and causes sunken bottoms followed by dried-up and fermented centers. In making creams and chewy centers, the cooking temperature may be increased 2 to 3 degrees in hot weather.

Wormy chocolates in hot weather are almost always caused by nuts. All nuts should be brushed free of dust before they are coated since worms hatch in the dust. Nuts should be fresh, ordered in small quantities, and kept in cold storage.

All nut and fruit and nut packages are always popular. There is a group of formulas in this chapter which gives the confectioner a potentially good all nut or fruit and nut assortment. Chocolates should be handled carefully when they are packed. Scratching should be avoided. Before it is placed in the box, every chocolate should be cupped. The bottom layer of the box should be given as much attention as the top layer. A hastily packed bottom in a box makes a poor impression on the consumer. When a two-layer box is packed, the partition and cardboard layer that separate the bottom from the top layer should be well fitted and supported by soft padding to avoid smashed chocolates. The top layer should be more tightly packed than the bottom one, since the box tends to stretch a little more at the top. One or two pieces foiled in fancy aluminum paper make the top layer more colorful. A few attractively colored, glossy Jordan almonds also help make the box more attractive. Partitions with gilt edges in the form of a zig-zag help give a classy appearance. When it is packed, the box should be checked for correct weight. A white soft cushion mat is placed on top, the wax-paper liners are folded over and closed, and the cover is placed on top of the box. If the box is wrapped by hand in No. 300 plain transparent cellophane paper, liquid glue is used to make the ends stick together. If machine wrapped, No. 300 moisture-proof cellophane is used. It is important to make certain that the box is the correct size. If it is too small the chocolates will be crowded in to meet the weight, and when the consumer opens the box, he will find crushed pieces. If the box is too large, the chocolates will be shaken in shipping, and will be scratched and stale looking when the box is opened.

Equipment for the chocolate department varies with the size and financial status of the concern. Where quality chocolates are made and coated by hand, they can be produced with a limited investment in equipment. The larger plant, however, may consistently spend thousands of dollars on up-to-date equipment and machinery.

The following equipment would constitute an up-to-date chocolate department in a smaller or medium shop. It should be equipped with a commercial air-conditioning unit capable of producing two to three tons of ice a day. All walls, windows, and doors in the chocolate department, including stock and storage rooms, are usually cork insulated. There should be two medium-sized chocolate-melting machines—one for dark bittersweet and sweet chocolate and another for milk chocolate. Six dipping tables should be installed to accommodate a shift of twelve hand coaters. Hard and chewy centers and firm starch-cast centers may be machine coated by a 24-inch enrober with stringing and packing attachments and a cooling tunnel. Bars and all medium-sized Easter eggs are also machine coated. Delicate cream centers, nuts, fruits, and cordials should be hand coated if a strictly quality product is desired. In exceptional cases (e.g., a wholesale manufacturer who produces for chain and syndicate stores that have also been moving toward quality products), an expert confectioner can develop some formulas for hand rolls which are of fair quality, have a firm finish when coming out of the cut roll machine, and will go through the enrober for the chocolate coating process without giving trouble. If well made, these centers will ripen in a few days to an excellent texture, smoothness, and softness.

All machine-coated chocolates are usually packed in stock boxes as they move along the belt after leaving the cooling tunnel. The number of packers required, as well as the entire servicing crew, is determined by the operator of the chocolate-enrobing machine.

Operating an enrober is an important job. Proficiency comes only with years of experience. When a new chocolate-enrobing machine is installed, a service man supervises its installation and helps the operator for a few days to familiarize him with the machine. Although all enrobers operate on the same principle, they differ somewhat in practice. The shop environment, the type of chocolate coating used, the type of centers to be coated, and the amount of chocolate to be put on the centers are the factors the machine operator must consider and adjust to in order to do a satisfactory job. An experienced operator who has handled machines of all sizes and makes and coated chocolates of all types and grades is a desirable employee.

When the melted chocolate coating is transferred from the melters to the chocolate tank in the enrobing machine, it must be cooled and adjusted so that the centers will be properly coated. The chocolate-machine operator should know when the chocolate is ready to go on the centers without even looking at the thermometer, since thermometers are not always reliable. When the coated centers travel to the cooling tunnel, the experienced operator can tell by keeping an eye on the belt whether the centers are coated too heavily or too lightly and can quickly make the necessary adjustments. The operator must see that the centers are at the proper temperature before they go on the machine for chocolate coating. Before starting the machine, he arranges his crew. The girls who string the centers before they go into the tunnel must be experienced chocolate coaters to do a good stringing job. The same applies to the chocolate packers. An experienced packer will move her fingers quickly and will not scratch the chocolates while packing them. The operator assigns each member of the crew to the job for which he or she is best fitted and shifts the crew whenever necessary.

For the various chocolate formulas, see the individual names, such as ITALIAN CREAM CHOCOLATES, PINEAPPLE-CREAM CHOCOLATES, BERNESE CHOCOLATES, ETC. *See also* COMPOSITION CHOCOLATES, LORRAINE CHOCOLATES, NORMANDY CHOCOLATES, OPERA CHOCOLATES, and SWISS FONDANT CHOCOLATES.

CHOP-SUEY BRITTLE

Sugar	12 lb
Corn Syrup	6 lb
Water	1½ qt
Salt	1 tablespoonful
Spanish Peanuts	5 lb
Sliced Coconut	2 lb
Bicarbonate of Soda	1 tablespoonful

The sugar, corn syrup, water, and salt are placed in a copper pan on the stove and stirred until dissolved. When the batch begins to boil, the peanuts are added and allowed to cook until the batch is quite thick. The sliced coconut is then added, while stirring constantly. The stirring and cooking process is continued until the peanuts are well roasted and begin to crack, and the candy breaks crisp when tested in cold water. The batch is then removed to a stand and the bicarbonate of soda is stirred in rapidly. The batch is then poured out on a greased slab and spread thinly with a palette knife. It is then loosened underneath with a long knife and stretched again as thinly as possible. After cooling on the slab, the candy is packed into tins.

This batch will yield approximately 22 pounds of candy.

CHOP-SUEY JELLIES: *See* SAVING WASTE.

CINNAMON BARLEY SUGAR CUTS: *See* BARLEY SUGAR CUTS, CINNAMON.

CINNAMON BUTTERSCOTCH WAFERS: *See* BUTTERSCOTCH WAFERS, CINNAMON.

CINNAMON CRIMPED RIBBON CANDY: *See* CRIMPED RIBBON CANDY, CINNAMON.

CINNAMON GUM DROPS: *See* GUM DROPS, CINNAMON.

CINNAMON IMPERIALS

Center

Sugar	25 lb
Corn Syrup	5 lb
Water	4 qt
Oil of Cassia	2 teaspoonfuls
Red Color	1½ teaspoonfuls

The sugar, corn syrup, and water are placed in a copper pan on the stove and stirred until dissolved. When the batch begins to boil, it is covered and allowed to steam for a few minutes. The sides are washed down and a thermometer is placed in the batch, which is cooked to 310°F. The batch is poured out on a cold, greased slab between iron bars, and the edges are folded together while the red color and the cinnamon flavor are stirred in. This folding process is continued until the batch is cool enough. Then it is taken over to the spinning table in front of the batch warmer and passed through a small heart roller machine (about 300 pieces to the pound). The small hearts are allowed to cool on the table and are then broken apart and put through a sieve. They are now ready for panning.

Panning

For this process, 20 pounds of sugar, 1 teaspoonful of cream of tartar, and 3 quarts of water are placed in a pan on the stove, stirred, and cooked to 242°F. About 1½ quarts of gum arabic solution are then stirred into the syrup, which is colored a deep bright red and flavored with 3 teaspoonfuls of oil of cassia. The red hearts are then placed in the revolving pan and processed with the red gummed syrup in the same manner as the SILVER DRAGEES (*see*).

Polishing

Polishing is also the same as in making SILVER DRAGEES (*see*). The polished candies are handled and finished as usual.

CINNAMON SQUARES

Sugar	18 lb
Water	3½ qt
Corn Syrup	7 lb
Red Color	1 teaspoonful
Oil of Cinnamon	3 teaspoonfuls

The water, sugar, and corn syrup are placed in a copper pan on the stove and stirred until dissolved. When the batch begins to boil, it is covered and allowed to steam for a few minutes. The sides are washed down and a thermometer is placed into the batch. When it reaches 280°F., the red color is added to the batch and cooking is continued to 290°F. The batch is then removed from the stove and the oil of cinnamon stirred in thoroughly. It is then poured on a greased slab with bars around, leaving just enough space to spread out the batch evenly to about ½ inch thickness. The batch is allowed to cool a few seconds and is then loosened by placing a large knife underneath. It is then cut with an adjustable cutter into 1-inch squares and allowed to cool off, at which time it is cut in sheets to the size desired and packed in boxes lined with wax paper between each sheet.

The batch will yield approximately 22½ pounds.

CLOVE BARLEY SUGAR CUTS: *See* BARLEY SUGAR CUTS, CLOVE.

COCONUT BISCUITS

Sugar	12 lb
Corn Syrup	8 lb
Water	1½ qt
Salt	1 teaspoonful
Macaroon Coconut	8 lb
Vanilla Flavor	2 teaspoonfuls

The sugar, water, corn syrup, and salt are placed in a clean copper pan on the stove and stirred until dissolved. When the batch begins to boil, half of the coconut is stirred in gradually. The batch is stirred and cooked until it forms a medium ball when tested in cold water. The balance of the coconut and the vanilla flavor are stirred in and mixed thoroughly. The batch is then turned out on a greased slab, dusted with flour, spread out thinly, and rolled out with a rolling pin. It is then cut out with cutting forms to the shape and size desired and placed in baking tins dusted with flour. The biscuits are allowed to set and they are then placed in a hot oven, which is heated to about 300°F, and allowed to remain until they are well browned. They are then removed, allowed to cool off, and packed in boxes.

This formula should yield approximately 30 pounds of candy.

COCONUT BRITTLE: *See* POPCORN COCONUT BRITTLE.

COCONUT BRITTLE, ROASTED

Sugar	12 lb
Corn Syrup	8 lb
Water	1½ qt
Long-Thread Coconut	5 lb
Oil of Lemon	1 teaspoonful

The sugar, water, and corn syrup are placed in a large copper pan on the stove and stirred until dissolved. When the batch begins to boil, the sides are washed down and a thermometer is placed in the batch, which is cooked to 240°F. The thermometer is then removed and the cooking is continued. The coconut is gradually added, while stirring the batch constantly. The stirring and cooking are continued until the coconut is toasted to a light brown color and the candy breaks crisp when tested in cold water. The batch is then removed to a stand, and the oil of lemon is stirred in. It is then poured out on a greased slab, spread thinly with a palette knife, and loosened underneath with a long knife. After it is stretched out quite thin, the batch is allowed to cool on the slab and then packed in tins.

This batch will yield approximately 25 pounds of candy.

COCONUT BUTTER CREAMS: *See* BUTTER CREAMS, COCONUT.

COCONUT BUTTER SQUARES

Sugar	20 lb
Corn Syrup	5 lb
Water	3 qt
Butter	1 lb
Salt	1 tablespoonful
Vanilla Flavor	3 teaspoonfuls
Fine Coconut	3½ lb

The sugar, water, and corn syrup are placed in a copper pan on the stove and stirred until dissolved. When the batch begins to boil, it is covered and allowed to steam for a few minutes. The sides are washed down and a thermometer is placed in the batch, which is cooked to 330°F. The heat is then shut off, while the butter, salt, coconut, and vanilla flavor are stirred in. Stirring is continued while the batch is allowed to cook again for about 2 minutes. It is then poured out on a greased slab between iron bars and spread out to about 1/8 inch thickness. A large knife is run under the batch to loosen it from the slab. Very quickly the batch is cut into small squares with a roller knife and allowed to cool on the slab. Then the squares are broken apart and packed into tins.

The batch will yield approximately 27 pounds.

COCONUT CANDIES

Since they stand up under any climate, coconut candies are popular the year round. Most Easter eggs, which sell so well during the first four months of the year, are made with coconut. Coconut kisses are popular in the summertime when chocolate sales dip down; so are iced and crystallized coconut dainties. In addition, coconut caramels make a good center for chocolates.

Years ago coconut candies of quality were usually made with fresh coconut processed in the candy factory, but modern developments have changed this situation. Because the labor and transportation needed to produce fresh coconut are almost prohibitive for the small and medium confectionery manufacturer, this work is left to large mechanized plants which specialize in processing coconut. Nowadays the manufacturing confectioner buys all his coconut in dry form, and the proficient candy maker can produce a quality product with dry coconut.

The equipment needed to make coconut candies is modest and simple. It includes open-fire or steam-cooking kettles, cooling slabs or large thick marbles, and cutting machines for production work. For smaller retail manufacturing establishments hand cutting is satisfactory.

Old-fashioned coconut pieces are the most popular candies in this group. It is important that they are made right. A few technical pointers on the making of these specialties follow. These suggestions should be adjusted to the confectioner's equipment and the size of his establishment.

Whether they are sugared, crystallized, iced, cream dipped, or chocolate coated, coconut candies are divided in two classes: chewy and creamy. It is important that chewies are made so that they will stay chewy a long time. For this reason, they must be "doctored" heavily. An extra pound or two of corn syrup in the formula will do that. When a formula calls for molasses, only the "open-kettle" type, which can be cooked on the open fire without scorching, should be used. When making white coconut candies, the batch must be cooked rapidly and stirred constantly so that its color does not darken. The same COLD-WATER TEST (*see*) that was used in making caramels is used here. The water must be cold when making this test.

Creamy coconut candies must be made so they will stay soft and creamy and will not ferment or dry out quickly. To prevent fermentation or drying out, the bonbon cream or French cream used in the formula must be smooth and not grainy. A small amount of glycerin mixed in with the coconut (about 1 ounce to every 20 pounds of candy) will help keep the batch soft. When spooning out kisses, the spoons should be dipped in cold water after every dip. The same applies to the ice-cream dipper when making patties. When dipping or spooning out these goods, the batch may feel dry toward the end. A few drops of water stirred in will help restore its natural gloss and color.

For the various coconut candy formulas, see the individual names, such as, COCONUT DIAMONDS, MAPLE COCONUT, ETC. *See also* COCONUT STARS and ICED COCONUT DAINTIES.

COCONUT CARAMELS, CHOCOLATE-COVERED: *See* CHOCOLATE-COVERED COCONUT CARAMELS.

COCONUT CORDIALS

The VANILLA CORDIALS (*see*) formula is followed, but the batch is flavored with ½ teaspoonful vanilla extract and 1½ teaspoonfuls of imitation coconut flavor. The coconut cordials are then finished in the usual manner.

The batch will yield approximately 20 pounds of cordials.

COCONUT-CREAM BARS, CHOCOLATE-COVERED: *See* CHOCOLATE-COVERED COCONUT CREAM BARS NO. 1.

COCONUT CREAM EGGS, CAST

RUNNING CREAM (*see*)	40 lb
Macaroon Coconut	5 lb
Vanilla Flavor	4 teaspoonfuls
Invertase	1 teaspoonful
Yellow Color	1 teaspoonful

The running cream is remelted in a steam-bath copper pan over slow heat as usual. The macaroon coconut is stirred in and the flavor and invertase are added. All of these ingredients are mixed thoroughly. About 5 pounds of this fondant are placed in a small pan and colored yellow. This portion of the batch is then cast in small yolk impressions molded in clean, dry starch and these are allowed to set until hard. They are then removed, and after they are brushed free of starch, they are ready for use.

For large cast eggs, deep starch boards (about four times the size of standard boards) are used. They are filled with clean dry starch and leveled as usual. They are then imprinted with large molds of easter egg halves in the various sizes desired. The remelted coconut cream is cast in the impressions of egg halves, and a yolk is placed in each half egg cast. Then for each board of eggs containing yolks, one board of eggs without the yolks is cast. All of the cast eggs are allowed to set overnight and they are then removed, brushed free of starch, and placed on boards.

About 5 pounds of running cream are remelted, and some of this cream is spread out on an egg half containing the yolk with a small flat knife. Then, working rapidly, a plain egg half is placed on top of the one with the yolk, and both halves are sealed. The egg is then placed on a clean board. This process is continued until all of the eggs have been sealed. They are then coated in pure, bittersweet chocolate and allowed to set until cool. The eggs are then ready for decorating.

The batch will yield approximately 60 pounds of large coconut cream eggs.

COCONUT CREAM EGGS, DECORATED, LARGE

The formula for COCONUT CREAM EGGS, HAND-ROLLED, SMALL (*see*) is followed, but the cooking temperature of the batch is changed to meet the following schedule. For 8 ounce to 1 pound eggs, the cooking temperature will be 260°F; for 1 to 3 pound eggs, the cooking temperature will be 264°F, and for 3 to 10 pound eggs, the cooking temperature will be 268°F. The usual creaming process is followed and the batch is then transferred to the work table, which has been dusted with flour. A small portion of the batch is colored yellow and is used as yolk. The white batch is kneaded smooth and spread out. The yellow batch is placed in the center of the white, and then folded and covered up with the white batch. The batch is then rolled out, cut, and each egg is weighed to the desired size. Each portion is then rolled round and allowed to set for a few minutes. It is then shaped in egg form and, after it hardens, the egg is coated with bittersweet chocolate. When cooled and set, the eggs are ready for decorating.

Note: When coating large eggs, the bottoms must be coated and cooled first. Then the entire egg is coated as usual.

All large coconut cream eggs have a tendency to expand after they are coated. In order to avoid this, as soon as the chocolate-covered egg begins to cool, it is scarred with a small pointed knife around its circumference. This allows some air to enter and prevents the cream from bursting and leaking through the chocolate.

The batch will yield approximately 60 pounds of large coconut-cream eggs.

COCONUT CREAM EGGS, HAND-ROLLED, SMALL

Sugar	35 lb
Water	7 qt
Cream of Tartar	2 teaspoonfuls
Melted Corn Syrup	8 lb
Macaroon Coconut (soaked in 3 cups of water)	5 lb
MARSHMALLOW WHIP (*see*)	4 lb
Vanilla Flavor	4 teaspoonfuls
Invertase	1½ teaspoonfuls
Yellow Color	1 teaspoonful

The sugar, water, and cream of tartar are placed in a clean copper pan on the stove and stirred until dissolved. When the batch begins to boil, it is covered and allowed to steam for a few minutes. The sides are washed down and a thermometer is placed in the batch, which is cooked to 254°F. The melted corn syrup is stirred in rapidly. The soaked coconut is spread out on the cream beater, which has been previously cooled. The batch is poured out on the cream beater and cooled rapidly for about 7 minutes. The marshmallow whip, flavor, and invertase are added, and the machine is set in motion. The creaming process is continued until the batch forms a lump. It is then transferred to the work table near the cut-roll machine. About 10 pounds of the coconut cream are colored a deep yellow. The white batch is kneaded until smooth and soft and is then placed in the cream-holding tank of the machine, leaving a space in the center of the tank, where the yellow cream is inserted. After the machine has been adjusted to cut the size desired, it is set in motion, and operation is continued until the entire batch is cut and rolled. The eggs are spread out on clean, wax-paper-lined trays and allowed to set until hard. They are then coated in a high-quality, dark, bittersweet chocolate. The eggs are allowed to set on the dipping trays and are then packed in boxes. The batch will yield approximately 70 pounds of small eggs.

Note: This type of coconut cream will give good results for eggs weighing 1/4 ounce to 4 ounces.

COCONUT CREAM LOAF

Sugar	30 lb
Water	6 qt
Cream of Tartar	2 teaspoonfuls
Melted Corn Syrup	5 lb
Macaroon Coconut	5 lb
MARSHMALLOW WHIP (*see*)	3 lb
Vanilla Flavor	4 teaspoonfuls
Invertase	1 teaspoonful

The coconut is dampened with 2 cups of water and spread out on the cream beater. The sugar, water, and cream of tartar are placed in a copper pan on the stove and stirred until dissolved. When the batch begins to boil, it is covered and allowed to steam for a few minutes. The sides are washed down and a thermometer is placed in the batch, which is cooked to 252°F. The melted corn syrup is stirred in rapidly and the batch is poured out on the cream beater and cooled rapidly for about 6 minutes. The marshmallow whip, flavor, and invertase are added and the beater is set in motion and operated until the batch creams to a lump. It is then kneaded until soft and smooth and shaped into rolls about 6 inches long, each roll weighing approximately 1 pound. When the rolls are cool, they are coated with bittersweet chocolate and packed as desired.

This formula will yield approximately 60 pounds of chocolate-covered coconut-cream loaves.

COCONUT CREAM SHERBETS

Sugar	30 lb
Water	6 qt
Cream of Tartar	2 teaspoonfuls
Melted Corn Syrup	5 lb
Macaroon Cocount (soaked in 1½ cups of water)	4 lb
Vanilla Flavor	3 teaspoonfuls
Melted Bitter Chocolate	1½ lb

The sugar, water, and cream of tartar are placed in a copper pan on the stove and stirred until dissolved. When the batch begins to boil, it is covered and allowed to steam for a few minutes. The sides are washed down and a thermometer is placed in the batch, which is cooked to 254°F. The melted corn syrup is stirred in rapidly and allowed to come back to a boil. The soaked coconut is spread out on the cooled cream beater, and then the batch is poured out on top of the coconut on the cream beater and allowed to cool for 5 minutes. The vanilla flavor is then added, the machine is set in motion and allowed to run until it creams the batch to a lump. The batch is allowed to sweat back for 20 minutes; then it is transferred to the work table which is dusted with flour. The batch is then kneaded and divided into two portions. The melted chocolate is mixed into one portion. The white portion is flattened out, and the chocolate portion is flattened out on top of the white. The batch is rolled round with the white covering the chocolate. It is then rolled out, cut in oblongs, and crystallized as usual (*see* CRYSTALLIZING).

This batch will yield approximately 55 pounds.

COCONUT CREAMS, HAND-ROLLED

The formula for VANILLA NUT CREAMS, HAND-ROLLED is followed with one exception: 4 pounds of macaroon coconut are dampened with 2 cups of water and added to the cooled batch instead of the nuts. The batch is then creamed, hand rolled, and cut as usual. The creams are then dipped in dark, vanilla semisweet chocolate with some macaroon coconut mixed in the coating. These centers are dipped with a rough finish.

Note: The average ratio for using chocolate coating for hand-rolled centers is approximately 20 percent for enrober coating and 30 percent for hand coating.

COCONUT DATE GOODIES

COCONUT CREAM (*see*)	6 lb
Chopped Soft Dates	4 lb
Crystal "A" Sugar	2 lb
Corn Syrup	3 lb
Coconut Butter	4 oz
Evaporated Milk	5 lb
Salt	1 teaspoonful
Maple Flavor	1 teaspoonful
Chopped Brazil Nuts	15 lb

The coconut cream is kneaded and mixed with the chopped dates. It is then rolled out into small balls on the table and allowed to set until dry and ready for the caramel dipping.

The sugar, corn syrup, salt, coconut butter, and half of the milk are placed on the stove and stirred until dissolved. When the batch begins to boil, the sides are washed down, and the stirring is continued. The balance of the milk is stirred in gradually, and the batch is cooked until it forms a soft ball when tested in cold water. The batch is then moved near the slab. The nuts are spread out on the slab, and the cream balls are dipped in the caramel and rolled in the nuts. They are allowed to cool on the slab and are then packed in boxes as usual.

This formula should yield approximately 35 pounds of candy.

COCONUT DIAMONDS

Sugar	8 lb
Corn Syrup	12 lb
Honey	1 qt
Water	1½ qt
Coconut Butter	8 oz
Macaroon Coconut	8 lb
Finely Ground Roasted Almonds	3 lb
Salt	1 teaspoonful
Vanilla Flavor	3 teaspoonfuls
Marshmallow Sheet	5 lb
MARSHMALLOW SPREAD (*see*)	

The sugar, corn syrup, honey, water, coconut butter, and salt are placed in a clean copper pan on the stove and stirred until dissolved. When the batch begins to boil, it is covered and allowed to steam for a few minutes. The sides are washed down and a thermometer is placed in the batch, which is cooked to 242°F. The coconut is stirred in rapidly, and the ground almonds and flavor are added and mixed thoroughly. The batch is turned out on a greased slab lined with iron bars and spread out with a palette knife to about ¼ inch thickness. One half of the batch is covered with marshmallow sheet and the other half is folded over the top of the marshmallow, rolled down evenly, cut into diamond shapes, and rolled in fine granulated sugar. It is then packed in trays lined with wax paper.

This formula should yield approximately 45 pounds of candy.

COCONUT FRUIT SQUARES

Sugar	12 lb
Corn Syrup	8 lb
Water	2 qt
Assorted Diced Fruits	2 lb
Chopped Glacéed Cherries	2 lb
Chopped Glacéed Pineapple	2 lb
Macaroon Coconut	8 lb
Vanilla Flavor	4 teaspoonfuls

The sugar, corn syrup, and water are placed in a copper pan on the stove and stirred until dissolved. When the batch begins to boil, it is covered and allowed to steam for a few minutes. The sides are washed down and a thermometer is placed in the batch, which is cooked to 242°F. The batch is moved near the slab and the diced fruits, cherries, and pineapple are stirred into the batch. The coconut is added and mixed thoroughly. The batch is then turned out on a greased slab lined with iron bars and is spread out evenly to about 1 inch thickness. It is allowed to set until hard, cut in small squares and rolled in fine granulated sugar. When the squares are dry, they are packed in boxes.

This formula will yield approximately 36 pounds of candy.

COCONUT NOUGAT, CHOCOLATE-COVERED: *See* CHOCOLATE-COVERED COCONUT NOUGAT.

COCONUT HAYSTACKS

Sugar	5 lb
Corn Syrup	5 lb
Molasses	1½ qt
Water	1 qt
Long-Thread Coconut	5 lb

The sugar, corn syrup, molasses and water are placed in a copper pan on the stove and stirred until dissolved. When the batch begins to boil, the sides are washed down and a thermometer is placed in the batch, which is cooked to 230°F. The coconut is then stirred in and mixed through thoroughly. The batch is turned out into a coarse sieve on a greased slab, drained until all the drippings have run out, and then formed into small haystacks, which are deposited on greased pans and allowed to cool until they are set. They are packed in boxes.

This batch will yield approximately 20 pounds of candy.

COCONUT KISSES, CHOCOLATE

The formula for COCONUT KISSES, VANILLA is followed, but the batch is cooked to 234°F. and 2 pounds of finely cut-up bitter chocolate are stirred in. Otherwise, the batch is finished as usual, and should yield approximately 32 pounds of candy.

COCONUT KISSES, CHOCOLATE-BOTTOMED

Sugar	8½ lb
Corn Syrup	2½ lb
Water	1½ qt
FRENCH CREAM (*see*)	6 lb
Macaroon Coconut	5 lb
Vanilla Flavor	2 teaspoonfuls
Glycerin	3 teaspoonfuls
Melted and Tempered Bittersweet Chocolate	6 lb

The sugar, water, and corn syrup are placed in a copper pan on the stove and stirred until dissolved. When the batch begins to boil, the sides are washed down and the batch is cooked to 236°F. It is then removed from the heat. The french cream is stirred in rapidly and remelted. Flavor and glycerin are added, and the coconut is stirred in gradually and mixed through well. The batch is then spooned out into small kisses and deposited on dipping trays lined with wax paper. The kisses are allowed to cool until they are dry. They are then coated half way in pure bittersweet chocolate. The white tops are left bare. The kisses are allowed to cool on the trays, and they are then packed in boxes in the usual manner.

This formula will yield approximately 30 pounds of candy.

COCONUT KISSES, STRAWBERRY

The formula for COCONUT KISSES, VANILLA (*see*) is followed, but before stirring in the coconut, the batch is colored a bright pink and flavored with strawberry. It is then finished as usual and should yield approximately 32 pounds of candy.

COCONUT KISSES, VANILLA

Sugar	10 lb
Corn Syrup	2 lb
Water	1½ qt
BONBON CREAM (*see*)	8 lb
Fine Macaroon Coconut	8 lb
Glycerin	4 teaspoonfuls
Vanilla Flavor	3 teaspoonfuls

The sugar, corn syrup, and water are placed in a copper pan on the stove and stirred until dissolved. When the batch begins to boil, it is covered and allowed to steam for a few minutes. The sides are washed down and a thermometer is placed in the batch, which is cooked to 236°F. The batch is then set on a stand near the slab, and the bonbon cream, glycerin, and flavor are stirred in. The stirring is continued until the cream is melted, and then the coconut is added and mixed thoroughly. The batch is then spooned out into small kisses, deposited on heavy wax paper on the slab, and allowed to cool until set. The candy is then packed in boxes.

This formula will yield approximately 32 pounds.

COCONUT NOUGAT

Sugar	12 lb
Corn Syrup	8 lb
Coconut Butter	1½ lb
Water	2 qt
MARSHMALLOW WHIP (*see*)	6 lb
Vanilla Flavor	3 teaspoonfuls
Macaroon Coconut	3½ lb

The sugar, corn syrup, coconut butter, and water are placed in a large copper pan on the stove and stirred until dissolved. When the batch begins to boil it is covered and allowed to steam for a few minutes. The sides are washed down and a thermometer is placed in the batch, which is cooked to 260°F. The batch is then placed on a stand, and the marshmallow whip is added, stirred, and beaten with a large spatula until thick. The flavor and coconut are then stirred in and mixed thoroughly. The batch is turned out on boards, which are lined and covered with wafer paper, and allowed to set overnight. It is then sliced down, cut in oblongs, and wrapped as usual.

This formula will yield approximately 36 pounds of candy.

COCONUT PATTIES, CHOCOLATE

The formula for COCONUT PATTIES, VANILLA (*see*) is followed. The batch is cooked to 236°F and 1½ pounds of finely cut bitter chocolate are stirred in. The batch is finished in the usual manner, and should yield approximately 25 pounds of candy.

COCONUT PATTIES, STRAWBERRY

The formula for COCONUT PATTIES, VANILLA (*see*) is followed. Before stirring in the French cream, the batch is colored a bright pink and flavored with strawberry. It is then finished as usual and should yield approximately 25 pounds of candy.

COCONUT PATTIES, VANILLA

Sugar	8 lb
Corn Syrup	2 lb
Water	1 qt
FRENCH CREAM (*see*)	8 lb
Glycerin	2 tablespoonfuls
Vanilla Flavor	2 teaspoonfuls
Fine Macaroon Coconut	4 lb

The sugar, corn syrup, and water are placed in a copper pan on the stove and stirred until dissolved. When the batch begins to boil, it is covered and allowed to steam for a few minutes. The sides are washed down and a thermometer is placed in the batch, which is cooked to 238°F. The batch is then moved to a stand near the slab, and the fondant (French cream), glycerin, and flavor are stirred in. The coconut is then added and mixed thoroughly. The candy is dipped out of the pan with an ice-cream dipper to the desired size and deposited on the slab, which is lined with heavy wax paper. It is allowed to cool on the slab until it comes off the paper clean. Each piece is then wrapped in moistureproof cellophane.

This formula should yield approximately 25 pounds of candy.

COCONUT SEA-FOAM KISSES: *See* SEA-FOAM KISSES, COCONUT.

COCONUT SQUARES, NEAPOLITAN TYPE

Sugar	15 lb
Corn Syrup	25 lb
Coconut Butter	1 lb
Water	2 qt
Macaroon Coconut	18 lb
Vanilla Flavor	2 tablespoonfuls
Bitter Chocolate (cut up)	2 lb

The sugar, corn syrup, coconut butter, and water are placed in a large copper pan on the stove and stirred until dissolved. When the batch begins to boil, it is covered and allowed to steam for a few minutes. The sides are washed down and a thermometer is placed in the batch, which is cooked to 234°F. The coconut and flavor are stirred in and mixed thoroughly. One third of the batch is poured out on a greased slab lined with iron bars and rolled out to about 1/4 inch thickness. Since this batch is not yet colored, it is referred to as a white batch. The remaining two thirds of the original batch is divided into two equal parts. The first third is placed in a separate pan, colored a bright pink, turned out on top of the white batch, and spread evenly. The bitter chocolate is stirred into the remaining third of the batch and mixed thoroughly. This last portion is turned out on top of the white and pink batch, which is spread out, sized down evenly and allowed to cool. It is then cut into squares, rolled in granulated sugar, and prepared for packing.

This formula should yield approximately 70 pounds of candy.

COCONUT SQUARES, RASPBERRY

Sugar	12 lb
Corn Syrup	12 lb
Water	2 qt
Raspberry Pulp	3 lb
Raspberry Flavor	3 teaspoonfuls
Red Color	2 teaspoonfuls
Macaroon Coconut	15 lb

The sugar, corn syrup, and water are placed in a large copper pan on the stove and stirred until dissolved. When the batch begins to boil it is covered and allowed to steam for a few minutes. The sides are washed down and a thermometer is placed in the batch, which is cooked to 236°F. The raspberry pulp is stirred in and the batch is colored a deep red, the flavor is added, and finally the macaroon coconut is stirred in gradually and mixed through thoroughly. The batch is then turned out on a greased slab between iron bars, spread out evenly with a palette knife to about 1 inch thickness and allowed to cool until hard. It is then sliced, cut into squares, rolled into fine granulated sugar, and packed in boxes.

This formula should yield approximately 50 pounds.

COCONUT STARS, CHERRY

The formula for COCONUT STARS, VANILLA (*see*) is followed, but the batch is colored a bright red and flavored with cherry. Also about 1½ pounds of chopped glacéed cherries are stirred into the batch with the coconut. The batch is then finished in the usual manner and should yield approximately 25 pounds of candy.

COCONUT STARS, CHOCOLATE

The formula for COCONUT STARS, VANILLA (*see*) is followed, with the addition of 1½ pounds of cut-up bitter chocolate, which is stirred in before the fondant (French cream) and vanilla flavor are added. Also, 1½ pounds of chopped glacéed pineapple is stirred in together with the coconut. The batch is then finished in the usual manner, and should yield approximately 28 pounds of candy.

COCONUT STARS, VANILLA

Sugar	8 lb
Corn Syrup	2 lb
Water	1 qt
FRENCH CREAM (*see*)	7 lb
Glycerin	2 tablespoonfuls
Vanilla Flavor	2 teaspoonfuls
Macaroon Coconut	3½ lb
Chopped Walnuts	1 lb

The sugar, water, and corn syrup are placed in a clean copper pan on the stove and stirred until dissolved. When the batch begins to boil, it is covered and allowed to steam for a few minutes. The sides are washed down and a thermometer is placed in the batch, which is cooked to 236°F. The batch is then moved to a stand near the slab, and the fondant, glycerin, and flavor are stirred in. The coconut is then added and mixed thoroughly, and the pan is set in a hot-water bath. For depositing, a baker's ornamenting bag with a star tube placed on the end is used. The bag is filled about three-quarters full with the coconut paste, which is squeezed out with the right hand and deposited on heavy wax paper in long shapes about 1½ inches each. After cooling on the slab, the candy is wrapped and packed.

This formula should yield approximately 25 pounds of candy.

COCONUT STRIPS NO. 1

Sugar	8 lb
Corn Syrup	12 lb
Molasses	4 qt
Evaporated Milk	4 lb
Salt	1 tablespoonful
Coconut Butter	1 lb
Fine Macaroon Coconut	15 lb
Vanilla Flavor	4 teaspoonfuls

All of the ingredients, excluding the coconut and vanilla flavor, are placed in a large copper pan on the stove and stirred until dissolved. When the batch begins to boil, the sides are washed down, the batch is stirred constantly and cooked until it forms a soft ball when tested in cold water. The coconut is then stirred in gradually, and the flavor is added and mixed through thoroughly. The batch is then turned out on a greased slab between iron bars and spread out evenly to about 1½ inch thickness. It is allowed to set on the slab until hard, when it is sliced down, cut into strips of desired size, rolled in granulated sugar, and finally packed in trays.

This formula should yield approximately 48 pounds of candy.

COCONUT STRIPS NO. 2: *See* SAVING WASTE.

COCONUT VARIETY CARAMELS: *See* VARIETY CARAMELS, COCONUT.

COFFEE CORDIALS

The VANILLA CORDIALS (*see*) formula is followed but for the vanilla flavor, ½ teaspoonful of coffee extract of the *mocha arabica* type is substituted. The cordials are then cast in starch, handled carefully, and finished the same way as the vanilla cordials.

The batch will yield approximately 20 pounds of cordials.

COFFEE ICE CREAM

Fresh 20% Cream	12 qt
Fresh 40% Cream	4 qt
Condensed Milk	4 qt
Sugar	12 lb
Pure Instant Coffee	6 oz

All of the ingredients are placed in a 40-quart ice-cream can and stirred with a long paddle until the sugar is dissolved. The mix is then strained into the freezer and processed in the usual manner. The same freezing process is used as in making VANILLA ICE CREAM NO. 1 (*see*).

The batch will yield approximately 38 quarts.

COFFEE PARFAIT

Fresh 20% Cream	16 qt
Powdered Instant Egg Yolks	1 lb
Sugar	15 lb
Pure Instant Coffee	6 oz

The egg yolks are mixed dry with the sugar and placed in the mixing can. The cream is poured on top of the sugar, and the mixture is stirred until all the sugar is dissolved. The instant coffee is stirred in and mixed thoroughly. The mix is then strained into the freezer and completed as under VANILLA ICE CREAM NO. 1.

The batch will yield approximately 35 quarts.

COFFEE SYRUP

Simple Syrup	4 qt
Instant Coffee	8 oz
Hot Water	1 qt
Glycerin	4 teaspoonfuls

The instant coffee is dissolved in the hot water. The glycerin is mixed into the coffee solution. This mixture is then stirred into the simple syrup and covered in a stone crock. The syrup is then ready for use.

The batch will yield approximately 5 quarts of coffee syrup.

COGNAC CORDIALS

The VANILLA CORDIALS (*see*) formula is followed. The batch is flavored with ½ teaspoonful of concentrated cognac extract. It is then processed in the same manner as the VANILLA CORDIALS (*see*).

The batch will yield approximately 20 pounds of cordials.

COGNAC TOFFEE

Sugar	12 lb
Corn Syrup	8 lb
Coconut Butter	4 oz
Salt	1 teaspoonful
Butter	1 lb
Fresh 18% Cream	2 qt
Evaporated Milk	2 lb
Concentrated Extract of Cognac	½ oz
Liquid Pink Color	1 teaspoonful

The sugar, corn syrup, salt, coconut butter, and cream are placed in a copper pan on the stove and stirred until dissolved. When the batch begins to boil, the evaporated milk is added and the stirring and cooking are continued while the butter is added. The batch is cooked to 254°F. It is then poured out on a cold greased slab between iron bars and the edges are folded together while stirring in the cognac extract. The batch is colored a light pink and cooled rapidly until firm enough to handle. It is then cut and wrapped as usual.

This batch will yield approximately 25 pounds of candy.

COLD-WATER TEST FOR CRISP HARD CANDY

A round stick is dipped first in cold water and then in the cooking batch. Some of the cooked candy is dipped out on the stick, which is placed back in the cold water. The candy is removed and cooled off properly. If the candy crunches in the mouth and does not stick to the teeth, it is properly cooked.

COLD-WATER TEST FOR SOFT CANDIES: *See* under COLONIAL NOUGAT, VANILLA.

COLONIAL NOUGAT, CHOCOLATE

The formula for COLONIAL NOUGAT, VANILLA (*see*) is followed with the following change: 3 pounds of chopped bitter chocolate are stirred into the cooked batch before the marshmallow whip is added. The batch is finished as usual and should yield approximately 35 pounds.

COLONIAL NOUGAT, VANILLA

Sugar	11 lb
Corn Syrup	9 lb
Salt	1 tablespoonful
Fresh 18% Cream	4 qt
Coconut Butter	1 lb
MARSHMALLOW WHIP (*see*)	5 lb
Chopped Brazil Nuts	6 lb
Vanilla Flavor	3 teaspoonfuls

The sugar, corn syrup, salt, cream, and coconut butter are placed in a large copper can on the stove and stirred until dissolved. Stirring is continued until the batch is cooked to a hard ball. It is tested in cold water. In making the COLD-WATER TEST a can of cold running water is placed on the table near the stove. When the batch looks nearly done a stick, about 8 inches long and 1 inch in diameter, is dipped first into the water and then into the cooking batch. A small sample of the batch is taken out on the stick, which is then put back into the cold water for a minute. The cool sample is taken out with the right hand and formed into a ball. The ball measures the consistency of the batch; it will feel soft, medium, or hard to the touch. After making the cold-water test the heat is shut off and the batch is removed to a stand near the slab where the marshmallow whip is added, stirred, and beaten with a large spatula until thick. The vanilla flavor is stirred in and the chopped Brazil nuts are added and mixed through. The batch is turned out on trays, which are lined and covered with wafer paper. It is allowed to set overnight, and is then cut and wrapped as usual.

This will yield approximately 35 pounds.

COMBINATION SUCKER (LOLLIPOP) AND HARD CANDY CUTTING MACHINE

This machine is used for the manufacture of lollipops as well as solid and filled hard candies. (Courtesy of the Vacuum Candy Machinery Co. and the Racine Confectioners' Machinery Co., New York, New York.)

COMMERCIAL CREAM COATING FOR COCONUT BONBONS

Sugar	50 lb
Corn Syrup	10 lb
Cream of Tartar	3 teaspoonfuls
Water	8 qt
Gelatin	1 oz

(150 bloom dissolved in 1 cup of water)

Melted 76% Coconut Butter	1 lb
Flavors and Liquid Colors	½ teaspoonful of each to each 10 lb of cream

The sugar, corn syrup, cream of tartar, and water are placed in a large copper pan on the stove and stirred until dissolved. When the batch begins to boil, it is covered and allowed to steam for a few minutes. The sides are then washed down and a thermometer is placed in the batch, which is cooked to 244°F. The batch is then poured out on a cooled, dampened cream beater and allowed to cool to about 110°F. The soaked gelatin is then added and the machine is set in motion. The flavor is added and then the melted coconut butter is mixed in. The machine is kept running until the batch creams to a lump. After sweating back for 20 minutes, the batch is ready to remelt.

For remelting, the cream is placed in a steam-bath copper pan on medium heat. Five pounds of invert sugar are added and stirred constantly until melted to the dipping consistency (not too thin).

The dipping process is somewhat similar to chocolate coating. Some cream is poured out on the marble. The flavor is stirred in while the cream is cooled to lukewarm. A handful of centers is covered with the cream at one end of the marble. They are then deposited on dipping trays lined with wax paper. The tops of the bonbons are touched with the finger tips to give them the desired rough appearance. The bonbons are allowed to cool until they set dry, when they are packed in the usual manner.

The honey coconut centers are dipped in white, yellow, pink, and green.

This formula should yield approximately 68 pounds of cream coating.

COMMERCIAL FUDGE, CHOCOLATE

The formula for COMMERCIAL FUDGE, VANILLA (*see*) is followed. When the batch is removed from the stove, 4 pounds of finely cut bitter chocolate are stirred in. When this is completely melted, the nuts and flavor are added and mixed thoroughly. The batch is finished in the usual manner and should yield 69 pounds of candy.

COMMERCIAL FUDGE, VANILLA

Sugar	22 lb
Corn Syrup	8 lb
Caramel Paste	5 lb
Salt	1 tablespoonful
Evaporated Milk	12 lb
Cream Fondant	16 lb
Chopped English Walnuts	4 lb
Vanilla Flavor	4 teaspoonfuls

The sugar, corn syrup, caramel paste, salt, and milk are placed in a copper pan on the stove and stirred until dissolved. When the batch begins to boil, the sides are washed down. Stirring is continued, and the batch is cooked until it forms a soft ball when tested in cold water. It is then moved to a stand near the slab and the fondant is stirred in rapidly. When the fondant is completely melted, the walnuts and the flavor are stirred in and mixed thoroughly. The batch is then turned out on boards lined with heavy wax paper and allowed to cool off. It is then marked and cut into squares and packed as usual.

This batch should yield 65 pounds.

COMPOSITION CHOCOLATES

Dark Vanilla Sweet Chocolate	10 lb
High-Quality Milk Chocolate	5 lb
Chopped Roasted Almonds	1 lb
Chopped Roasted Filberts	1 lb
Chopped Pecans	1 lb
Chopped Pistachio Nuts	2 lb

The dark and milk chocolate are melted and cooled as usual. When they are cool enough for dipping, the almonds, filberts, and pecans are placed in a deep crystal pan and mixed thoroughly. Some colored aluminum cups are laid out in flat pans, and the mixed nuts are mixed in with the chocolate coating in the same way as for clusters. Small tidbits of the chocolate composition are deposited into the cups and topped with chopped pistachio nuts. The chocolates are cooled and packed in the usual manner.

The yield of this batch should be exactly 20 pounds.

Note: These chocolates are used to decorate the top layer of all nut and fruit and nut packages.

CONTINUOUS HARD-CANDY CUTTER

This machine automatically cuts all sorts of chips, lumps, pillows, butter cups, and kisses in 3/8, 3/4 and 1 1/2 inch lengths. (Courtesy of the Savage Brothers Co., Chicago, Illinois.)

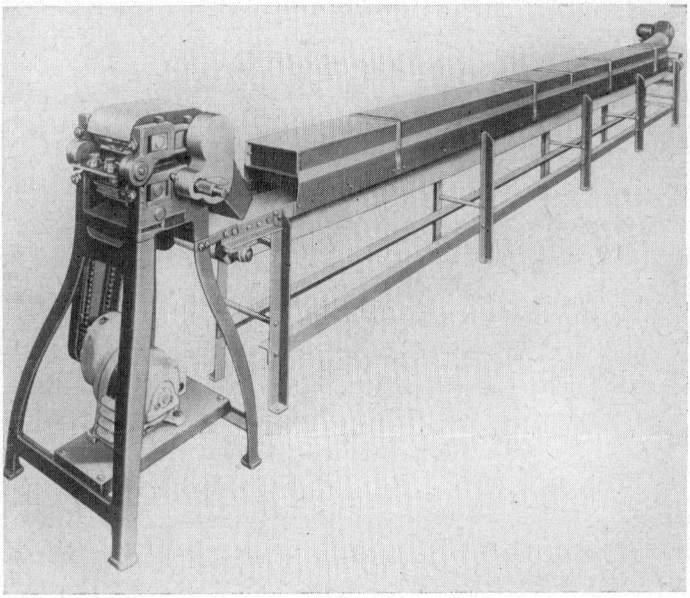

CORDIALS

In this encyclopedia four different types of cordials are described: (1) fruits hand dipped in cream, (2) starch-cast liquid fruits, (3) cordials formed in metal molds, and (4) starch-cast liquid syrup cordials. Since cordials are the most expensive chocolates to make, the cautioning statements which supplement the formulas should be carefully noted.

Where the fruits are first hand dipped in the fondant and then chocolate coated, it is important to remember that they must be well drained. There is enough juice inside the canned fruit to melt part of the cream and liquefy the fruit after it is chocolate covered. If the fruit is not thoroughly drained, there is a possibility that excessive juice remaining outside the fruit will cause the chocolates to ferment shortly after they are coated. When the cream is remelted, it should not be overheated. If the fruit is dipped in overheated cream, excessive heat will make it sweat too much juice and the cream may melt away before the dipped fruit is cool enough to be chocolate covered. The fruits should be chocolate coated rapidly as soon as they cool; this prevents the centers from liquefying before they are chocolate covered.

Hand-dipped liquid fruits are a high-quality product. They are made by hand every step of the way, usually by retail manufacturers who can command a high price. Starch-cast liquid fruits are usually made by wholesale manufacturers as a competitive item. Their *raison d'être* is economy. Although they do not compare in quality with hand-dipped liquid fruits, they can be made so that they stay liquid and are fairly good eating. In order to produce this type of cordials, the following suggestions should be noted. The starch for casting must be clean and dry. It should be kept in a warm room overnight. This will prevent the starch from sticking to the cordials. For the same reason, the fruit must be well drained before it is deposited in the molds. The remelted cream must not be overheated. The usual remelting temperature for all starch-cast cream centers also holds for starch-cast liquid fruits. The production schedule is usually arranged so that these goods are cast, cleaned of starch, and chocolate covered on the same day. Permitting the centers to stay in starch too long will cause the liquid to leak into the starch and make the starch stick to the centers. This will affect eating quality. Also, if the brushed-off centers remain uncoated for any length of time, they will melt and stick together. That is why making starch-cast liquid fruits should be a one-day process. If this suggestion is followed, the fruits will stay liquid.

In making the third type of cordial, the metal molds should first be warmed; not too warm, just enough to take the chill off. The chocolate should be cooled in the same way as it is for hand coating, but it should not be too cold. When the first chocolate shell is deposited, the molds are chilled to permit the shell to harden; then the centers are put in. The centers should be cooled before their tops are covered with cooled chocolate, since they have to be at the same temperature as the shells before chocolate ceilings are put on them. If the centers were to remain warm, they would cause the chocolate ceilings to turn gray. The chocolates must cool to the hardening point before they are removed. Removing chocolates from molds prematurely causes excessive breakage and produces dull chocolates. When they are removed at the hardening point, they have a nice gloss.

Liquid-syrup cordials are difficult to make; so difficult that most confectioners do not attempt to make them. The formulas in this chapter were developed by the writer after several years of experiment. The citrus pectin in these formulas stabilizes and liquefies at the same time. The result is firm cordials that are not hard to handle, cordials that will be 90 percent liquid in 3 to 4 days after they are coated.

Liquid-syrup cordials are of a low cook (234°F). Theoretically, the liquid state is achieved in the following manner: When the crystal syrup is cast in warm starch, the moderate heat of the starch and the warm room cause the syrup to crystallize and form a shell around the molded piece. Some syrup remains inside the molded piece and it becomes a cordial.

This method has not proved satisfactory because it does not work systematically and uniformly. Some of the molded cordials have too much syrup and some have none. This means excessive breakage and a nonuniform product. The pectin acts as a stabilizer and binder so that the cast centers are all bound together and act alike. They set quickly and do not absorb any starch. This means they are completely free of starch when brushed off and easy to handle. All centers have the same amount of cordial in them.

The confectioner should follow the liquid syrup formulas with special care.

No extra equipment is needed to make chocolate cordials. The necessary equipment is usually found in the cream and chocolate departments.

For the various cordial formulas, *see* the individual names, such as, LIQUID CHERRIES, LIQUID PINEAPPLES, LIQUID STRAWBERRIES, ETC. *See also* GENEVA CHOCOLATE CORDIALS.

COSMOPOLITAN CHOCOLATES

Dark Sweet Vanilla Chocolate	20 lb
High-Quality Milk Chocolate	20 lb

A good, varied set of metal molds in assorted fancy shapes is needed to make this type of chocolates. At first, the molds are warmed a little, enough to take the chill off. The chocolate is then melted as usual, stirred, and cooled to about 84°F. It is then cooled again and tempered in the same way as regular chocolate coating. The molds are filled with the chocolate, and then a small palette knife is run over the top so that the edges are trimmed off. The filled molds are given a slight tap on the table, so that the tops are smoothed. The molds are then placed in a cool room at 64°F, where they remain until the chocolate is thoroughly cool and feels crisp. The molds are then tapped flat on the table and the chocolates are carefully removed and packed assorted.

The batch will yield 40 pounds of chocolates.

COUGH DROPS

Sugar	25 lb
Water	3½ qt
Cream of Tartar	2½ teaspoonfuls
Licorice Paste	1 oz
Charcoal Color	1 tablespoonful
Oil of Wintergreen	10 g
Oil of Sassafras	10 g
Paregoric	3 g
Oil of Aniseed	5 g

The sugar, water, and cream of tartar are placed in a copper pan on the stove and stirred until dissolved. When the batch begins to boil, it is covered and allowed to steam for a few minutes. The sides are washed down and a thermometer is placed in the batch, which is cooked to 336°F. The batch is poured out on a greased slab and colored with licorice paste and charcoal color. The edges are folded together, while stirring in all four extracts prescribed. Folding is continued until the batch is cool enough, when it is taken over to the spinning table in front of the batch warmer. It is then passed through a flat or round roller machine and allowed to cool on the table. The drops are broken apart and blended with a mixture of magnesia powder and powdered sugar (half and half). They are then put through a sieve and packed as desired.

The batch will yield approximately 23½ pounds.

COUNTER GOODS

Counter goods, or miscellaneous candies, usually get more display space than chocolates in the average sweet shop. They differ from chocolates in several ways. Since chocolate centers are covered with black or milky coating, the consumer cannot tell how the chocolates will taste merely by glancing at them. However, when counter goods are displayed, their many colors and flavors appeal directly to the eye. The consumer comes closer to tasting counter goods when he looks at them. Whereas chocolates are usually gift purchases, counter goods are usually bought for home consumption. Attractive appearance and display are important factors in building a consistent demand for these goods. Whereas chocolates have no protection against hot weather, there are some crystallized soft candies, hard candies, and chewy candies wrapped in moistureproof cellophane which can stand up under any weather conditions. Some of these, e.g., lime and peppermint candies and jellies, and crystallized creams, have flavors that are particularly refreshing in hot weather.

When the sale of chocolates declines during the summer months, the sale of counter goods usually picks up because they are more adaptable to hot weather. It is a good practice to introduce a special summer package consisting of about 8 to 10 popular candy types picked from the general counter-goods line. Usually a one-layer box is used. The color scheme of the box usually suggests the summer scene. The one-pound box is usually wrapped in moistureproof cellophane. There can also be a half-pound box with an open-top cellophane display.

Several suggestions for making counter goods that will stand up under all weather conditions follow. Hard candies cannot be cooked any higher than the formulas specify. The only precaution that can be taken against hot weather is in wrapping them. The hard candies should be wrapped immediately before they get sticky. They should be wrapped either in wax paper or in moistureproof cellophane. All chewy candies should be cooked 2 to 4 degrees higher. The amount of coconut butter that goes into the batches should be increased about 10 percent. These candies are also wrapped immediately after they are made, and wax paper or moistureproof cellophane is best for wrapping. Where wrapping machines are used, hard cocoabutter is rubbed on the surface of the batch to prevent it from sticking to the cutting knife and tearing up the wrappers. When making gums and jellies, the cook cannot be changed. Therefore, the method of crystallizing is changed slightly. The crystal is cooked 1 to 2 degrees higher. The centers to be crystallized must be completely dried before they are placed in the crystal pans. The crystal pans should be only one third instead of one half filled. When the gums and jellies are drained of the crystal, the drying is prolonged until they are perfectly dry. After they are crystallized and dried, these candies should be packed with extra care.

When making crystallized creams and bonbons, the fondant should be cooked 2 to 4 degrees higher. The crystal should be handled as it is in making gums and jellies. When making hand-dipped bonbons, the fondants for both the centers and the dipping cream are cooked 2 to 4 degrees higher. All hand-dipped bonbons should be packed in waxed-paper cups.

With some exceptions, counter goods do not present equipment and machinery problems. Most of these are hand-made products. Almost every department in the busy factory contributes a few counter-goods items. The investment involved in making Vienna-filled plastic hard candies is considerable. A technician from the factory supervises the installation of the machinery and directs its operation.

For the various counter-goods formulas, *see* the individual names, such as ALMOND-CRISP PATTIES, BUTTERSCOTCH SUCKERS, LOG-NUT ROLL, ETC. *See also* SEA-FOAM KISSES.

CREAM COCONUT SQUARES, MAPLE

The formula for CREAM COCONUT SQUARES, VANILLA is followed, but the batch is flavored and colored with 4 teaspoonfuls of maple flavor. It is then finished in the usual manner, and should yield approximately 22 pounds of candy.

CREAM COCONUT SQUARES, STRAWBERRY

The formula for CREAM COCONUT SQUARES, VANILLA is followed, but the batch is colored pink and flavored with strawberry. It is then finished in the usual manner, and should yield approximately 22 pounds of candy.

CREAM COCONUT SQUARES, VANILLA

Sugar	6 lb
Corn Syrup	4 lb
Water	2 qt
RUNNING CREAM (*see*)	5 lb
Sliced Coconut	4 lb
Vanilla Flavor	2 teaspoonfuls
Glycerin	2 teaspoonfuls

The sugar, corn syrup, and water are placed in a copper pan on the stove and stirred until dissolved. When the batch begins to boil, it is covered and allowed to steam for a few minutes. The sides are washed down and a thermometer is placed in the batch, which is cooked to 244°F. The batch is then moved near the slab, and the running cream is added and stirred until dissolved. The coconut is mixed and the batch is stirred as the flavor and glycerin are added. The batch is then turned out on boards lined with wax paper and spread out to about ½ inch thickness. This is allowed to set overnight. The boards are then turned upside down on the table, the paper is pulled off, and the coconut candy is cut into 1½ inch squares. After the squares are dry, they are packed in the usual manner.

This formula should yield approximately 22 pounds of candy.

CREAM CUTS: *See Individual Cream Cuts Formulas,* e.g., ALMOND CREAM CUTS and PECAN CREAM CUTS.

CREAM DATE KISSES

Sugar	8 lb
Cream of Tartar	1 teaspoonful
Water	1 qt
FRENCH CREAM (*see*)	6 lb
Orange Emulsion	1 teaspoonful
Chopped Dates	3 lb
Glycerin	2 teaspoonfuls

The sugar, water, and cream of tartar are placed in a small copper pan on the stove and stirred until dissolved. When the batch begins to boil, it is covered and allowed to steam for a few minutes. The sides are washed down, and a thermometer is placed in the batch, which is cooked to 238°F. The batch is then removed from the heat, and the cream is stirred in until it is melted. The flavor and glycerin are then added, and the chopped dates are stirred in and mixed through. The batch is then spooned out into small kisses and deposited on wax paper-lined trays. The kisses are allowed to set until dry and are then packed in boxes.

This formula should yield approximately 18 pounds of candy.

CREAMED ALMONDS

Gum Arabic (dissolved in 2 qt of water)	1 lb
Cream of Tartar	5 teaspoonfuls
Sugar	60 lb
Water (enough to dissolve the sugar)	
White Vanilla Flavor	1½ oz
Medium-Size Almonds	25 lb

The gum arabic solution is heated and strained. The almonds are placed in a warm revolving pan and allowed to run until they are fairly warm. The gum arabic solution is then applied to the almonds and allowed to revolve until the almonds are fairly coated with the gum arabic. They are then covered with about 2 pounds of sifted powdered sugar. Revolving is continued until the almonds are dry and roll freely in the pan. They are then ready for the first coating.

To make the coating, 20 pounds of sugar, 3 quarts of water, and 2 teaspoonfuls of cream of tartar are placed in a copper pan on the stove and stirred until dissolved. The sides are washed down and the batch is cooked to 242°F. This syrup is then applied gradually to the almonds while the pan is in motion and about one third of the syrup is used up. A wooden paddle is used to stir the almonds and syrup to keep them from sticking together. The almonds will crystallize and a second charge of syrup is then applied. This process is continued until all the syrup is used up, and then the nuts are allowed to run freely for a few minutes. They are now ready for the second coating.

This consists of 30 pounds of sugar, 4 quarts of water and 3 teaspoonfuls of cream of tartar which are cooked to 244°F. Again, the almonds are treated with this syrup as in the first coating.

The syrup must be added in a slow stream to prevent the nuts from expanding and breaking through the coating. The third and final coating is then prepared as follows:

Ten pounds of sugar, 3 pints of water, and ½ teaspoonful of cream of tartar are cooked to 248°F. The vanilla flavor is stirred in, and the almonds are finally treated with this syrup as in the first coating operation. The nuts are allowed to run until they are dry. Then they are removed, placed in crystal pans, and crystallized with a warm dry crystal for approximately 6 hours (*see* CRYSTALLIZING). They are then drained, allowed to dry, and packed in boxes.

The batch will yield approximately 175 pounds.

CREAMED BARS: *See Individual Creamed Bars Formulas,* e.g., ORANGE CREAMED BARS and RASPBERRY CREAMED BARS.

CREAM FILBERTS

Raw Large Filberts	20 lb
Sugar	100 lb
Water	12 qt
Gum Arabic	6 lb
(dissolved in 6 qt of water)	
Corn Syrup	20 lb

The gum arabic solution is placed in a copper pan on the stove over low heat and stirred until dissolved. It is then put through a fine sieve and the corn syrup is stirred into the gum and mixed through thoroughly. The gum arabic solution is then ready for use.

The raw filberts are placed into the revolving pan, which is set in motion. The filberts are sprinkled with the gum arabic solution and dried with tapioca flour. They are then removed from the pan, spread out on clean trays, and allowed to dry overnight.

The sugar and the water are then placed in a large copper pan on the stove and stirred until dissolved. When the batch begins to boil, it is covered and allowed to steam for a few minutes. The sides are washed down and a thermometer is placed in the batch, which is cooked to 246°F.

The syrup is allowed to cool about 20 minutes, and then the dry filberts are placed back into the revolving pan and started up. The syrup is sprinkled on the filberts, a little at a time, until one third of the syrup is used up. The filberts in the pan are stirred through with one hand to prevent them from sticking together. The nuts are allowed to run in the pan until they are dry, and then the second portion of the syrup is applied to the filberts in the same way as the first time. They are allowed to dry again. The last third of the syrup is flavored with 1½ ounces of vanilla flavor and then the last operation is completed following the same method. The filberts are then removed and spread out on clean trays to dry. They are then placed in crystal pans and crystallized in the usual manner with a short crystal (*see* CRYSTALLIZING). The nuts are then drained, dried, and packed in boxes.

This batch should yield approximately 175 pounds of candy.

CREAM FONDANT FOR FUDGE

Sugar	40 lb
Water	8 qt
Cream of Tartar	3 teaspoonfuls
MARSHMALLOW WHIP (*see*)	5 lb
Invertase	2 teaspoonfuls

The sugar, water, and cream of tartar are placed in a copper pan on the stove and stirred until dissolved. When the batch begins to boil, it is covered and allowed to steam for a few minutes. The sides are washed down and a thermometer is placed in the batch. When it reaches 240°F, it is reduced with 1 quart of water, and it is finished cooking at 244°F. The batch is then poured into a dampened cream-beating machine and sprinkled with a few drops of cold water on top. The batch is cooled rapidly to lukewarm, and the invertase and marshmallow whip are added. The power is then turned on and the beating process begins and continues until the batch becomes a white lump. It is then placed in a metal container, covered, and used when wanted. This batch should yield 50 pounds of cream fondant.

CREAM MINTS, ASSORTED

Sugar	25 lb
Water	3 qt
Cream of Tartar	2 teaspoonfuls
Oil of Peppermint	3 teaspoonfuls
Glycerin	1 oz

The sugar, water, and cream of tartar are placed in a copper pan on the stove and stirred until dissolved. When the batch begins to boil, it is covered and allowed to steam for a few minutes. The sides are then washed down and a thermometer is placed in the batch, which is cooked to 270°F. The batch is then poured out on a cold, greased slab and the edges are folded together while the oil of peppermint is added and mixed through. The batch is cooled rapidly until it is hard enough; then it is taken over to the pulling machine. The glycerin is added while the batch is being pulled. The pulling is continued for 7 minutes, and then the batch is taken over to the "long taffy" wrapping machine, where it is cut and wrapped in the usual manner. The mints are kept overnight in a warm place and are then creamed and ready for sale. Creamed mints are usually made in assorted colors of yellow, green, and pink. They are all flavored mint and the color is added to the batch on the pulling machine.

This batch should yield approximately 25 pounds of candy.

CREAM MINTS, CAST

The formula for VANILLA CREAM, CAST is used. The cream is flavored with 4 teaspoonfuls of the best oil of peppermint. The flat patty mold is used. The creams are finished in the usual way. Thirty pounds of cream centers will yield about 40 pounds of chocolates when chocolate coated by enrobing machine, or 45 pounds when hand dipped.

CREAM WAFERS: *See* STRAWBERRY-CREAM WAFERS.

CREAMS

Fondants and creams are the most delicate candy specialties. When chocolate covered and correctly flavored, smooth cream centers are a favorite. When the flavors are correctly blended, crystallized French creams in pastel colors are delicious.

When the fondant is the main ingredient, smoothness is most important. Some important pointers in making fondant follow. Cream fondant is always cooked 2 to 4 degrees higher in hot humid weather; cooked in this fashion, it will stand up more firmly and absorb more moisture. Water is an important ingredient in cream fondant. More water kills the grain, inverts more of the syrup, and gives a smoother fondant. Another important point: The batch should be well cooled before creaming is started. This will help prevent grainy batches, since cool batches prolong the beating time of the cream before creaming occurs. Prolonged beating causes more crystals to break up, and this means a smoother cream.

The fondant should be remelted on low heat and the consistency of the cream should be reduced with some simple syrup. Invertase is used as a softener. When making fondant that will be remelted for cast centers, it is good practice to have the fondant rather firm with more body to it (i.e., not too flat). When a firm cream is remelted with the proper amount of simple syrup, it does not require too much heating before it gets to the proper consistency for casting. It also sets more rapidly after it has been cast in starch, so that when the creams are removed from the starch boards and brushed free of starch, they are rather firm and stand up well. This makes them more suitable for a good job of chocolate coating.

However, when a flat cream is remelted, more applied heat is necessary to enable the creams to set after they are cast in starch. More heat gives them a rough finish and kills the smoothness. If not enough heat is applied in remelting flat creams, the setting is slow and the result is soft creams with a rough finish that are flat and sticky and very difficult to coat with chocolate. The finished product is of poor quality.

To obtain a firm cream, the batch should be cooked 1 or 2 degrees higher. The water that is poured on the cream beater to dampen it should be watched very carefully; never should more be used than ½ cup of water for a smaller batch and 1 cupful for a large batch. Also, no more than a few drops of cold water should be sprinkled on top of the batch after it has been poured on the cream beater. Although it has been stated that the batch should be well cooled before starting the creaming process, excessive cooling should be avoided. The batch should never be cooler than 90°F. If these suggestions are followed, the batch will cream up perfectly.

After being coated or dipped, cream centers undergo a chemical change. Formulas must, therefore, be followed carefully. Invertase, which is used as a softener, causes a chemical change in cream centers after they are chocolate covered or crystallized. If softening is to occur properly, i.e., in 2 or 3 days after the cream centers are chocolate covered or crystallized, the addition of invertase should be carried out with great care. There are two ways of adding invertase to the creams: on the cream beater before creaming begins (as is noted in most of the formulas for cream centers); or, when making remelted creams, during the remelting process. The first method, i.e., adding invertase on the cream beater, is the only one that can be used when making hand-rolled creams. There is no problem here, because there is no remelting to be done. The hand rolling is the final stage. When making remelted creams, however, it should be emphasized that no matter how the invertase is added (whether during remelting or on the cream beater before the creaming process begins) excessive heat in remelting will cancel out the softening effect of the invertase afterward.

Copper pans, machines, and all tools that are used in the cream department should be spotlessly clean. This will prevent the finished product from fermenting and thus safeguard its shelf life.

All cream batches should be poured into clean, cooled cream beaters. The water should be kept running under the machine while the syrup is cooling. Graining should be carefully watched. Whenever grain appears, the batch should be sprinkled with cold water. The water may be shut off while the machine is in motion.

In remelting creams, low heat is used. The cream is stirred gradually until it is melted and then it is immediately removed from heat. The remelting point is usually 120°F for cast creams or 130°F for bonbon dipping. The cream may be reduced with a little water if it thickens during the dipping process.

When a batch running on the cream beater takes too long to cream up, the situation is usually remedied by shutting the machine off for a few minutes. This gives the batch a chance to crystallize partially and causes it to cream more rapidly.

Citric acid solution, which is used in combination with fruit flavors, is prepared by dissolving 1 pound of citric acid in 1 pound of water and stirring until the acid is dissolved. The solution is then poured into clean glass bottles, which are covered, and is ready to use when needed. When used in combination with fruit flavors and occasional fruit jams to flavor cream centers, e.g., strawberry, raspberry, lemon, and orange, the citric acid solution helps bring out the flavor more strongly. It also affects invertase to a certain extent and helps produce softer and semifluid centers.

In former years, the basic fondant for cream centers was usually made of sugar and corn syrup and then remelted with a cooked bob syrup. The bob syrup was considered a labor-saving ingredient; the all-sugar fondant was considered impractical for many reasons. When invertase was still in its infancy, the all-sugar cream was considered a risk because the confectioner was not fully informed on how to use it. Its use was still something of a gamble and without this softener the cream centers made with all-sugar cream dried up in a short time, fermented, and were a total loss. That is why most confectioners considered a corn-syrup cream formula more practicable. The bob syrup was used with the corn-syrup-doctored cream, because it was easier to make than a remelted cream and saved labor. Although this method was, and still may be, effective, creams made in this fashion will produce chocolates of second quality. Cream centers of this type lack the smooth creamy texture. The centers are either rough or too fluid.

The author has recently developed a basic all-sugar cream formula for cream centers cast in starch which works satisfactorily and gives outstanding quality (*see* RUNNING CREAM). The all-sugar cream with cream-of-tartar "doctor" has the following merits: (1) It is a sweet center; (2) it is easy to remelt with low heat and does not require any bob syrup to help its remelting; (3) it produces firm centers that are easily brushed free of starch and easily blended with all flavors (either spice or fruit) and all types of fruit jams; and (4) the invertase which is used in the formula works effectively as a softener and the amount of invertase used can be regulated by the confectioner to produce centers as soft and as fluid as he wants them to be.

Many confectioners prefer cast cream centers because they lend themselves to a variety of shapes. The shapes are practically uniform throughout the candy-eating world. The vanilla cream comes in a round mold with a pointed top. The maple cream is a flat oblong shape. The lemon cream usually comes in a fruit-slice mold. The orange cream is diamond-shaped, and the raspberry heart is an old standard.

Most retail manufacturers prefer hand-rolled creams, because they are smooth and tender. Until approximately 30 years ago, hand rolls were hard to make because of fermentation and the resultant short shelf life of the product. However, with the advent of invertase, this problem was solved and the life of hand rolls extended. The author has developed a group of original hand-rolled formulas that have given satisfactory results. (See the individual formulas for creams and chocolates.)

Hand-rolled centers are usually made with invert sugar and invertase. Although this method gives a full sweet center, the excessive acid usually found in invert sugar produces two unfavorable reactions: (1) It darkens the color of the batch and cannot produce a pure white cream, and (2) if it is used more freely, i.e., 1 or 2 pounds more to a 50-pound batch of cream, it may cause a small amount of fermentation, particularly in fruit centers.

In the method developed by the author, invert sugar is replaced by corn syrup, which is more economical. The corn syrup is not cooked with the rest of the batch, but is kept ready melted and stirred into the batch after it has finished cooking. Crystallization and fermentation of the cream centers are completely eliminated. For example: 40 pounds of sugar, 2 gallons of water, and 3 teaspoonfuls of cream of tartar "doctor" may be cooked as usual. The batch is cooked 5 degrees higher (i.e., at 250°F instead of 245°F). The melted corn syrup is then added and stirred in rapidly. The batch is poured on a cooled cream beater in the usual way. It can be cooled prematurely to 100°F before creaming. The marshmallow whip, invertase, and all necessary flavoring can be added to the batch, which then is subjected to the usual creaming. After it creams to a lump, the batch is allowed to sweat back for about 15 to 20 minutes. It is then cut and rolled as customary and when the centers cool off, they are chocolate covered as usual.

Other advantages of this method are as follows: The batch is firmer after creaming and yields firm centers that are not sweaty or sticky. Vanilla or mint creams have an unusually white color—not a yellowish white. After the cream centers emerge from the hand-roll machine, they are firm rather than sticky and are suitable for chocolate coating either by hand or by the chocolate-enrobing machine. They can be machine coated without difficulty. After the chocolates are coated and placed in the storeroom for ripening, they will ripen in 3 to 4 days and may be kept in the regular stockroom for many months. They are secure against fermentation. The author kept small Easter eggs made in this fashion for 1 year and found them still edible at the end of that time.

Although the cream department is as important as the chocolate department, its quota of equipment and machinery is more modest. In the smaller candy factory the cooking is done mostly by open fire on gas stoves. In the medium shop, the steam-jacketed kettle is used. The open-type, round Ball cream beater is more satisfactory than the enclosed continuous cream beater, because it is so easy to keep clean. Its continuous water-running arrangement for cooling also improves quality because it cools the batch off rapidly and prevents grainy batches, the most persistent enemy of silky, smooth fondants.

In the smaller shop, the cream is remelted in the steam-bath copper pan; in the medium factory, it is remelted in the cream-breaker steam kettle. Either equipment will give good results if handled with experience and care.

The casting of cream centers into starch-molded impressions is usually done through the funnel by hand in the smaller retail shop. The depositing machine is used for this purpose in the larger candy factory. Both methods give good results. The main point is not to cast the cream with too much heat.

Removing and cleaning the centers of starch are also handled in two different ways. Although more expensive from the labor-cost standpoint, the hand method with the sieve and brush is the best way, particularly when a top-quality soft center is required. It is used mostly by retail manufacturers who are interested in quality. However, in the medium-sized wholesale candy factory where labor saving is a problem, firm cream centers can be made and the standard starch buck can remove them and brush off the starch.

In making hand-rolled creams, the batches are cooked and creamed as they are in making cast creams. Smaller shops use the hand method for cutting and rolling the centers, whereas the medium and larger shops use the cut-roll machine. Whatever method is used, the main thing is to follow the instructions in the formulas.

For the various cream formulas, see the individual names, such as, PECAN CREAM CUTS, BONBON CREAM, ALMOND PASTE BONBONS, etc.

CREMO FUDGE, CHOCOLATE

The formula for CREMO FUDGE, VANILLA (*see*) is followed, but before stirring in the fondant, 3 pounds of chopped bitter chocolate are stirred in. The batch is finished in the usual manner and should yield 43 pounds of candy.

CREMO FUDGE, CHOCOLATE NUT

The formula for CREMO FUDGE, VANILLA (*see*) is followed, but the batch is flavored with 3 pounds of bitter chocolate. Three pounds of chopped pecans are stirred into the batch while the vanilla is added. The batch is finished in the usual manner and should yield 46 pounds.

CREMO FUDGE, VANILLA

White Sugar	12 lb
Brown Sugar	3 lb
Corn Syrup	4 lb
Salt	1 tablespoonful
Fresh Light Cream	4 qt
Evaporated Milk	4 lb
Cream Fondant	12 lb
Vanilla Flavor	3 teaspoonfuls

Both sugars, the corn syrup, salt, and cream are placed in a large copper pan on the stove and stirred until dissolved. When the batch begins to boil, the sides are washed down and the evaporated milk is stirred in gradually. Stirring is continued and the batch is cooked until it forms a soft ball when tested in cold water. The batch is then moved to a stand near the slab, and the cream fondant is stirred in and mixed through gradually while adding the flavor. The batch is then poured out on greased pans and allowed to set overnight. It is then ready for use as desired.

This batch should yield 40 pounds of candy.

CRIMPED CHOCOLATES

Dark Vanilla Sweet Chocolate	10 lb
Pitted Dates	5 lb
Whole Brazil Nuts	5 lb

The dates and Brazil nuts are coated with milk chocolate in the usual manner. They are allowed to cool on the dipping trays. The dark chocolate is then cooled in small portions as for hand dipping. A crimping tube is inserted on the end of a baker's ornamental rubber bag. The bag is half filled with the cooled processed chocolate. The chocolate from the bag is squeezed out with the fingers of the right hand and applied on top of the covered dates and Brazil nuts. These are allowed to cool and are then packed as usual.

The yield of this batch should be approximately 45 pounds of chocolates.

Note: These pieces are used to decorate the top layers of assorted chocolates.

CRIMPED RIBBON CANDY, CINNAMON

The formula for CRIMPED RIBBON CANDY, LEMON is followed. A small white pulled stripe is prepared and reserved. The main batch is colored a deep bright red and flavored with oil of cinnamon. The batch is finished according to the original formula. It will yield approximately 28½ pounds.

CRIMPED RIBBON CANDY, LEMON

Sugar	30 lb
Water	5 qt
Cream of Tartar	3 teaspoonfuls
Lemon Flavor	3 teaspoonfuls
Color	1 teaspoonful

The sugar, water, and cream of tartar are placed in a clean copper pan on the stove and stirred until dissolved. When the batch begins to boil, it is covered and allowed to steam for a few minutes. The sides are washed down and a thermometer is placed in the batch, which is cooked to 336°F. The batch is poured out on a greased slab, and the edges are folded together while stirring in the oil of lemon. A small piece (about 1 pound) is taken off the batch for a stripe. It is pulled white and placed in front of the batch warmer to keep soft, while the main batch is folded and kneaded until it is cool enough to handle. The main batch is then transferred to the spinning table in front of the batch warmer and formed flat. It is then striped down the center on both sides with the pulled white. The flat batch is given four fine, even stripes down the center; then it is turned over and striped down the center on the other side. The batch is then spun into flat, almost paper-thin ribbons and passed through a candy crimper. The finished candy is allowed to cool on the table in front of an electric fan. It is then packed in tins.

The batch will give about 28½ pounds.

CRIMPED RIBBON CANDY, LIME

The formula for CRIMPED RIBBON CANDY, LEMON (*see*) is used. While the edges are folded together a small piece is taken off and pulled white for a stripe. The main batch is colored a delicate green and flavored with oil of lime. It is finished in the usual manner and will yield approximately 28½ pounds.

CRIMPED RIBBON CANDY, PEPPERMINT

The formula for CRIMPED RIBBON CANDY, LEMON (*see*) is followed. When the batch is poured on the greased slab, a small piece is taken off and colored a deep red, then prepared for the stripe. While the edges are folded together, the oil of peppermint is stirred into the main batch. Kneading and folding are continued until the batch is cool enough to handle. It is then pulled partly on the hook and twisted to a glossy white. The batch is then transferred to the spinning table in front of the batch warmer, where it is formed, striped, spun, and finished as usual.

It will yield about 28 pounds.

CRYSTAL SYRUP

Sugar	50 lb
Water	10 qt

The sugar and water are placed in a large copper pan on the stove and stirred until dissolved. When the batch begins to boil, it is chilled with a little cold water. Then the scum is removed, the sides are washed down, and the batch is cooked to 34 degrees on the syrup gage, which is the equivalent of 218°F. The syrup is then moved to a cool place, where it should not be disturbed in any way. The confectioner should particularly refrain from moving the copper pan or doing anything that will shake the floor. A crust will gradually form on top of the syrup, and when it is about lukewarm it is ready to use.

The soft candies to be crystallized are placed in crystal pans, which are never more than half filled. Now the syrup is poured over the soft candies. A dipper, large enough to fill one of these pans, is used. This eliminates unnecessary stirring and disturbing of the crystal. A pan is placed under the dipper to prevent the used syrup from dripping back into the unused syrup in the copper pan. A sheet of wax paper, with a hole punched in its center, is placed on top of each crystal pan to prevent a heavy crust from forming on top of the pan. Stacked from two to four pans high, the crystal pans are placed in a warm room and allowed to set until the next morning, when the syrup is drained. To facilitate draining, these pans are placed on a long, V-shaped rack which is on a large table. A receptacle for the drained syrup is placed underneath the rack on one end. The crystal pans are half tilted for about 10 minutes until the syrup starts to drain into the receptacle. The pans are then stood up on their edges and arranged so that every two adjacent pans cross each other and form a V. This arrangement permits the draining to be completed. When the crystallized goods are almost dry, the pans are emptied out on boards which are lined with wax paper. The candy is allowed to set until it is completely dry and then it is packed in boxes. The boxes are lined with wax paper; a cardboard and wax paper are used between each layer, and waxed dividers are usually used to separate each row on the top layer.

When cold crystal syrup is called for, it is cooked in the morning and put on the candies before closing time. The candies are drained the next morning.

In gum and jelly work and other cases where a hot crystal is needed, the candies must be perfectly dry. Otherwise the crystal will peel off and give a poor finish. There are two ways of making hot crystal. One is the overnight crystal method and the other is called the short-day crystal method. The overnight crystal turns out more brilliantly than the short-day crystal, but the second is preferred when making goods of a more delicate nature because it crystallizes much faster. When making the short-day crystal, the syrup is cooked 2 degrees higher on the syrup gage (i.e., 36 degrees). It is only allowed to cool for 2 hours, and is then poured on top of the candies in the pans. The pans are covered with sheets of wax paper and, as soon as a light crust forms on top of the pans, they are drained and handled in the usual manner. The same procedure is followed in making the overnight hot crystal except that the syrup is cooked 2 hours before quitting time, it is cooked 2 degrees lower (i.e., 34 degrees) on the syrup gage, is applied to the candies just before quitting time, and is drained the following morning.

Automatic machinery for doing crystal work has not, as yet, been developed, though units which save some time and labor are available. However, the principle and system of crystallizing described are the same.

CRYSTAL SYRUP: *See also* CRYSTALLIZING.

CRYSTALLIZED FRENCH CREAMS

Sugar	50 lb
Water	8 qt
Cream of Tartar	4 teaspoonfuls
Melted Corn Syrup	10 lb
Liquid Colors and Flavors	to suit
Invertase	3 teaspoonfuls

The sugar, water, and cream of tartar are placed in a large copper pan on the stove and stirred until dissolved. When the batch begins to boil, it is covered and allowed to steam for a few minutes. The sides are washed down, and a thermometer is placed in the batch, which is cooked to 252°F. The melted corn syrup is then stirred in rapidly. The batch is poured out on a dampened cooled cream beater and allowed to cool for about 8 minutes. The invertase is then added to the batch and the machine is set in motion and allowed to run until the batch creams to a lump. It is then allowed to sweat back for about 20 minutes. The cream is then remelted in a steam-bath copper pan and divided into four parts as follows: The white is flavored mint; the second part is colored a delicate yellow and flavored lemon; the third part is colored a pastel green and flavored pistachio, and the fourth part is colored a delicate pink and flavored with oil of rose. The melted cream is cast into fancy-shaped impressions, molded in starch, and allowed to set overnight. The creams are then removed, brushed free from starch, and placed in crystal pans, which are only half filled. They are crystallized as usual with a cool overnight crystal syrup (*see* CRYSTALLIZING). They are then drained, dried, and packed as usual.

This batch will yield approximately 75 pounds.

Note: One and one-half teaspoonfuls of mint flavor is the dose for the mint creams. One-half teaspoonful of liquid yellow color and 1½ teaspoonfuls of lemon flavor are the proper amounts for the yellow creams. One-half teaspoonful of liquid pink color and 1½ teaspoonfuls of strawberry flavor are the doses for the pink creams; and ¼ teaspoonful of liquid green color and 1 teaspoonful of oil of lime are the proper amounts for the green creams.

CRYSTALLIZED MINTS, CHOCOLATE

The formula for CRYSTALLIZED MINTS, SPARKLING (*see*) is followed, but the batch is cooked to 270°F and then 2 pounds of finely cut up bitter chocolate are stirred in together with 8 ounces of corn syrup. The batch is stirred and cooked for another minute. It is then poured out on a cold greased slab and finished in the usual manner. When all the colored mints are drained of crystal and dry, they are packed assorted in airtight cans.

This batch should yield approximately 30 pounds of candy.

CRYSTALLIZED MINTS, SPARKLING

Sugar	30 lb
Cream of Tartar	3 teaspoonfuls
Water	5 qt
Oil of Peppermint	3 teaspoonfuls
Glycerin	1 oz

The sugar, water, and cream of tartar are placed in a copper pan on the stove and stirred until dissolved. When the batch begins to boil, it is covered and allowed to steam for a few minutes. The sides are washed down, and a thermometer is placed in the batch, which is cooked to 272°F. The batch is then poured out on a cold greased slab, and the edges are folded together while the oil of peppermint is added and mixed through. The batch is cooled rapidly until it is hard enough; it is then taken over and placed on the pulling machine. The glycerin is added to the pulled batch and pulling is continued for 7 minutes. The batch is then transferred to the spinning table and passed through the mint-cutting machine. It is then placed on boards lined with wax paper and dusted with powdered sugar. The boards are placed in a warm room overnight. When they are creamed, the mints are broken apart and put in crystal pans. The pans are only half filled to avoid breakage. The mints are crystallized (*see* CRYSTALLIZING) overnight as usual. This formula is suitable for yellow, green, and pink colors. Excessive coloring should be avoided, since the ideal is a delicately colored mint.

This formula should yield approximately 30 pounds of candy.

CRYSTALLIZED OPERA CREAMS, *See* OPERA CREAMS, CRYSTALLIZED.

CRYSTALLIZED PINEAPPLE HEARTS

Canned Pineapple Hearts (slices)	20 lb
Crystal "A" Sugar	20 lb
Water	3 qt

The pineapple is drained on a fine wire sieve. A pan is placed underneath to receive the juice. The pineapple is allowed to drain for 1 hour and is then placed in earthenware crocks. Each crock is only half filled. The sugar and water are placed in a copper pan on the stove and stirred until dissolved. Then the solution is cooked to 216°F. The syrup is cooled until it is lukewarm and is then poured over the pineapple so that it is well covered. This is allowed to set overnight.

Some of the drained pineapple juice and 5 pounds of sugar are added to the syrup. It is again cooked to 218°F and again cooled. The pineapple is covered with this syrup for the second time and allowed to set overnight. The following morning the same process with the syrup is repeated. This time the syrup is cooked to 220°F, and the pineapple is covered for the third time and allowed to set overnight. Finally, on the following morning, the pineapple and syrup are placed in a copper pan over low heat, allowed to come to a boil, and cooked for 5 minutes. The pineapple is then poured back into the crock with the syrup and kept covered. It is kept in a cool place for use when wanted.

This formula should yield approximately 30 pounds of slices.

CRYSTALLIZING

Crystallization is applied when making various soft candies to protect them from drying out and sticking together. Soft goods that are properly crystallized will stay edible and salable for 6 to 8 months.

Crystallizing is a tedious process, which must be handled with considerable care from beginning to end. The first step is to cook the crystal syrup. In making this syrup, crystal "A" sugar is preferred. It is necessary to use just enough water to dissolve the sugar before boiling and no more: 4 quarts of water to each 20 pounds of sugar is the correct measure. Excessive water will weaken the crystal and discolor it. The crystal syrup should be cooked rapidly, 25 to 30 minutes for a 50-pound batch. It may be recooked once by adding fresh sugar and water, but one recooking is the limit. Then syrup can be used up in making fondants and creams.

CUSTARD BUTTER CREAMS: *See* BUTTER CREAMS, CUSTARD.

CUTTING AND WRAPPING MACHINE, THE FORGROVE 42B PLASTIC

This machine cuts and fold-wraps plastic candies, caramels, and toffee. (Courtesy of the Package Machinery Co., Springfield, Massachusetts.)

D

DAINTIES: *See Individual Dainties formulas,* e.g., ALMOND DAINTIES and CHOCOLATE-COVERED DATE DAINTIES.

DAIRY CREAM CHOCOLATES: *See* JERSEY DAIRY CREAM CHOCOLATES.

DAIRY MAID KISSES

Sugar	8 lb
Corn Syrup	2 lb
Salt	1 teaspoonful
Fresh 18% Light Cream	3 qt
Butter	1 lb
FRENCH CREAM (*see*)	8 lb
Vanilla Flavor	2 teaspoonfuls
Glycerin	3 teaspoonfuls
Chopped Walnuts	3 lb

The sugar, corn syrup, salt, and cream are placed in a copper pan on the stove and stirred until dissolved. When the batch begins to boil, the sides are washed down; the butter is stirred in and the batch is stirred constantly and cooked until it forms a soft ball when tested in cold water (or to about 238°F). The batch is then removed from the heat and the French cream is added and stirred until it is melted. The flavor and glycerin are stirred in, and the chopped walnuts are added and mixed through. The batch is then spooned into small kisses and deposited on wax-paper-lined trays. These are allowed to cool until set and dry, and they are then packed in boxes as usual.

This formula should yield approximately 25 pounds of candy.

DAISY ROLLIES

The formula for ICELAND ROLLIES is followed for batch No. 1. Red color is added to the chocolate batch to give it a red-brown color. When the batch is cooled on the marble, it is formed into long thin rolls as usual. For batch No. 2, again the same formula is followed until the cooking is finished. The batch is then moved to the marble and 1 pound of French cream and 1 teaspoonful of vanilla flavor are stirred in. The rolls are dipped in the sauce and rolled in green-colored coconut which has been spread out on the marble. They are then finished in the usual manner.

This formula should yield approximately 38 pounds of rollies.

DATE DAINTIES, CHOCOLATE-COVERED: *See* CHOCOLATE-COVERED DATE DAINTIES.

DATE VARIETY CARAMELS: *See* VARIETY CARAMELS, DATE.

DATES, CHOCOLATE-COVERED: *See* CHOCOLATE-COVERED DATES.

DECORATED COCONUT CREAM EGGS: *See* COCONUT CREAM EGGS, DECORATED.

DECORATIONS: *See* EASTER EGGS.

DE LUXE CARAMELS, CHOCOLATE

Sugar	11 lb
Corn Syrup	15 lb
Coconut Butter	1 lb
Salt	1 tablespoonful
Fresh 18% Cream	4 qt
Evaporated Milk	14 lb
Vanilla Flavor	2 teaspoonfuls
Bitter Chocolate (cut up finely)	4 lb

The sugar, corn syrup, coconut butter, salt, and cream are placed in a large copper pan on the stove and stirred until dissolved. Stirring is continued while the batch is cooking. The evaporated milk is added gradually, and the batch is cooked until it forms a medium ball when tested in cold water. The batch is then removed to a stand, where the chocolate and flavor are stirred in, and the batch is then poured on a greased slab between iron bars, spread out evenly with a palette knife, and allowed to cool until hard. It is then passed through a sizing machine, cut in square caramels, and packed in bonbon cups.

This batch will yield approximately 45 pounds of candy.

DE LUXE CARAMELS, CHOCOLATE NUT

The formula for DE LUXE CARAMELS, CHOCOLATE (*see*) is followed, but 4 pounds of chopped Brazil nuts are stirred into the batch before pouring. It is finished in the usual manner. This batch will yield approximately 45 pounds of candy.

DE LUXE CARAMELS, VANILLA

Sugar	11 lb
Corn Syrup	14 lb
Coconut Butter	1½ lb
Salt	1 tablespoonful
Fresh 18% Cream	4 qt
Evaporated Milk	14 lb
Vanilla Flavor	4 teaspoonfuls

The sugar, corn syrup, coconut butter, salt, and cream are placed in a large copper pan on the stove and stirred until dissolved. When the batch begins to boil, stirring is continued while the evaporated milk is added gradually. The batch is cooked to a medium-firm ball when tested in cold water. The batch is then removed to a stand, and the vanilla flavor is stirred in. It is then poured out on a greased slab between iron bars, spread out evenly with a palette knife, and allowed to cool until hard. It is then put through a sizing machine, cut into square caramels, and packed in bonbon cups.

This batch should yield approximately 40 pounds.

DE LUXE CARAMELS, VANILLA NUT

The formula for DE LUXE CARAMELS, VANILLA (*see*) is followed, but 3½ pounds of walnut pieces are stirred into the batch before pouring. The batch is finished in the usual manner and should yield approximately 40 pounds of candy.

DEPOSITOR D-100

This machine automatically deposits pastes, jellies, gums, and creams into impressions molded in starch trays. It fills a full tray in one operation. (Courtesy of the National Equipment Corporation, New York, New York.)

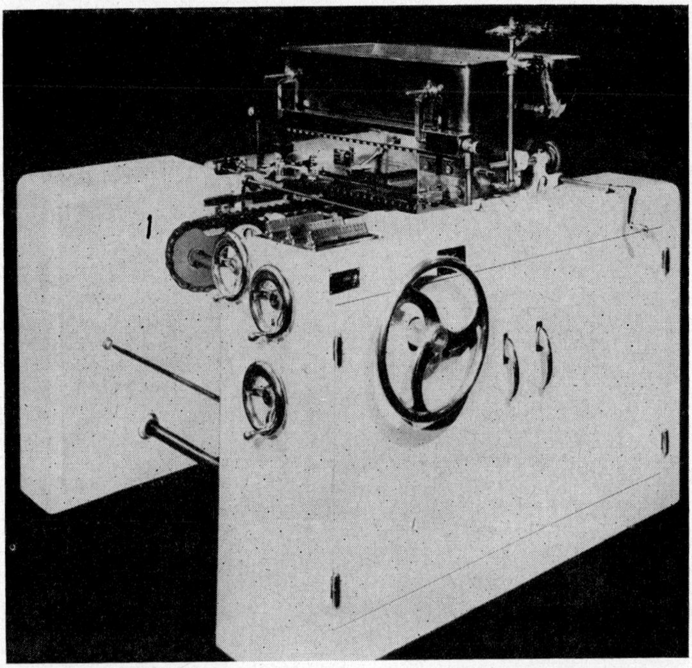

DIPPING CARAMELS: *See* LOG-CABIN PECAN ROLL, LOG NUT ROLL.

DOCTORING CANDIES: *See* HARD CANDY, CHEWING CANDIES.

DRESDEN CHOCOLATES

The molds are prepared in the same way as for COSMOPOLITAN CHOCOLATES (*see*). For the center RUNNING CREAM (*see*) is used. Small portions of this cream are placed on a marble and flavored vanilla, mint, coffee, etc. The flavor is mixed in thoroughly and the fondant is ready to use.

The molds are filled with chocolate rapidly and then turned

upside-down instantly so that a shell of chocolate is formed in the mold. The molds are moved to the cool room and left there until the chocolate shell hardens. They are then placed on the work table, and the flavored cream is placed in a baker's decorating bag equipped with an ordinary tube. A small amount of cream is deposited in the center of each mold and then the mold is filled with tempered chocolate and placed in the cool room again. When the chocolates harden, they are removed from the molds and packed in the usual manner.

The yield on Dresden chocolates will be exactly pound per pound of the ingredients used in its making.

DUTCH HOPJES CARAMELS

Crystal "A" Sugar	15 lb
Brown Sugar	7 lb
Corn Syrup	18 lb
Fresh 18% Cream	10 qt
Salt	2 teaspoonfuls
Butter	6 lb
Hopjes Concentrated Flavor	2½ oz

The sugars, salt, corn syrup, and cream are placed in a large copper pan on the stove and stirred until dissolved. The sides are washed down and stirring is continued until the batch is cooked to a hard crisp or about 290°F. The butter is then stirred in and allowed to cook for about 2 minutes. The batch is then poured out on a greased slab lined with bars and spread evenly to approximately ¾ inch thickness. After cooling for a few minutes, it is cut into small squares with an adjustable cutter. The cutting is done rapidly in order to avoid scrap. The candies are then loosened, cooled, broken apart, and wrapped in fancy tin foil papers.

This formula should yield approximately 50 pounds of caramels.

DUTCH HOPJES TOFFEE

Sugar	12 lb
Corn Syrup	8 lb
Coconut Butter	4 oz
Salt	1 teaspoonful

Fresh Cream	2 qt
Evaporated Milk	2 lb
Butter	2 lb
Concentrated Hopjes Flavor	4 teaspoonfuls

The sugar, corn syrup, coconut butter, salt, and fresh cream are placed in a copper pan on the stove and stirred until dissolved. When the batch begins to boil, the evaporated milk is stirred in, and stirring is continued while the butter is added. The batch is cooked until it forms a hard ball when tested in cold water (*see* COLD-WATER TEST). The batch is then poured out on a greased slab between iron bars, and the edges are folded together while stirring in the Hopjes flavor, which is mixed through. The batch is cooled rapidly and, when hard enough to handle, it is cut to the desired size and wrapped with double foil on the inside and moistureproof cellophane on the outside.

This batch will yield approximately 25 pounds of candy.

E

EASTER EGGS

Small Easter eggs and novelties wrapped in cellophane and tinfoil are in constant demand around Easter time. Since most Easter eggs are really chocolates in a larger form, they should be given the same care as expensive chocolates.

Most manufacturers begin making Easter eggs early in January and stock the warehouse with large quantities of these goods. Since the goods do not reach the consumer until Easter time, their shelf life is an important problem. If the confectioner will follow formulas accurately and use the best ingredients, he will safeguard his Easter eggs against early deterioration.

In each Easter egg formula, invertase is listed as an ingredient. Invertase is necessary because it acts as a softener and preserves the life of the cream center, but it is up to the confectioner to adjust the amount of invertase used to suit his particular trade. Some customers prefer a soft and fluffy cream and some prefer a more solid cream. Since the invertase begins to act about a week after the cream centers are made, they gradually become softer as they lie on the shelves. The confectioner should use the proper amount of invertase, so that when the eggs finally reach the consumer, they will meet his requirements. In the following formulas enough invertase is given to produce a medium soft cream.

From 2 to 3 ounces of chocolate to each pound of cream is the usual proportion in Easter eggs of fairly good quality. The more chocolate is put on the egg, the higher its quality.

The icing should be made so that the decorations do not break easily. The icing formula calls for gum arabic solution and some cream of tartar. Gum arabic solution makes the icing stick to the egg without falling off. Cream of tartar keeps the icing more plastic so that it does not crumble or break. These two ingredients should not be omitted or cut down in quantity.

Some decorations, like the borders around the circumference of the egg, large roses, and flowers, require a thicker icing. To thicken the icing more powdered sugar is mixed in. Each time more sugar is added the icing should be beaten for a few minutes to mix in the sugar thoroughly. In writing names, drawing lines, or etching small flowers and leaves, thin icing is needed. A few drops of water mixed through the icing will reduce it and make it thin. Each time the icing is reduced, it should be beaten so the water will be mixed in thoroughly.

Most decorators use a baker's ornamental decorating bag with tubes of various designs. For putting borders on eggs, the bag is half filled with icing and a star tube is connected to its end. For decorations like flowers and roses, the icing is colored pink, yellow, red, and green. There are special tubes for drawing roses and flowers. Drawing birds and animals is particularly difficult for the beginner, who is advised to take a few lessons from a professional decorator. A confectioner with some sort of art background will usually make a good decorator.

Decorations are fragile. Even when they are sold locally, the eggs should be carefully packed. Enough shredded wax paper should be placed on the bottom, sides, and top of the box. Before the box is closed, it should be tightly packed with enough shredded wax paper to keep the egg from shaking.

When Easter eggs are shipped out of town, they require special attention. For this purpose, the Easter-egg boxes should be packed in single layers in a cardboard carton; never should one box be placed on top of the other when packing the cardboard cartons (sometimes called outside packers). The outside packers are then placed in a heavy shipping carton, which should never exceed more than 50 or 60 pounds in weight. The shipping cartons should be large enough to absorb the outside packers without leaving too much vacant space between them.

Candy factories that operate chocolate and cream departments are equipped to make Easter eggs. The regular cream department has the equipment needed to produce hand-rolled or starch-cast cream centers. Smaller Easter eggs are usually cut and rolled by the average cut-roll machine, with the usual adjustments for the various sizes. Larger hand-rolled eggs are strictly a hand-made product. As for chocolate coating, the average enrober will coat smaller eggs up to 1 pound with a slight adjustment. Whether hand rolled or starch cast, all large eggs should be hand coated.

Metal molds for Easter eggs, shells, rabbits, chicks, etc., are an extra investment which usually pays. The larger the set of molds, the more goods will be produced. Cream rabbits, chicks, and other Easter animals made with plaster of Paris molds are old favorites. Although they are difficult to make, the formula found under CHOCOLATE HOLLOW EASTER EGGS should give good results if followed closely.

There are machines that turn out hollow chocolate Easter eggs and novelties in large quantities. However, these machines are very expensive and large investments are involved. If the manufacturer's demand can be met by hand production methods, the investment is not warranted.

For the various Easter egg formulas, see the individual names, such as COCONUT CREAM EGGS, FRUIT AND NUT EGGS, etc.

Note: All small Easter eggs may be cut and rolled by hand where a cut-roll machine is not available. The method is simple: the cream is kneaded and rolled in long rolls on the table and then cut, weighed, and shaped into eggs.

EASTER MOLDS: *See* CHOCOLATE RABBITS, CHICKS, etc.

EGG CUTTING: *See* COCONUT CREAM EGGS, HAND-ROLLED, SMALL and PEANUT-BUTTER EGGS.

EGG POLISHING: *See* PANNED PASTEL COCONUT EGGS.

EGG SCARRING: *See* COCONUT-CREAM EGGS, DECORATED, LARGE.

EGGNOG CREAMS, HAND-ROLLED

The formula for HAND ROLLED CREAM CENTERS, VANILLA NUT is followed up to the cooling point. The marshmallow whip and invertase are added to the batch together with 5 pounds of selected diced assorted fruits. The machine is then set in motion and, while it is running, the batch is colored straw yellow and flavored with 5 teaspoonfuls of eggnog flavor. The batch is then finished in the usual way.

Note: The average ratio for using chocolate coating for hand-rolled centers is approximately 20 percent for enrober coating and 30 percent for hand coating.

EGGNOG PARFAIT

Fresh 20% Cream	16 qt
Instant Egg Yolks	1 lb
Sugar	15 lb
Diced Assorted Fruits	5 lb
Yellow Color	½ teaspoonful
Eggnog Flavor	2½ oz

The egg yolks are mixed with the sugar and placed in the mixing can. The cream is poured on top of the sugar and stirred until the sugar is dissolved. The color and flavor are stirred in, and the mix is strained into the freezer. When the cream freezes about halfway, the diced fruits are added. The parfait is completed in the same manner as VANILLA ICE CREAM NO. 1.

The batch will yield approximately 38 quarts.

EMPLOYEES: *See* SAVING WASTE.

ENGLISH WALNUT BRITTLE

Sugar	6 lb
Corn Syrup	6 lb
Water	1 qt
Salt	2 tablespoonfuls
English Walnut Pieces	8 lb

The sugar, corn syrup, water, and salt are placed in a copper pan on the stove and stirred until dissolved. When the batch begins to boil, it is covered and allowed to steam for a few minutes. The sides are washed down and a thermometer is placed in the batch, which is cooked to 294°F. The nuts are added gradually, stirred, and cooked on a slow flame for another few minutes. The batch is then poured out on a greased slab, the edges are folded together, and the batch is kept in a soft lump in the center of the slab. One half of the batch is rolled out in a thin sheet with a rolling pin; then turned on the opposite side and stretched quite thin. It is moved over to the end of the slab to cool, and the same process is done with the remaining part of the batch. The candy is then packed into tins.

This batch will yield approximately 20 pounds of candy.

ENGLISH WALNUT PATTIES

Sugar	6 lb
Corn Syrup	4 lb
Water	1 qt
Salt	2 teaspoonfuls
English Walnuts	10 lb

The sugar, corn syrup, water, and salt are placed in a copper pan on the stove and stirred until dissolved. When the batch begins to boil, it is covered and allowed to steam for a few minutes. The sides are washed down and a thermometer is placed in the batch, which is cooked to 290°F. The walnuts are then stirred in rapidly and the batch is allowed to cook. The stirring is continued for another 2 minutes, and the batch is turned out on a warm, greased slab. The edges are folded together until the batch is partly cool. It is then kept folded in a warm spot, and a few small pieces, about the size of a miniature ball, are pulled from the batch and placed in round fritter molds on the slab. The process is continued, working rapidly, until the entire batch is molded into patties. After the patties are cool, they are wrapped in moistureproof cellophane and packed as usual.

This formula should yield approximately 19 pounds of patties.

ENROBING MACHINE: *See* CHOCOLATE ENROBER.

F

FIG PASTE, CHOCOLATE-COVERED: *See* CHOCOLATE-COVERED, FIG PASTE.

FIG PASTE, TURKISH: *See* TURKISH FIG PASTE.

FIG TAFFY CRISP

Sugar	5 lb
Corn Syrup	3 lb
Water	½ qt
Coconut Butter	2 oz
Pulled Figs (cut in quarters)	6 lb

The sugar, corn syrup, water, and coconut butter are placed in a copper pan on the stove and stirred until dissolved. When the batch begins to boil, it is covered and allowed to steam for a few minutes. The sides are washed down and a thermometer is placed in the batch, which is cooked to 270°F. The figs are gradually stirred in, and the batch is then poured out on a greased slab. It is spread out with a palette knife, stretched as thinly as possible, then cut into square sheets. After cooling, it is packed into tins.

This batch will yield approximately 15 pounds.

FIG VARIETY CARAMELS: *See* VARIETY CARAMELS, FIG.

FIGS, CHOCOLATE-COVERED: *See* CHOCOLATE-COVERED FIGS.

FILBERT BRITTLE

The ENGLISH WALNUT BRITTLE (*see*) formula is duplicated, using filberts instead of English walnuts.

The batch will yield approximately 20 pounds.

FILBERT PASTE BONBON CENTERS

Sugar	25 lb
Cream of Tartar	2 teaspoonfuls
Water	5 qt
Melted Corn Syrup	4 lb
Ground Filberts	5 lb
Honey Flavor	3 teaspoonfuls
Invertase	1 teaspoonful
Glycerin	1 teaspoonful to every 5 lb of bonbon cream

The sugar, water, and cream of tartar are placed in a copper pan on the stove and stirred until dissolved. When the batch begins to boil, it is covered and allowed to steam for a few minutes. The sides are then washed down and a thermometer is placed in the batch, which is cooked to 247°F. The melted corn syrup is stirred in rapidly and the batch is poured out on a cooled, dampened cream beater and allowed to cool for about 7 minutes. The nuts, flavor, and invertase are added, and the beater is set in motion and allowed to run until the batch creams to a lump. The batch is then allowed to sweat back for 15 minutes; then it is transferred to the working table. It is kneaded, rolled, and cut into small oblongs, which are ready as centers for orange and pink bonbons.

The batch should yield approximately 40 pounds of centers.

FILBERT-PASTE BONBONS: *See* BONBONS, FILBERT-PASTE.

FILBERTS, CHOCOLATE-COVERED: *See* CHOCOLATE-COVERED FILBERTS.

FILBERTS, SALTED: *See* SALTED FILBERTS.

FILLED BUTTER CUPS, LEMON

The formula for FILLED BUTTER CUPS, VANILLA is followed, but roasted ground almonds are used in the center instead of peanuts. The outside batch is colored yellow and flavored lemon. It is finished in the usual manner and will yield approximately 38 pounds.

FILLED BUTTER CUPS, PISTACHIO

The formula for FILLED BUTTER CUPS, VANILLA is used, but chopped pecans are used in the center instead of peanuts. The outside batch is colored a delicate green and flavored with pistachio. The batch is finished as usual and will yield about 38 pounds.

FILLED BUTTER CUPS, ROSE FILBERT

The formula for FILLED BUTTER CUPS, VANILLA is followed, but roasted chopped filberts are used for the center and the outside batch is colored a delicate pink and flavored with 2 teaspoonfuls of oil of rose. The batch is finished in the same manner as the original and will yield approximately 38 pounds.

FILLED BUTTER CUPS, VANILLA

Center

Sugar	4 lb
Corn Syrup	5 lb
Water	½ qt
Ground Roasted Peanuts	5 lb
Salt	1 tablespoonful

The sugar, water, corn syrup, and salt are placed in a copper pan on the stove and stirred until dissoved. When the batch begins to boil, the sides are washed down and a thermometer is placed in the batch, which is cooked to 234°F. The peanuts are stirred in to make a stiff paste. This center is kept warm until ready for use.

Outside Batch

Sugar	25 lb
Water	3½ qt
Cream of Tartar	2½ teaspoonfuls
Vanilla Flavor	3 teaspoonfuls

The sugar, water, and cream of tartar are placed in a copper pan on the stove and stirred until dissolved. When the batch begins to boil, it is covered and allowed to steam for a few minutes. The sides are washed down and a thermometer is placed in the batch, which is cooked to 336°F. The batch is then poured out on a greased slab, and the edges are folded together while the vanilla flavor is stirred in. The folding and kneading are continued until the batch is cool enough to handle; then it is pulled partly on the hook and twisted to a glossy white. It is then transferred to the spinning table in front of the batch warmer, kneaded again, and formed into a flat batch. The warm peanut center is then placed in the middle of the flat batch, the edges are wrapped around, and the ends closed up. The batch is formed round, spun, and passed through a cutting machine to a butter-cup size. It is cooled on the table at the other end and then packed in tins.

The batch will yield approximately 38 pounds.

FILLED BUTTER CUPS, VIOLET

The formula for FILLED BUTTER CUPS, VANILLA is followed, but chopped walnuts are used for the center, and the outside batch is colored a bright purple and flavored with 2 teaspoonfuls of violet flavor. It is finished as usual and will yield approximately 38 pounds.

FONDANT MARZIPAN

Sugar	20 lb
Corn Syrup	4 lb
Water	4 qt
Invertase	3 teaspoonfuls
ALMOND PASTE (*see*)	18 lb

The sugar, corn syrup, and water are placed in a copper pan on the stove and stirred until dissolved. When the batch begins to boil, it is covered and allowed to steam for a few minutes. The sides are washed down, and a thermometer is placed in the batch, which is cooked to 248°F. The batch is poured out on a clean, dampened cream beater and sprinkled on top with ½ cup of cold water. The batch is cooled rapidly to about 100°F and the almond paste and invertase are then added. The machine is set in motion and allowed to run until the batch creams to a lump. It is then placed in clean cans and kept covered until it is needed.

The batch will yield approximately 45 pounds of fondant marzipan.

FONDANTS: *See* CREAMS.

FOUNTAIN SYRUPS

Soda-fountain syrups include simple and flavored syrups for plain sodas, chocolate syrups for sodas and ice-cream sundaes, sauces, such as caramel, butterscotch, fudge, and mint, for ice-cream sundaes, and nut toppings (nuts preserved in syrup) for ice-cream sundaes. These syrups, sauces, and nut toppings are not particularly difficult to make. No machinery is involved. Blending and cooking the ingredients is all there is to it.

However, problems sometimes do develop. Crystallization and fermentation are the two most frequent ones. Syrups may crystallize for several reasons. If the syrup is cooked too rapidly (the formulas specify a low flame), not all of the sugar crystals will dissolve. These crystals will tend to settle on the inner surface of the container and fermentation will probably follow. In simple syrup, water acts as a "doctor." The amount specified in the formulas should never be decreased. If the confectioner is overly hurried and cuts down the amount of water in order to speed up the cooking, the syrup will crystallize and then ferment. If the syrup is kept in uncovered containers, some of it will stick to the sides of the container and form crystals when it is dipped out. As the syrup is dispensed in successive dips, more of it will crystallize because the already crystallized syrup will be mixed with the liquid syrup. Ultimately the uncovered syrup will ferment. If excessive stocks of syrup are kept on hand, the very fact that the syrup is lying around will make it susceptible to crystallization. Finally, all syrups should be kept in a cool room, for an overheated room will cause the syrup to crystallize even if it is well covered.

To prevent crystallization cooking kettles and other utensils must be spotlessly clean. Fine granulated sugar and clean water should be used in all the formulas. Crystal "A" sugar is harder to dissolve. The batch should be stirred until all the sugar dissolves. Before the batch starts to cook, the sides of the batch should be washed down, because, as the batch is heated and stirred, many crystals accumulate on the sides of the cooking pan. Washing down the sides of the batch dissolves these crystals and keeps them out of the syrup. The syrups, which are always cooked on low heat, should always be cooled before they are poured into containers because, if they are stored while still warm, they are likely to crystallize. Stone crocks with clean covers are the best containers for all syrups and sauces.

Nut and syrup toppings should be kept in ½-gallon jars and covered with clean, airtight lids. No more than 1 week's supply of all syrups and sauces should be made at a time.

Once the syrup crystallizes, it should be scrapped. All syrup and sauce scrap can be used in making chocolate-covered fudge.

Fermentation usually follows crystallization. The two processes are closely related. If crystallization is prevented, there is less chance of fermentation. If the syrup or sauce crystallizes and is held over for any length of time, it will ferment and turn sour. Exposed syrup containers which attract bacteria are another cause of fermentation. Fermented syrups and sauces should immediately be scrapped.

Chocolate and other syrups which contain butterfat cannot be held for more than 2 weeks because they will eventually ferment. A soybean product, lecithin, prevents the butterfat from separating from the other ingredients and prolongs the usefulness of these syrups. However, lecithin will not eliminate fermentation. There are several chocolate-syrup formulas in this chapter. The confectioner should select the formula that suits his purpose best and follow it carefully. Chocolate syrup should be made in small quantities, i.e., it should be freshly made every week. Chocolate syrups and sauces are cooked higher (about 218°F) to prevent the butterfat from turning rancid too quickly. Because these products are cooked higher, they have more body and may prove too thick. If the chocolate syrup is too thick, it can be diluted to the right consistency with some simple syrup (i.e., sugar and water). This also applies to all sauces for soda-fountain use.

MARSHMALLOW WHIP (*see*) is suitable for soda-fountain use. If it is too thick, it can also be diluted with some simple syrup. If marshmallow whip is to be shipped wholesale for the fountain trade, it should be vacuum packed in 5-pound tin cans. This also applies to chocolate syrups and other sauces that contain butterfat. Nut toppings should be packed in airtight ½-gallon jars.

For the various fountain-syrup formulas, see the individual names, such as, CHERRY SYRUP, CHOCOLATE SYRUP, etc.

FOUR-UNIT PLANT, THE HANSELLA PLASTIC

This machine is used for center filling, batch forming, rope sizing, and die cutting. A two-way cooling unit is attached for high-speed production of Vienna filled hard candies. (Courtesy of the Package Machinery Co., Springfield, Massachusetts.)

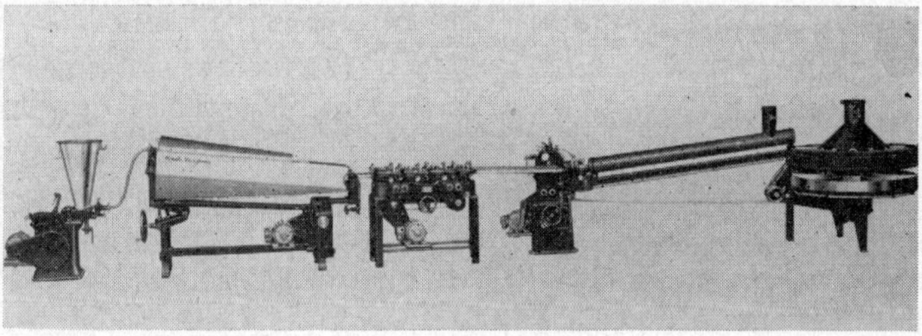

FRENCH CHEWING TAFFY, CHOCOLATE

The formula for FRENCH CHEWING TAFFY, VANILLA is followed. The batch is cooked at 252°F and 1½ pounds of chopped bitter chocolate are stirred into the batch while gelatin and marshmallow whip are added.

The batch will yield approximately 25 pounds of candy.

FRENCH CHEWING TAFFY, STRAWBERRY

The formula for FRENCH CHEWING TAFFY, VANILLA is followed, but the batch is colored a light pink and flavored strawberry. It is finished as usual.

The batch should yield approximately 25 pounds of candy.

FRENCH CHEWING TAFFY, VANILLA

Coconut Butter	8 oz
Sugar	12 lb
Corn Syrup	8 lb
Water	1½ qt
Salt	1 teaspoonful
Evaporated Milk	1½ lb
Silver Sheet Gelatin (dissolved in ½ cup of water)	1 oz
Vanilla Flavor	3 teaspoonfuls
Marshmallow Whip	1 lb

The sugar, corn syrup, water, coconut butter, and salt are placed in a copper pan on the stove and stirred until dissolved. When the batch begins to boil, it is covered and allowed to steam for a few minutes. The sides are washed down and a thermometer is placed in the batch, which is cooked to 260°F. The evaporated milk is stirred in, and the batch is reduced to 254°F. The heat is then shut off, the gelatin and marshmallow are stirred in, and the batch is poured out on a greased cold slab between iron bars. The edges are folded together while stirring in the vanilla extract. The batch is cooled rapidly, and when cool enough, it is taken to the pulling machine and pulled for approximately 7 minutes. It is then transferred to the spinning table, formed flat, stretched out to about 3-inch width, and cut in bars about 4 inches long. The candy is allowed to cool on the table and wrapped in moistureproof cellophane.

The batch will yield approximately 25 pounds of candy.

FRENCH CREAMS, CRYSTALLIZED: *See* CRYSTALLIZED FRENCH CREAMS.

FRENCH CRISP, CHOCOLATE

The same formula is followed as for FRENCH CRISP, VANILLA (*see*), but before the batch is poured on the slab it is flavored with 1½ pounds of cut-up bitter chocolate, which is mixed in thoroughly.

The batch will yield approximately 20½ pounds.

FRENCH CRISP, STRAWBERRY

The same formula is followed as for FRENCH CRISP, VANILLA, with the following changes: The batch is colored a light pink and flavored with 1 ounce of strawberry fruit flavor. It is finished as usual.

The batch will yield approximately 19 pounds.

FRENCH CRISP, VANILLA

Sugar	20 lb
Water	3 qt
Cream of Tartar	2 level teaspoonfuls
Vanilla Extract	½ oz

The sugar, water and cream of tartar are placed in a copper pan on the stove and stirred until dissolved. When the batch begins to boil, it is covered and allowed to steam for a few minutes. The sides are washed down and a thermometer is placed in the batch, which is cooked to 320°F. The batch is poured on a greased slab and flavored with the vanilla extract, while the edges are folded together. Folding and kneading are continued until the batch is cool enough for pulling. It is then pulled on the hook until light and fluffy. Then it is taken over to the spinning table and formed in the shape of a triangle and spun out in long sticks about 1 inch thick. The candy is then allowed to cool on the table, cut in 4 inch lengths, and packed in tins.

The batch will yield approximately 19 pounds.

FRENCH FRUITS: *See* GLACÉED FRENCH FRUITS.

FRENCH NOUGAT

Batch No. 1

Corn Syrup	6 lb
Instant Egg Albumen	4 oz
Water	½ qt

The egg albumen, water, and corn syrup are placed in a medium-sized, open-type beater on medium speed, started and beaten until thick.

Batch No. 2

Sugar	12 lb
Corn Syrup	6 lb
Coconut Butter	12 oz
Water	1½ qt
Salt	1 teaspoonful
Vanilla Flavor	4 teaspoonfuls
Granulated Gelatin	4 oz
Water	½ qt
Roasted Cashew Nuts	4 lb

The sugar, corn syrup, coconut butter, water, and salt are placed in a copper pan on the stove and stirred until dissolved. When the batch begins to boil, it is covered and allowed to steam for a few minutes. The sides are washed down and a thermometer is placed in the batch, which is cooked to 264°F. The batch is removed to a stand, and the gelatin, which has been previously soaked in water, is now stirred into it. This batch is then poured slowly into batch No. 1 in the beater and is beaten at high speed until thick. The machine is then reduced to low speed, and the nuts and flavor are stirred in. The batch is then turned out on trays, which are lined and covered with wafer paper, and allowed to set overnight. It is then cut and wrapped as usual.

This batch should yield approximately 25 pounds.

FRENCH VANILLA ICE CREAM NO. 3

Fresh Eggs	2 doz
Cooking Cornstarch	1 lb
Fresh Milk	3 qt
Fresh 20% Cream	6 qt
Condensed Milk	4 qt
Fresh 40% Cream	3 qt
Sugar	12 lb
Pure Vanilla Extract	1½ oz

The milk is placed in a double boiler on the stove and heated thoroughly. The eggs and starch are placed in a mixing bowl and mixed to a paste, which is stirred slowly into the hot milk. This custard is stirred and cooked for 15 to 20 minutes and is then set aside. The creams, condensed milk, and sugar are placed in a large ice-cream can and stirred with a large paddle until the sugar is dissolved. The custard and vanilla flavor are stirred in and mixed thoroughly. The mixture is then strained and allowed to cool for about 1 hour. It is then poured into the freezer and finished as usual. The same freezing process is used as for VANILLA ICE CREAM NO. 1 (*see*).

The batch will yield approximately 40 quarts.

FROZEN CONFECTIONS

Ice cream was first made in the early part of the 19th century. It made its first appearance in wealthy homes where a French chef was usually employed. It was later developed by caterers and confectioners and bakers who sold it at retail. The commercial production of ice cream on a wholesale scale began in the United States about the early part of the 20th century.

Today ice cream is an important food. About 6 percent of the total milk production in the United States is used up in the manufacture of ice cream and frozen delicacies.

Ice cream is made of butterfat, which is provided by heavy cream or light cream and evaporated milk. Nonfat milk solids are furnished by fresh skimmed milk. Sugar and stabilizers are used in ice-cream formulas, and some ice creams also contain egg-yolk solids.

Most state laws provide for an 8 to 14 percent butterfat content for ice cream. The standards for fruit, nut, and chocolate ice cream often permit a slightly lower percentage of butterfat than is required for plain ice cream. Some state laws also provide for a minimum weight per quart of ice cream.

When ice cream is first made, the mix is held back several hours before freezing. During the freezing process, some air is whipped into it, bringing about an increase in volume (overrun). Ice cream usually has about 100 percent overrun, which means that the air whipped into the mix has increased the final product to twice the volume of the original mix.

An average ice-cream mix contains about 12 percent of butterfat, 11 percent nonfat milk solids, and about 15 percent sugar.

With the addition of a stabilizer and flavoring ingredients, it contains about 38 percent total solids.

An excessive overrun in ice cream will affect the texture and the nutritive value of the ice cream. Good ice cream should have a fresh flavor. The body should be firm, not fluffy. The texture should be smooth, neither coarse nor icy.

In this encyclopedia more than six hundred formulas, describing every kind of candy, are presented. However, retail manufacturers who maintain soda-fountains and ice-cream departments will also be interested in ice cream, frozen confections, and fountain syrups.

There are four kinds of frozen confections: (1) ice cream in a wide variety of flavors for the bulk and package trade; (2) parfaits in a more limited variety of flavors; (3) fruit ices; and (4) sherbets, which are a cross between ice cream and fruit ice and particularly suitable for customers who are allergic to ice cream with a high butterfat content.

Each of these frozen confections offers its own production problems. However, flavor, quality, and smoothness are problems common to all of them. The flavors should be distinct and pronounced so that every flavor identifies itself. Vanilla should taste like vanilla and not leave an anonymous taste. Pronounced flavors are achieved by avoiding flavor substitutes and using the best ingredients. Years ago genuine vanilla beans were used in this way: 1 pound of vanilla beans was ground fine with 1 pound of granulated sugar. This was reground several times until it was pulverized. The pulverized vanilla powder was used for vanilla flavoring in quality ice cream. Although it produced an excellent flavor, it had a shortcoming. It made the vanilla ice cream somewhat sandy and prevented perfect smoothness. Nowadays reputable flavoring houses produce excellent pure vanilla extracts, and the ice-cream manufacturer should select the best pure-vanilla extract to flavor his ice cream. For only the best extract will give an excellent flavor without affecting smoothness.

Most fruit ice creams are flavored with fresh fruit in season. When fresh fruit is not available, canned fruit is used. In either case the flavoring of the ice cream should be supported by some pure flavors and extracts. Fruit emulsions produce the best results when used for this purpose. They may also be used in the other frozen confections.

Smoothness is achieved by following instructions carefully. The ice-cream mix should be well stirred until all the sugar is dissolved and strained through a clean, fine sieve. If sugar crystals, or other foreign matter, e.g., lumps of pectin, find their way into the freezer, the result will be a rough product. Stabilizers and emulsifiers are important in producing smoothness. Not only do they give the ice cream body, but they also help prevent separation. When a formula calls for 6 or 8 ingredients and one or more of these ingredients does not blend with the others, it is said to separate. Sometimes the fruit, sometimes one of the butterfat ingredients separates. If the fruit flavor starts to separate from the rest of the ice cream, it gives a bleeding effect. When the consumer eats the ice cream, some parts have more fruit than the others. In the same way if the evaporated milk does not mix well with the fresh cream, the consumer may taste ice cream with only evaporated milk in it. Most ice-cream manufacturers use gelatins or similar powders as stabilizers, but the ice-cream formulas that follow call for 100 grade citrus pectin. This ingredient is more expensive than gelatin, but it gives the ice cream unparalleled smoothness. Gelatins sometimes give the ice cream too much body, i.e., they make it too solid, and the consumer is conscious of a rough texture. Unlike gelatin, pectin is derived from pure citrus fruits and has a fruity texture. Not only is it a powerful binder and emulsifier and successul in preventing separation, but it also brings out the flavor in all its fragrance and is more successful in liquefying the solid ingredients.

Since parfait is a luxury product, the second group of formulas will be of interest to quality firms only. Parfait gets egg yolk as an extra ingredient. Since this serves partially as a stabilizer, no further stabilizer is needed. All parfait formulas are nonfruit formulas, e.g., spiced flavors. Pectin works best with fruit flavors, but in parfait the egg yolk blends particularly well with the spiced flavors and gives the desired delicacy.

Fruit ices can never take the place of ice cream, but some people prefer an occasional change. It is important to follow these formulas with particular care. The following phrase, which recurs frequently, is very important: ". . . until all the sugar is dissolved and the mixture registers 20 degrees on the syrup gage." When the formula is balanced according to instructions, this degree is usually achieved, i.e., the ice is not too watery and not too solid. However, if it does not, and some adjustment is needed, the following procedure is recommended: If the mix is above 20 percent, more water is added until the syrup gage registers 20 percent. If it is below 20 percent, more sugar is added until the syrup gage registers 20 percent. The sugar is then stirred again until it dissolves.

Fruit is the main ingredient in fruit ices. Therefore, it is important to concentrate on this ingredient. The fruit-ice formulas that follow include a few original numbers, e.g., honeydew ice, loganberry ice. It is a good idea to make these novel ices when the fresh fruit is in season.

Sherbet is an ice cream with a low butterfat content. Pectin is also used as a stabilizer in making sherbet. The pectin compensates for the low butterfat content. It cannot bring out a rich cream with a lot of butterfat, but it does bring out an unusually rich fruit flavor.

In all ice-cream, parfait, fruit-ice, and sherbet mixes, only fine granulated sugar is used because this dissolves easily whereas crystal "A" sugar is hard to dissolve in cold mixes.

Although the products made in accordance with these formulas are of top quality, only modest equipment is needed for their manufacture. Whenever a formula calls for warming up some mixes, an ordinary gas stove is used. The cold mixing is done in an ordinary 40-quart metal ice-cream can. The freezer and storage cabinet usually operate on the frigidaire system. There should be space in the storage cabinet for fifteen to twenty ice-cream cans.

Most homemade-sweets shops usually use fancy molds for special occasions, e.g., ice-cream cakes for birthday parties, and ice-cream baskets for Easter. The metal molds for these specialties are purchased in equipment and supply houses. The standard formulas are used. The special handling and decorating involved in making these products is a challenge to the ice-cream manufacturer's ingenuity.

For the various frozen confection formulas, *see* the individual names, such as VANILLA ICE CREAM, CHERRY VANILLA ICE CREAM, LIME ICE, etc.

FRUIT ACID SOLUTION

Powdered Citric Acid	1 lb
Water	1 qt

The water is heated and removed from the stove. The citric acid is added to the hot water and stirred until dissolved. The solution is filtered and cooled and then it is ready for use.

This batch will yield approximately $1\frac{1}{2}$ quarts of fruit-acid solution.

FRUIT AND NUT BARS

Sugar	7 lb
Corn Syrup	3 lb
Water	1 qt
Raw Almonds	1 lb
Dates	1 lb
Figs	1 lb
Citron	1 lb
Glazed Cherries	1 lb
Glazed Pineapple	1 lb
Sliced Coconut	8 oz
Cinnamon Pearls	2 oz

The slab is greased and lined with iron bars 3 feet long, 4 inches wide, and $1\frac{1}{4}$ inches high. The fruits and nuts and the coconut are mixed and filled into the spaces between the bars. The red pearls are sprinkled on top.

The sugar, corn syrup, and water are placed in a copper pan on the stove and stirred until dissolved. When the batch begins to boil, it is covered and allowed to steam for a few minutes. The sides are washed down and a thermometer is placed in the batch, which is cooked to 290°F. The candy is then poured out evenly on the fruits and nuts on the slab and allowed to cool for a few minutes. It is then loosened with a large knife, sliced with a sharp knife, and, after cooling, wrapped in moistureproof cellophane. It is then packed into tins.

This batch will yield approximately 20 pounds of candy.

FRUIT AND NUT EGGS

The PINEAPPLE EGGS (*see*) formula is followed up to the point where the batch is poured out on the cream beater. While the batch is cooling for 7 minutes, 3 pounds of chopped mixed pineapple, cherries, and citron are added; also 3 pounds of pecan pieces. The beater is set in motion and 3 teaspoonfuls of Roman punch flavor, 1 teaspoonful invertase, and 3 pounds of MARSHMALLOW WHIP (*see*) are added. After the batch has creamed to a lump, it is kneaded soft and put through the cut-roll machine to the desired size. After the eggs have cooled on boards, they are coated and finished roughly in a high-quality milk chocolate. They are then allowed to cool on trays and are finally packed in boxes.

The batch will yield approximately 70 pounds of fruit and nut eggs.

FRUIT AND NUT PILLOWS, LEMON

Center

Sugar	10 lb
Corn Syrup	15 lb
Water	2 oz
Raisins	5 lb
Diced Assorted Fruits	5 lb
Roasted Peanuts	5 lb
Walnut Pieces	5 lb
Fine Coconut	5 lb
Vanilla Flavor	2 qt

The water, sugar, and corn syrup are placed in a copper pan on the stove and stirred until dissolved. When the batch begins to boil, it is covered and allowed to steam for a few minutes. The sides are washed down and a thermometer is placed in the batch, which is cooked to 232°F. All the nuts and fruits are added to the batch; also the vanilla flavor. It is then stirred until well mixed and kept warm until ready for use.

Jacket

Sugar	20 lb
Water	3½ qt
Cream of Tartar	2 teaspoonfuls
Oil of Lemon	3 teaspoonfuls

The water, sugar, and cream of tartar are placed in a copper pan on the stove and stirred until dissolved. When the batch begins to boil, it is covered and allowed to steam for a few minutes. The sides are washed down and a thermometer is placed in the batch, which is cooked to 338°F. The batch is then poured out on a greased slab and colored to a bright yellow. The edges are folded together while stirring in the oil of lemon. Folding and kneading are continued until the batch is cool enough; then it is taken over to the hook, pulled partly, and twisted to a glossy bright yellow. It is then transferred to the spinning table in front of the batch warmer and stretched and formed into a flat batch large enough to take the center. Twelve pounds of the warm center are then placed in the center of the jacket. The jacket is wrapped over and the ends are closed. The batch is then shaped round, spun, and cut in small pillows on a cutting machine. After cooling on the table, it is packed in tins.

This batch will yield approximately 31 pounds.

FRUIT AND NUT PILLOWS, LIME

The formula for FRUIT AND NUT PILLOWS, LEMON is followed, but the batch is colored a light green and flavored with oil of limes. It is finished as usual.

The batch will yield approximately 31 pounds.

FRUIT AND NUT PILLOWS, ORANGE

The formula for FRUIT AND NUT PILLOWS, LEMON is followed, but the batch is colored orange and flavored with oil of orange. It is finished as usual.

The batch will yield approximately 31 pounds.

FRUIT AND NUT PILLOWS, STRAWBERRY

The formula for FRUIT AND NUT PILLOWS, LEMON is followed, but the batch is colored a light pink and flavored with strawberry flavor.

The batch will yield approximately 31 pounds.

For the complete assortment, all four colors are spread out on the packing table, mixed, and packed in tins.

FRUIT AND NUT SLICES, ENGLISH STYLE

Batch No. 1

Sugar	4 lb
Corn Syrup	3 lb
Water	1 qt
Diced Fruits	1 lb
Glacéed Cherries	1 lb
Glacéed Pineapple	1 lb
Raw Chopped Almonds	1 lb
Chopped Pecans	8 oz
Chopped Pistachio Nuts	8 oz
Vanilla Flavor	3 teaspoonfuls

The sugar, corn syrup, and water are placed in a copper pan on the stove and stirred until dissolved. When the batch begins to boil, it is covered and allowed to steam for a few minutes. The sides are washed down and a thermometer is placed in the batch, which is cooked to 244°F. All of the fruits and nuts are then stirred into the batch gradually. The flavor is added and stirred in thoroughly. The batch is then turned out on a greased slab and allowed to cool until firm. It is then formed into rolls about 8 inches long and 1½ inches thick. The rolls are now ready for caramel dipping in batch No. 2.

Batch No. 2

Sugar	7 lb
Corn Syrup	5 lb
Coconut Butter	8 oz
Salt	2 teaspoonfuls
Evaporated Milk	6 lb
Butter	1 lb
Vanilla Flavor	3 teaspoonfuls
English Walnut Pieces	20 lb

The sugar, corn syrup, coconut butter, salt, and milk are placed in a copper pan on the stove and stirred until dissolved. The sides are washed down and the batch is stirred and cooked until it forms a soft ball when tested in cold water. The butter is stirred in until dissolved and the batch is moved near the slab.

The vanilla flavor is stirred in and the fruit and nut rolls are then dipped in the caramel (using two dipping forks) and rolled in the walnuts. This process is continued until all the rolls are dipped. They are shaped round and cooled. They are now ready for slicing.

This formula should yield approximately 50 pounds of slices.

FRUIT AND NUT TAFFY CRISP

Sugar	8 lb
Corn Syrup	4 lb
Water	1 qt
Walnuts	1 lb
Roasted Almonds	1 lb
Chopped Dates	1 lb
Glaceéd Cherries	1 lb
Chopped Citron	1 lb
Glacéed Cut Pineapple	1 lb

The sugar, water, and corn syrup are placed in a copper pan on the stove and stirred until dissolved. When the batch begins to boil, it is covered and allowed to steam for a few minutes. The sides are washed down and a thermometer is placed in the batch, which is cooked to 290°F. All the fruits and nuts are gradually stirred in and the batch is cooked about 2 or 3 minutes longer. The batch is then poured out on a greased slab and spread out quite thin with a palette knife. It is cut into squares, allowed to cool on the slab, and then packed into tins.

This batch will yield approximately 20 pounds.

FRUIT CARAMELS, RUSSIAN TYPE, LEMON

Center, Apple Paste

Ripe Apples	20 lb
Sugar	20 lb

The apples are placed in a large copper pan, covered with water, and then placed on the stove under moderate heat. The pan is covered with a metal or wooden cover and the apples are heated until they come to a boil. They are then strained, crushed, and put though a meat-chopping machine. The pulp is returned to the copper pan on the stove and the sugar is added. The batch is stirred continuously and cooked to a paste or a firm jelly. The following test may be applied: A spoonful of jelly is spread out on wax paper and placed in a cool spot for a few minutes. The jelly should set and come off the wax paper clean. The apple paste is now ready for use as centers and should be kept moderately hot until wanted.

This batch will yield approximately 35 pounds of apple paste.

Jacket

Sugar	16 lb
Corn Syrup	4 lb
Water	2½ qt
Powdered Citric Acid	2 teaspoonfuls
Oil of Lemon	2 teaspoonfuls

The sugar, corn syrup, and water are placed in a copper pan on the stove and stirred until dissolved. When the batch begins to boil, it is covered and allowed to steam for a few minutes. The sides are washed down and a thermometer is placed in the batch, which is cooked to 300°F. The batch is poured out on a cooled greased slab between iron bars, and the edges are folded together while the acid and flavor are stirred and mixed in thoroughly. Folding is continued until the batch is cool enough to handle. One third of the batch is then kept soft in front of the batch warmer, and the balance of the batch is pulled on the hook until it turns white. This portion is then taken over to the spinning table in front of the batch warmer and kneaded until it is smooth. The clear, soft portion of the batch previously placed in front of the batch warmer is also kneaded smooth and spread out flat, large enough to hold the center. Approximately 10 pounds of the moderately hot apple paste are placed in the center of the clear flat batch. The batch is then wrapped over and closed all around. It is kept in shape while the white kneaded batch is spread out flat and wrapped over the entire center batch. The batch is then formed into a bottle shape and kept warm in front of the batch warmer. The batch is then spun with gloved hands into medium-sized sticks, and it is then passed through a pillow-cutting machine. After the candies have cooled on the table, they are double-wrapped in fancy lithographed papers, which are lined with plain wax paper. Fancy printed moistureproof cellophane or printed tin foil wraps may be used instead of the double wrappers. These candies are usually wrapped by twisting the ends of the wraps into points instead of folding them under. The fruit caramels are then packed in boxes.

This batch will yield approximately 28 pounds of fruit caramels.

FRUIT CARAMELS, RUSSIAN TYPE, ORANGE

The formula for FRUIT CARAMELS, RUSSIAN STYLE, LEMON (*see*) is used with the following exception: The hard-candy batch is colored and flavored orange. The batch is then finished in the usual manner and should yield approximately 28 pounds of fruit caramels.

FRUIT CARAMELS, RUSSIAN TYPE, PINEAPPLE

The formula for FRUIT CARAMELS, RUSSIAN TYPE, LEMON (*see*) is followed all the way through with the following exception: The outside batch is colored a light yellow and flavored pineapple. It is finished in the same manner as the lemon batch and should yield approximately 28 pounds of fruit caramels.

FRUIT CARAMELS, RUSSIAN TYPE, RASPBERRY

The formula for FRUIT CARAMELS, RUSSIAN STYLE, LEMON (*see*) is used with the following exception: The jacket batch is colored red and flavored raspberry. The batch is finished as usual and should yield approximately 28 pounds of fruit caramels.

FRUIT CUTS, LEMON

Sugar	25 lb
Water	3½ qt
Cream of Tartar	2½ teaspoonfuls
Oil of Lemon	3 teaspoonfuls
Powdered Citric Acid	1 oz
Color	

The sugar, water, and cream of tartar are placed in a copper pan on the stove and stirred until dissolved. When the batch begins to boil, it is covered and allowed to steam for a few minutes. The sides are washed down and a thermometer is placed in the batch, which is cooked to 338°F. The batch is put on a greased slab and the edges are folded together while the oil of lemon and the citric acid are stirred in. A small portion (about 3 pounds) is taken off and pulled white and kept warm for a jacket. The main batch is kneaded until cool enough, when it is taken to the spinning table in front of the batch warmer. It is then formed into a square batch. The white is then formed flat and stretched large enough to wrap over the square. The batch is kept square and spun into square long sticks in the length of the table. They are usually kept square between two long iron bars and allowed to cool on the table. Cutting is usually done as follows: A short iron bar is placed on top of a sieve. Then the sticks are chopped with the end of the shears until they are all cut up. The candy is then placed in tins.

The batch will yield approximately 23½ pounds of candy.

FRUIT CUTS, LICORICE

The formula for FRUIT CUTS, LEMON (*see*) is followed, but the citric acid is omitted and enough black charcoal color is used to give a satisfactory color. Oil of anise is added for flavor.

The batch will yield approximately 23½ pounds.

After the entire assortment of fruit cuts is completed, all the candies are mixed on the packing table and packed in tins.

FRUIT CUTS, LIME

The formula for FRUIT CUTS, LEMON (*see*) is followed, but the batch is colored a delicate green and flavored with oil of lime. It is handled and finished as usual.

The batch will yield approximately 23½ pounds.

FRUIT CUTS, ORANGE

The formula for FRUIT CUTS, LEMON (*see*) is followed, but orange flavor, color, and acid are stirred in while the batch is folded and completed in the same manner as the lemon formula.

This batch will yield approximately 23½ pounds.

FRUIT CUTS, RASPBERRY

The formula for FRUIT CUTS, LEMON (*see*) is followed, but raspberry flavor, red color, and the citric acid are stirred in while the edges are folded together and the batch is finished as usual.

It will yield approximately 23½ pounds.

FRUIT ICES: *See* FROZEN CONFECTIONS and INDIVIDUAL FRUIT ICE FORMULAS, e.g., LOGANBERRY ICE, RASPBERRY ICE, etc.

FRUIT MARMALADE, RUSSIAN TYPE, APPLE

Jap Gelatin (Agar-Agar)	8 oz
Water	8 qt
Sugar	24 lb
Corn Syrup	12 lb
APPLE PASTE (*see*)	12 lb
Pure Apple Flavor	4 teaspoonfuls
Citric Acid Solution	4 teaspoonfuls

The water and agar-agar are placed in a large copper pan on the stove and stirred until the agar is completely dissolved. When the batch begins to boil, the sugar is stirred in gradually. The batch is allowed to boil again and the corn syrup is stirred in gradually. It is then cooked until it forms a good heavy string when tested between fingers (or at about 224°F). The batch is then moved to a cool spot and transferred to another clean pan. The apple paste is stirred in, and the batch is allowed to cool until all the heat has left it (to about 90°F). The flavor and citric acid are then added and mixed through well. The batch is then cast into impressions of assorted fruit shapes, which are molded in clean dry starch. The jellies are allowed to set overnight, and they are then removed, brushed free of starch, and dampened with a little simple syrup. They are then rolled in fine granulated sugar and allowed to dry. The fruit marmalades are then placed in crystal pans for an overnight crystal (*see* CRYSTALLIZING). They are then finished in the usual way.

This formula should yield approximately 60 pounds.

FRUIT MARMALADE, RUSSIAN TYPE, CHERRY

The formula for FRUIT MARMALADE, RUSSIAN TYPE, APPLE (*see*) is followed, but the batch is colored cherry red and flavored with 2 pounds of pure cherry jam and 4 teaspoonfuls of pure wild cherry extract. The batch is then finished in the usual manner and should yield approximately 60 pounds of marmalade.

FRUIT MARMALADE, RUSSIAN TYPE, GRAPE

The formula for FRUIT MARMALADE, RUSSIAN TYPE, APPLE (*see*) is followed, but the batch is colored a bright purple and flavored with 2 pounds of pure grape jam and 4 teaspoonfuls of pure grape extract. Citric acid solution must be used with each of the fruit flavors, and the proper time to stir it in is when the batch is sufficiently cool. The batch is then finished in the usual way and should yield approximately 60 pounds of marmalade.

FRUIT MARMALADE, RUSSIAN TYPE, LEMON

The formula for FRUIT MARMALADE, RUSSIAN STYLE, APPLE (*see*) is followed all the way through except for the color and flavor. The batch is colored a pale yellow and flavored with pure oil of lemon. Otherwise it is finished in the usual manner and should yield approximately 60 pounds of marmalade.

FRUIT MARMALADE, RUSSIAN TYPE, LIME

The formula for FRUIT MARMALADE, RUSSIAN STYLE, APPLE (*see*) is followed with the following exception: The batch is colored a delicate green and it is flavored with pure oil of limes. It is finished in the usual way and should yield approximately 60 pounds of candy.

FRUIT MARMALADE, RUSSIAN TYPE, ORANGE

The formula for FRUIT MARMALADE, RUSSIAN TYPE, APPLE (*see*) is used with the following exceptions: The batch is colored a bright orange and flavored with pure oil of orange. Also 2 pounds of pure orange jam are mixed through the batch. It is important to remember that before the citric acid solution is stirred in, all the heat must be out of the batch. This batch is finished in the usual manner and should yield approximately 60 pounds of marmalade.

FRUIT MARMALADE, RUSSIAN TYPE, PINEAPPLE

The formula for FRUIT MARMALADE, RUSSIAN TYPE, APPLE (*see*) is followed, but the batch is colored a straw yellow and flavored with 2 pounds of pure pineapple jam and 4 teaspoonfuls of pure pineapple extract. The batch is then finished in the usual manner and should yield approximately 60 pounds of marmalade.

FRUIT MARMALADE, RUSSIAN TYPE, RASPBERRY

The formula for FRUIT MARMALADE, RUSSIAN TYPE, APPLE (*see*) is used. The batch is colored a deep red and flavored with 4 teaspoonfuls of pure raspberry flavor and 2 pounds of pure raspberry jam. Otherwise, the batch is finished as usual and should yield approximately 60 pounds of marmalade.

FRUIT MARMALADE, RUSSIAN TYPE, STRAWBERRY

The formula for FRUIT MARMALADE, RUSSIAN TYPE, APPLE (*see*) is used. The batch is colored a delicate pink and flavored with 2 pounds of strawberry jam and 4 teaspoonfuls of pure strawberry flavor. The batch is then finished in the usual manner and should yield approximately 60 pounds of marmalade.

Note: When the assortment of Russian fruit-marmalade flavors has been completed, they are packed assorted in plywood boxes. Waxed dividers are used to separate the rows, and soft padding mats are placed between the layers and on top of the box.

FRUIT MELLOWS, CHOCOLATE

The formula for FRUIT MELLOWS, LEMON is followed, but the batch is cooked to 266°F and 2 pounds of finely cut-up bitter chocolate are stirred in. The cooking is continued for 1 minute longer. The batch is then poured out on a cold greased slab, cooled rapidly, and then pulled, handled, and finished as usual.

This batch should yield approximately 25 pounds of candy.

Note: The fruit mellows may be packed assorted or in separate flavors.

FRUIT MELLOWS, LEMON

Sugar	25 lb
Water	3½ qt
Cream of Tartar	2 teaspoonfuls
Yellow Color	1 teaspoonful
Powdered Citric Acid	2 oz
Oil of Lemon	4 teaspoonfuls

The sugar, water, and cream of tartar are placed in a copper pan on the stove and stirred until dissolved. When the batch begins to boil, it is covered and allowed to steam for a few minutes. The sides are then washed down and a thermometer is placed in the batch, which is cooked to 270°F. The batch is poured out on a cold, greased slab between iron bars, and the edges are folded together while the color, flavor, and citric acid are added and mixed through. The batch is cooled rapidly and, when hard enough, it is taken over to the pulling machine and pulled for 7 minutes. It is then taken over to the spinning table and formed flat, after which it is passed through the mint-cutting machine. It is then placed on boards lined with wax paper and dusted with powdered sugar. The boards are placed in a warm room overnight. When creamed, the mints are broken apart and packed in airtight tins.

This batch should yield approximately 25 pounds of candy.

FRUIT MELLOWS, LIME

The formula for FRUIT MELLOWS, LEMON (*see*) is followed. The batch is colored a bright green and flavored with oil of lime while the citric acid is mixed in. The batch is then completed in the usual manner, and should yield approximately 25 pounds of candy.

FRUIT MELLOWS, ORANGE

The formula for FRUIT MELLOWS, LEMON is followed, but the batch is colored a deep orange and flavored with oil of orange while the citric acid is added. It is then finished in the same manner as the lemon batch and should yield approximately 25 pounds of candy.

FRUIT MELLOWS, RASPBERRY

The formula for FRUIT MELLOWS, LEMON is used, but the batch is colored a deep red and flavored with raspberry while the citric acid is added. It is then finished in the usual manner and should yield aproximately 25 pounds of candy.

FRUIT NOUGAT: *See* SAVING WASTE.

FRUIT NOUGAT, TURKISH: *See* TURKISH FRUIT NOUGAT.

FRUIT-PUNCH CHOCOLATES

Sugar	30 lb
Honey	2 lb
Fresh 18% Light Cream	8 qt
Salt	2 teaspoonfuls
Melted Corn Syrup	3 lb
MARSHMALLOW WHIP (*see*)	3 lb
Melted Bitter Chocolate	3 lb
Assorted Diced Fruits	2 lb
Chopped Pecans	2 lb
Roman Punch Flavor	3 teaspoonfuls
Invertase	2 teaspoonfuls

The sugar, honey, cream, and salt are placed in a copper pan on the stove and stirred until dissolved. The sides are washed down and stirring is continued. The batch is cooked until it forms a rather soft ball, or until it registers 242°F. The melted corn syrup is stirred in rapidly and the batch is poured out on a clean, cooled, dampened cream beater and cooled down rapidly to about 100°F. The marshmallow and chocolate are added, the beater is set in motion, and the fruits, nuts, flavor, and invertase are added. Operation of the beater is continued until the batch is creamed to a lump. It is then allowed to sweat back for 10 minutes and then kneaded until soft and put through a cut-roll machine (or cut and rolled by hand). The small centers are coated with high-quality milk chocolate and after they are set, they are packed in the usual manner.

This formula will yield approximately 55 pounds of centers, or 75 pounds of hand-dipped chocolates.

FRUIT PUNCH CREAMS, CHOCOLATE-COVERED: *See* CHOCOLATE-COVERED FRUIT PUNCH CREAMS.

FUDGE

In the summer, when chocolates start melting, fudge makes an ideal substitute. Fudge can be made in many varieties and combinations, depending on the confectioner's resources and his imagination. Fruits are abundant in the summer, so that it is practicable to make fudge with all kinds of fruits, e.g., cherries, pineapple, strawberries, etc. As in making all candy, it is necessary to follow formulas closely and use the best ingredients.

Although it is usually one of the busiest units in the candy factory, the fudge department does not require much equipment. Cooking kettles, marbles and slabs, boards, and pans are all the equipment needed. Cooking must be watched carefully. The batch should be stirred constantly to prevent scorching. Evaporated milk should be added gradually in a slow stream to prevent curtailing and give a smooth texture. To preserve most of the butter flavor, butter is always added when the batch is almost done.

When making fondant for fudge, considerable care is necessary. A smooth cream will produce a smooth fudge whereas grainy cream will produce a rough, grainy, sandy fudge. The suggestions that follow are offered to prevent the cream from graining. While the batch is being cooked, it is better to allow for a little extra water than to underestimate the amount of water needed. When the batch is first started, it must be stirred constantly until it dissolves. As stated under the formulas, the batch must be covered before boiling and the sides must be washed down while boiling. While the batch is being cooled on the cream beater, its top should be sprinkled with a few drops of cold water after it is poured out. The water valve underneath the cream beater should be opened before the batch is poured, and the water should continue to run until the machine is started for the creaming. Creaming should never be interrupted until it is finished. If these suggestions are followed, a smooth batch of cream will result.

Pouring the fudge is another delicate step. To prevent spotting, it must not be too hot when poured. However, if the fudge is poured when it is too cold (which sometimes happens when the candymaker is careless or moves too slowly), it will lose its gloss and acquire a dull finish. In order to achieve a happy medium, the confectioner should cool the batch first and then, if he finds that it is too thick, reduce it to a liquid consistency with a little cream before he pours it.

The confectioner should study his equipment, ingredients, and surroundings and act accordingly.

For the various fudge formulas, *see* the individual names, such as SEA FOAM FUDGE, BOARDWALK FUDGE, etc. *See also* COMMERCIAL FUDGE, CREMO FUDGE, and VARIETY FUDGE.

G

GENEVA CHOCOLATE CORDIALS, GRAPE

The formula for GENEVA CHOCOLATE CORDIALS, RUM RAISIN (*see*) is followed, but milk chocolate is used for the shell and the cream for the center is flavored with grape extract. Drained processed grapes are used for the cordial center. The grape cordials are processed and finished in the usual manner.

The yield equals the weight of the ingredients used.

GENEVA CHOCOLATE CORDIALS, PEACH

The formula for GENEVA CHOCOLATE CORDIALS, RUM RAISIN (*see*) is followed, but dark vanilla chocolate is used for the shell, drained processed peach cubes are used for the fruit cordial, and the cream for the center is flavored with peach extract. The peach cordials are then processed and finished in the usual manner.

The yield equals the weight of ingredients.

GENEVA CHOCOLATE CORDIALS, RASPBERRY

The formula for GENEVA CHOCOLATE CORDIALS, RUM RAISIN is followed, but dark vanilla chocolate is used for the shell and drained processed raspberries for the cordial. The cream for the center is flavored with raspberry flavor. The raspberry cordials are then processed and finished in the usual manner.

The yield equals the weight of the ingredients used.

GENEVA CHOCOLATE CORDIALS, RUM RAISIN

Swiss Processed Milk Chocolate	20 lb
Large Processed Raisins in Syrup	5 lb
RUNNING CREAM (*see*)	10 lb
Roman Punch Flavor	2 teaspoonfuls

The raisins are allowed to drain for 2 hours. The assorted fruit-shaped molds are warmed slightly for about 30 minutes. The chocolate is melted in the usual manner for coating and then cooled as for hand coating. The molds are filled with the cool chocolate and drained immediately so that only a thin shell of chocolate remains in them. A raisin is then placed in each mold. The running cream is remelted as usual and flavored with the Roman punch extract. It is then deposited in small bits with a funnel on top of the raisins. This is allowed to cool for about 10 minutes and the molds are finally filled with cooled chocolate. The edges of the molds are trimmed with a small palette knife; they are then placed in the cool room where they remain until they set and cool. The chocolates are then carefully removed by tapping the molds lightly on the table so that they are easily removed. The cordials are then wrapped in colored aluminum foil and packed in boxes.

The yield of this batch will equal the weight of all the ingredients used in its making.

GINGER DAINTY KISSES

Sugar	8 lb
Water	1½ qt
Cream of Tartar	1 teaspoonful
FRENCH CREAM (*see*)	6 lb
Chopped Drained Ginger	4 lb
Vanilla Flavor	1 teaspoonful
Glycerin	2 teaspoonfuls

The sugar, water, and cream of tartar are placed in a copper pan on the stove and stirred until dissolved. When the batch begins to boil, it is covered and allowed to steam for a few minutes. The sides are washed down and a thermometer is placed in the batch, which is cooked to 238°F. The batch is removed from the heat and the French cream is stirred in rapidly. The vanilla flavor and glycerin are added, the chopped ginger is stirred in, and all the ingredients are mixed through well. The batch is then spooned out into small kisses and deposited on wax-paper-lined trays. They are allowed to cool until set and dry, and are then packed in boxes.

This formula should yield approximately 20 pounds of candy.

GLACÉED FRENCH FRUITS

Assorted French Fruits	15 lb
Crystal "A" Sugar	20 lb
Water	3 qt
White Vanilla Extract	2 teaspoonfuls

The sugar and water are placed in a copper pan on the stove and stirred until dissolved. The batch is cooked to 232°F and is then moved near the table. The flavor is stirred in. The French fruits are placed into the syrup and spread out evenly. The pan is then placed on the stove over low heat and warmed for about 1 minute. It is then moved back near the table. The syrup is rubbed against the sides of the pan for a few minutes, and the fruits are mixed through the syrup carefully. This is continued until the syrup begins to look milky. The fruits are then deposited on a wire screen over a metal table. They are carefully separated and allowed to dry. The fruits are then packed in wax-paper-lined boxes, with dividers placed between each line.

This formula should yield approximately 25 pounds of French fruits.

GLACÉED NUTS AND FRUITS

Glacéed Pineapple (cut into small pieces)	1 lb
Glacéed Cherries	1 lb
Brazil Nuts	1 lb
English Walnuts	1 lb
Large Filberts	1 lb
Large Raw Almonds	1 lb
Crystal "A" Sugar	8 lb
Corn Syrup	1 lb
Cream of Tartar	½ teaspoonful
Water	1 qt

The nuts and fruits are all prepared in separate boxes, and made ready for dipping. The sugar, corn syrup, cream of tartar, and water are placed in a small copper pan on the stove and stirred until dissolved. When the batch begins to boil, it is covered and allowed to steam for a few minutes. The sides are washed down and a thermometer is placed in the batch, which is cooked to 296°F. The batch is then placed on a round pan on the edge of the greased slab. The fruits and nuts are dipped into this mixture singly with a dipping fork and are deposited rapidly on the greased slab. This process is continued as long as the cooked glacée remains liquid enough for dipping. The glacéed candies are loosened with a clean knife and packed in jars, or they may be coated in a dark sweet vanilla chocolate. This formula should yield approximately 14 pounds of ordinary glacéed nuts and fruits, or 25 pounds of chocolate-coated glacéed nuts and fruits.

GLACÉED PINEAPPLE

Crystallized Pineapple Slices	25 lb
Crystal "A" Sugar	20 lb
Water	3 qt

The crystallized pineapple (slices) described under CRYSTALLIZED PINEAPPLE HEARTS *(See)*, together with the syrup, is placed in a copper pan on the stove over low heat. It is heated to the boiling point. The pineapple is then placed on a coarse sieve under a pan to drain. The sugar and water are placed in a copper pan on the stove, stirred until dissolved, and cooked to 221°F. The drained pineapple is then put into the warm syrup and spread out evenly. The syrup is rubbed against the sides of the copper pan until it shows a grain. The pineapple is then stirred gently through this syrup over and over again until it absorbs some of the heavy syrup and the syrup begins to look milky. The glacéed pineapple is then placed on a screen which rests on a lined metal table. It is carefully separated, allowed to dry, and finally packed.

This formula should yield approximately 40 pounds of glacéed pineapple.

Glaze A for Fudge and Other Candies

This glaze will comply with all pure food laws and will give fudge a glossy finish. It is used mostly for commercial fudge and fudge bars.

Granulated Gum Arabic (soaked in 2 qt of water)	2 lb
Sugar	4 lb
Water	1 qt
Instant Egg Albumen (soaked in ½ pt of water)	2 oz
Pure Alcohol	24 oz

The soaked gum arabic is melted in a double boiler. The sugar and water are placed in a clean, small copper pan on the stove and stirred until dissolved. When boiling, the sides are washed down and the batch is cooked to 236°F. It is then removed to cool. The melted gum arabic is then stirred into the syrup, and the soaked egg albumen is then partly beaten and mixed with the alcohol. This alcoholic mass is then stirred into the gummed syrup, put through a fine sieve to drain, and allowed to set for 3 hours, when it is ready for use. The glaze is applied on the fudge with a soft, flat brush, running the brush in one direction. When the glaze is dry, the fudge is ready for the usual packing.

This formula should yield 8½ pounds of *Glaze A*.

GLAZED BUTTER CREAMS: *See* BUTTER CREAMS, GLAZED.

GLOSS GOODS: *See* HARD CANDY.

GLYCERIN THROAT PASTILLES

Powdered Gelatin (soaked in 2 qt of water)	2 lb
Sugar	25 lb
Corn Syrup	10 lb
Water	1½ qt
Glycerin	16 oz
Tar Extract	3 oz

The sugar, corn syrup, and water are placed in a copper pan on the stove and stirred until dissolved. When the batch begins to boil, the sides are washed down and a thermometer is placed in the batch, which is cooked to 256°F. The batch is removed and the soaked gelatin is stirred in rapidly. The batch is cooled for about 15 minutes, the scum is removed, and the glycerin and tar extract are them stirred in and mixed through thoroughly. The batch is then cast in small tablet shapes which are molded in starch and allowed to set overnight. The pastilles are then removed, sifted free of starch, and put in a large clean pan. Gum arabic solution is sprinkled and rubbed through the pastilles until they are all damp and clear bright. They are then spread out on boards lined with wax paper to dry and packed as desired.

This batch will yield about 42 pounds of candy.

Note: The gum arabic solution is made of 1 ounce of powdered gum arabic dissolved in ½ quart of water.

GOLD CHOCOLATE STRINGS

The formula for SILVER CHOCOLATE STRINGS (*see*) is followed, but the batch is flavored and colored with molasses and cooked to 330°F. It is finished by the same method as the silver chocolate strings.

It will yield approximately 16 pounds.

GOLDEN ROD

Sugar	20 lb
Water	3½ qt
Cream of Tartar	2 teaspoonfuls
Molasses	½ qt
Bonbon Cream	10 lb
Fine Coconut	2 lb
Salt	1 tablespoonful
Vanilla	½ oz

The center is first prepared. The bonbon cream is melted over low heat while the coconut is being mixed in. This center is kept warm until ready for use. For the jacket, the sugar, water, and cream of tartar are placed in a copper pan on the stove and stirred until dissolved. When the batch begins to boil, it is covered and allowed to steam for a few minutes. The sides are washed down and a thermometer is placed in the batch, which is cooked to 340°F. While the molasses and salt are being stirred in, the batch is poured out on a greased slab. The edges are folded together while the vanilla is stirred in. Folding and kneading are continued until the batch is cool enough to handle, when it is partly pulled on the hook and twisted until it shows a gloss and a bright golden color. It is then taken over to the spinning table in front of the batch warmer, kneaded, and formed into a flat batch. The warm coconut-cream center is then placed in the center of the jacket, wrapped around, and the ends are closed. The batch is formed and kept in triangle shape, spun out, and twisted in long sticks almost the length of the spinning table. When cool enough, it is cut with shears into 4 inch long sticks. The candy is then packed in tins.

This batch will yield approximately 29 pounds.

GOODIES: See Individual Goodies Formulas (e.g. Apricot Pecan Goodies, Cocoanut Date Goodies)

GRAPE JELLY

Canned Grapes	10 lb
Sugar	8 lb
Agar-Agar	4 oz
Water	8 qt
Sugar	10 lb
Corn Syrup	3 lb
Grape Flavor	4 teaspoonfuls

The canned grapes and the 8 pounds of sugar are placed in a copper pan on the stove, stirred constantly, cooked to a firm jelly (about 223°F), and set aside.

The agar-agar and water are placed in a copper pan on the stove and stirred constantly. When the batch begins to boil the sugar is stirred in gradually, and the stirring is continued while the corn syrup is added. The batch is cooked to 224°F and is then moved to a stand. The grape fruit and flavor are stirred in and mixed through well. The jelly is run out into fancy impressions cast in starch and allowed to set over night. The candy is then removed, cleaned of starch, dampened, and sugared. When dry, it is placed in crystal pans and crystallized (*see* CRYSTALLIZING) as usual.

This batch will yield approximately 45 pounds of candy.

GRAPE SALT WATER TAFFY: See Salt Water Taffy, Grape.

GREEN BONBONS: *See* BONBONS, GREEN.

GUM ARABIC MARSHMALLOWS

Slow-Set Pectin	3 oz
Sugar	30 lb
Corn Syrup	20 lb
Water	6 qt
Instant Egg Albumen (dissolved in 2 qt of water)	1 lb
Powdered Gum Arabic	1 lb
Pure Vanilla Flavor	1½ oz

The pectin is mixed dry with 5 pounds of the sugar and placed in a large copper pan on the stove with 2 quarts of the water and stirred until dissolved. The balance of the sugar and water are then placed into the pan and stirred until dissolved. The sides of the pan are washed down and the batch is cooked to 234°F. The powdered gum arabic is stirred into the corn syrup, and this mixture is warmed up and added to the pectinized sugar batch. The batch is then poured into the beater and the dissolved egg albumen is added. The beater is started at low speed for about 5 minutes and the speed is shifted to high while the vanilla flavor is added. The beating is continued until the marshmallow weighs about 1 pound to the quart. The marshmallow is then placed in a warm depositor and cast into oval shapes molded in dry clean starch. These oval pieces are placed in the hot room of 100°F overnight. The marshmallows are then removed, brushed free of starch, and packed as usual.

This batch should yield approximately 65 pounds of marshmallow.

GUM ARABIC SOLUTION: *See* HONEY TABLETS.

GUMDROPS: *See* SOFT GOODS and *Individual Gumdrop Formulas.*

GUMDROPS, ANISE

(Gum Arabic Type)

Batch No. 1

Powdered Gum Arabic	10 lb
Water	12 qt
Corn Syrup	10 lb

Batch No. 2

Sugar	45 lb
Corn Syrup	25 lb
Water	8 qt
Oil of Anise	6 teaspoonfuls

For batch No. 1, the water and the gum arabic are placed in a copper pan on the stove over low heat and stirred constantly until the gum is completely dissolved. The corn syrup is then added and stirred until hot. This melted gum mixture is then poured through a fine brass-wire sieve into another copper pan and set aside.

For batch No. 2 the sugar, water, and corn syrup are placed in a large copper pan on the stove and stirred until dissolved. When the batch begins to boil it is covered and allowed to steam for a few minutes. The sides are washed down and a thermometer is placed in the batch, which is cooked to 246°F. The batch is then moved to a stand and the melted gum arabic mixture is stirred in and mixed thoroughly. This is allowed to set for 30 minutes and the scum is then removed from the top and the flavor is stirred in. It is then cast in the usual gumdrop impressions, which are molded in warm starch. The filled starch boards are placed in the hot room overnight. They are then removed, the drops are brushed free of starch, dampened, rolled in sugar, dried, and then placed in crystal pans and crystallized (*see* CRYSTALLIZING) as usual.

This batch should yield approximately 125 pounds of candy.

GUMDROPS, CINNAMON

(Gum Arabic Type)

The formula for GUMDROPS, ANISE (*see*) is followed, but the batch is colored a deep red and flavored with oil of cassia. It is then finished in the usual manner and should yield approximately 125 pounds of candy.

GUMDROPS, LEMON

(Exel-o Type)

Water	12 qt
Sugar	30 lb
Corn Syrup	30 lb
Cornstarch (best cooking grade, dissolved in 12 qt of water)	6 lb
Tartaric Acid (dissolved in water)	1 oz
Cream of Tartar (dissolved in water)	3 teaspoonfuls
Oil of Lemon	5 teaspoonfuls

The water, sugar, and corn syrup are placed in a large steam-cooking kettle and stirred until dissolved. When the batch begins to boil, the dissolved cornstarch is stirred in gradually. Stirring is continued while the tartaric acid and cream of tartar are stirred in. The batch is cooked to a firm jelly string or to about 226°F. It is then drawn out into a large, clean, copper pan, and the flavor is stirred in and mixed through. The batch is then poured into a depositing machine, heated to about 140°F and cast in the usual gumdrop impressions, which are round, flat-topped molds. The impressions are molded in warm starch that has been kept in the hot room for 24 hours. The cast gumdrops are placed in a hot room of 150°F and allowed to set for about 2 days. The gumdrops are then removed, brushed free of starch, dampened, sugared, and allowed to dry. They are then placed in crystal pans and crystallized (*see* CRYSTALLIZING) in the usual manner.

The batch will yield approximately 100 pounds of candy.

GUMDROPS, LIME

(Exel-o Type)

The formula for GUMDROPS, LEMON (*see*) is followed, but the batch is colored a bright green and flavored with oil of lime. It is then completed in the usual manner and will yield approximately 100 pounds of candy.

GUMDROPS, ORANGE

(Exel-o Type)

The formula for GUMDROPS, LEMON (*see*) is followed, but the batch is colored a bright orange and flavored with oil of orange. It is then completed in the usual manner and will yield approximately 100 pounds of candy.

GUMDROPS, RASPBERRY

(Exel-o Type)

The formula for GUMDROPS, LEMON (*see*) is followed, but the batch is colored deep red and flavored with 3 pounds of raspberry pulp and 3 teaspoonfuls of raspberry flavor. When all four flavors are crystallized and dried, they are assorted and packed in boxes.

The raspberry batch should yield approximately 100 pounds.

GUMDROPS, SPEARMINT

(Gum Arabic Type)

The formula for GUMDROPS, ANISE is followed, but the batch is colored a bright green and flavored with oil of spearmint. It is then finished in the usual manner and should yield approximately 125 pounds of candy.

GUMDROPS, VIOLET

(Gum Arabic Type)

The formula for GUMDROPS, ANISE is followed, but the batch is colored a bright violet and flavored with violet flavor. It is then completed in the usual manner and should yield approximately 125 pounds of candy.

H

HAND-ROLLED CHOCOLATE CREAMS: *See* CHOCOLATE CREAMS, HAND-ROLLED.

HAND-ROLLED COCONUT-CREAM EGGS: *See* COCONUT-CREAM EGGS, HAND-ROLLED.

HAND-ROLLED COCONUT CREAMS: *See* COCONUT CREAMS, HAND-ROLLED.

HAND-ROLLED EGG-NOG CREAMS: *See* EGG-NOG CREAMS, HAND-ROLLED.

HAND-ROLLED HONEY FILBERT CREAMS: *See* HONEY FILBERT CREAMS, HAND-ROLLED.

HAND-ROLLED MAPLE PECAN CREAM CENTERS: *See* MAPLE PECAN CREAM CENTERS, HAND-ROLLED.

HAND-ROLLED ORANGE CREAMS: *See* ORANGE CREAMS, HAND-ROLLED.

HAND-ROLLED PINEAPPLE CREAMS: *See* PINEAPPLE CREAMS, HAND-ROLLED.

HAND-ROLLED RUM RAISIN CREAMS: *See* RUM RAISIN CREAMS, HAND-ROLLED.

HAND-ROLLED STRAWBERRY CREAMS: *See* STRAWBERRY CREAMS, HAND-ROLLED.

HAND-ROLLED VANILLA NUT CREAMS: *See* VANILLA NUT CREAMS, HAND-ROLLED.

HARD CANDY

The main ingredients that go into the making of hard candies are water, sugar, and corn syrup. Water is the only ingredient that costs nothing, but if it goes in free, it also comes out free. It cooks out entirely and does not increase the weight of the batch. This is reminiscent of an old Polish confectioner. While cooking his hard candy batch, he kept pouring quarts of water into it, thinking that this would make the batch larger. He finally decided that water has magical qualities. The more water he poured into the batch the longer it took to cook, but it did not weigh an ounce more.

Concerning sugar, white granulated sugar is satisfactory, but the crystal "A" type is preferred because it cooks more clearly and produces a far better clear, white, hard candy. Corn syrup is used as a "doctor." The "doctor" kills the grain and keeps the hard candy plastic. Without it, the candy would crystallize and turn into a lump of cloudy sugar shortly after it was poured on the slab. When the all-sugar type of hard candy is made, cream of tartar may be used as a "doctor" instead of corn syrup.

Many other ingredients go into hard candies. Each one has a specific function. In all-molasses hard-candy mixtures, molasses is used as coloring and flavoring. It is also used in making chips and sponge for chocolate centers. Honey is used as a flavor in some hard candies and also in some filled centers. Chocolate is used for flavoring and coloring. Fruits and nuts are used in filled centers. In coloring the candies, paste colors are used. Great care is required in making bright colors neither too dark nor too pale. Special flavors and oils are made for hard-candy manufacture. It takes a special highly concentrated flavor to produce good results since some of the flavor evaporates while it is stirred into the batch.

In cutting filled hard candies or punching out drops and tablets, small hand-turned machines or power-driven medium-sized rollers and cutting machines are used. The rollers are used for solid hard candies and the cutting machines for filled or solid hard candies. An electric fan blows on the candies; this cools them off and keeps them glossy. These machines are available in many forms and designs. The more machinery the confectioner uses the more varied will be his hard candy assortment.

Hard candies are very good when made right. Packed in air-tight tins or glass jars, they will keep in good condition for a long time. High-grade hard candies, known as gloss goods, are made of pure sugar "doctored" with cream of tartar. Some candy-makers believe that a corn-syrup "doctor" will keep them longer. The confectioner will get good results with any formula if he follows it carefully.

In the average candy factory, the hard candy department is usually the busiest unit. Aside from the regular line of hard-candy specialties, many hard specialties serve as centers for chocolate coating. Two kinds of hard candy are made. The cheaper kind is usually made for variety stores and wholesale distributors. This line of candy is produced in huge quantities and sold on price, with a small margin of profit. It should, therefore, be made on a production basis. The cooking is usually done in the vacuum pan by steam. The batches range from 500 to 1000 pounds and over. The cooked batches are poured on water-cooled slabs. A mixing and kneading machine is used to get the batch ready for spinning. Large pulling machines are used for pulling. The batch-roller machine keeps rolling the batch while the confectioner spins it. For cutting in various forms and shapes, the automatic cutting machine is used. In the summer time, these goods are usually sugared to keep them from sticking; a revolving pan does the sugaring.

In making high-priced, quality hard candy, the cooking is usually done on the open-fire gas stove. As in the production of cheaper hard candy, the batches are poured on water-cooled slabs, but mixing and kneading are done by hand. Spinning is carried out mostly from the flat table, but where a satin finish or gloss goods are wanted, the batch roller is sometimes used. Most of the pulling is done on a hook.

Special production problems: Hard candy is "doctored" to keep it plastic and avoid crystallization. To "doctor" pure-sugar hard candy, 1 even teaspoonful of cream of tartar is used for every 10 pounds of crystal "A" sugar. 1½ quarts of water are enough to dissolve 10 pounds of sugar. The cooking temperature ranges from 330 to 340°F, depending on the kind of hard candy desired and the climate of the place where the shop is located.

When a corn-syrup "doctor" is preferred, 1½ pounds of pure refined corn syrup for every 10 pounds of sugar is the general rule. One quart of water is sufficient to dissolve the sugar.

The cooking temperature ranges from 310 to 330°F. From 20 to 25 minutes is the proper cooking time for a 20 to 25 pound batch of hard candy. It should be remembered that slow cooking will invert some of the sugar and give a sticky batch of hard candy. To avoid grainy batches, each batch should be covered before it starts to boil; then the sides of the batch should be washed down before a thermometer is inserted.

If, as sometimes happens, the confectioner has forgotten whether or not he has "doctored" the batch, he should stir one half of the usual amount of "doctor" into the batch before he finishes cooking it.

Getting the gloss in hard candy is a problem. It should be remembered that a lower-cooked batch may be easier to work, but will have no gloss. An overcooked batch, however, will yield sticky hard candies.

Pulling the batch by hand or machine is another problem. When penny goods are made and light weight is desired, the pulling should begin when the batch is still soft and continue until it gets light and fluffy. For gloss goods, pulling should begin when the batch is firmer and continue for 3 to 4 minutes. Kneading should begin while the batch is being formed and continue until a glossy texture is achieved.

Color, flavor, and acid are usually stirred in while the batch is on the slab and has partially cooled along the edges. These elements are mixed in while the batch is being folded together. Paste colors are best to use. For flavoring hard candies, oils, extracts, and concentrated flavors are preferred. In blending colors, yellow and red will give an orange shade; pink and blue will give a purple or violet color. Blending flavors is an art, and only an expert should try it.

When making filled hard candy, it is important that the center and outside batch are kept at the same temperature. Excessive heat and cold should be avoided. Excessive cold will harden the batch, and excessive heat will split the batch open. In either case, the result will be a total loss.

Spinning hard candy uniformly is a matter of experience. The more the operator works at it, the more efficiently and uniformly he will spin.

Although hard candy is an all-year-round confection, it is very susceptible to dampness. When it sweats, a thin syrup forms on the outer surface. If the candy is dry or closely "doctored," the sweating will be followed first by drying and later by graining. If the candy is overcooked or too highly "doctored," the sweating becomes a real menace, for it will continue until the candy liquefies. The best solution to the sweating problem is to "doctor" the hard candy closely so that if it does start to sweat, it will become grainy—but will still be edible. From the point of view of the storekeeper and buyer, this is a better state of affairs than a damp, sticky mess.

When the outer surface of hard candy is covered with fine granulated sugar, it is given a protective coating which makes it decidedly moisture resistant. More corn syrup can be used where a cheaper product is wanted. Before applying the sugar, the candy is steamed or dampened. After the sugar is applied, the candy should be allowed to dry before it is packed.

All hard candy scrap that does not contain acid can be cooked over again. Most of it can go into horehound drops. Introducing this kind of scrap need not affect the flavor because the burned sugar for the coloring and the extract of horehound (both found in HOREHOUND DROPS, *see*) will offset all foreign flavors. Some scrap can also be used in molasses chewing candies, but great care must be taken not to overdo it. Only a small amount of scrap can be used so that it blends into each batch without injuring the flavor. Extra flavor is always used in scrap-cooked batches. All scrap that does contain acid must be used up the same day. It cannot be cooked, but must be kneaded in from one batch to the other.

All hard-candy spinning must be done with gloved hands. Canvas or leather gloves may be used. Spinning with bare hands kills the gloss and makes the hard candy sweat and stick.

For the various hard-candy formulas, see the individual names, such as, LEMON TABLETS, SPEARMINT LEAVES, etc. *See also* BARLEY-SUGAR CUTS, CANDY CURLS, CRIMPED RIBBON CANDY, FILLED BUTTER CUPS, FRENCH CRISP, FRUIT CUTS, STICK CANDY, WALNUT PILLOWS, and YELLOW JACK.

HARD CANDY DEPARTMENT: *See* SAVING WASTE.

HARD CANDY SCRAP: *See* SAVING WASTE.

HARD CANDY, VACUUM-COOKED

Sugar	200 lb
Corn Syrup	100 lb
Water	40 qt

The sugar, water, and corn syrup are placed in the attached open-steam kettle with the steam turned on, and stirred until dissolved. When the batch begins to boil, the sides are washed down and a thermometer is placed in the batch, which is cooked to 280°F. The batch is then drawn into a vacuum pan after the vacuum has been applied to 28 inches for 6 minutes. The batch is poured on greased slabs, colored and flavored as desired, and finished as usual.

It will yield about 280 pounds.

HAZELNUT BRITTLE

Sugar	12 lb
Corn Syrup	8 lb
Water	1½ qt
Salt	1 tablespoonful
Hazelnuts	12 lb

The nuts are kept warm near the stove. The sugar, corn syrup, water, and salt are placed in a copper pan on the stove and stirred until dissolved. When the batch begins to boil, it is covered and allowed to steam for a few minutes. The sides are washed down and a thermometer is placed in the batch, which is cooked to 300°F. It is then removed to a stand, and the nuts are all stirred in and mixed through thoroughly. The batch is then poured out on a greased slab, and the edges are folded together while rolling with a rolling pin. The batch is formed square between iron bars and sized down to approximately 1-inch thickness. While the batch is still soft, it is sliced and cut into squares. After cooling on the slab, it is packed into tins.

This batch will yield approximately 30 pounds.

HAZELNUT CORDIALS

Sugar	10 lb
Water	2 qt
Cream of Tartar	1 teaspoonful
Slow-Set Citrus Pectin	1 oz
Filbert Paste (dissolved in ½ cup of hot milk and honey)	2 oz
Vanilla Flavor	½ teaspoonful
Invertase	½ teaspoonful

The pectin is mixed dry in 3 pounds of the sugar and then placed on the stove in a small copper pan which contains 1 quart of water and the cream of tartar. This mixture is stirred until it is completely dissolved. The balance of the sugar and water are then added and stirring continued until they are dissolved. When the batch begins to boil, it is chilled with a cupful of cold water. The scum is removed and then the dissolved filbert paste is added. When the batch has cooked to 232°F, it is removed from the stove and cooled for 5 minutes. It is then cast in bottle-shaped impressions molded in clean, warm, dry starch. The impressions are placed in a warm room overnight. The cordials are then removed, brushed carefully, and coated with heavy dark vanilla sweet chocolate. They are allowed to set and are then wrapped in colored aluminum foil and packed in boxes.

The batch will yield approximately 20 pounds of cordials.

HAVANA BANANA PASTE

Slow Set Citrus Pectin	4 oz
Water	6 qt
Sugar	12 lb
Corn Syrup	8 lb
Sodium Citrate (dissolved in water)	1 oz
Citric Acid (dissolved in water)	2 oz
Banana Pulp	6 lb
Banana Flavor	3 teaspoonfuls

The water is placed in a clean copper pan on the stove and heated to the boiling point. The pectin is mixed dry with 3 pounds of the sugar and stirred in rapidly. This mixture is then stirred constantly and allowed to boil again. The dissolved sodium and half of the dissolved citric acid are then stirred in. Stirring is continued while the balance of the sugar and the corn syrup are added. When the batch has reached about 224°F, the banana pulp is stirred in and the batch is cooked to 231°F. It is then removed from the stove and cooled for a while. The flavor and the balance of the citric acid solution are then stirred in and the batch is poured out on heavily-dusted starch boards and allowed to set overnight. It is then sliced, cut into small squares, and rolled in a mixture of powdered sugar and cornstarch. When dry, it is packed in boxes.

This will yield about 35 pounds of candy.

HONEY ALMOND TOFFEE

Sugar	10 lb
Corn Syrup	10 lb
Coconut Butter	8 oz
Pure Clover Honey	1 qt
Fresh 18% Cream	2 qt
Butter	1 lb
Chopped Roasted Almonds	2 lb
Evaporated Milk	2 lb
Oil of Bitter Almond	2 teaspoonfuls

The sugar, corn syrup, coconut butter, and fresh cream are placed in a copper pan on the stove and stirred until dissolved. When the batch begins to boil, the evaporated milk is added gradually and the batch is stirred constantly. When boiling thick, the honey and butter are stirred in gradually. Stirring and cooking are continued until the batch forms a medium ball when tested in cold water (*see* COLD-WATER TEST). The chopped almonds are mixed in, and the batch is then poured out on a greased, cold slab. The flavor is stirred in while the edges are folded together. The batch is cooled rapidly and then cut to the desired size and wrapped.

This batch will yield approximately 25 pounds of candy.

HONEY COCONUT

Sugar	8 lb
Corn Syrup	7 lb
Pure Clover Honey	1 qt
Water	1 qt
Long-Thread Coconut	8 lb
Vanilla Flavor	2 teaspoonfuls

The sugar, corn syrup, honey, and water are placed in a copper pan on the stove and stirred until dissolved. When the batch begins to boil, it is covered and allowed to steam for a few minutes. The sides are washed down and a thermometer is placed in the batch, which is cooked to 242°F. The coconut is then stirred in gradually and cooked for about 2 minutes longer. The batch is turned out on a greased slab, stretched out thin, and allowed to cool. It is then cut into large squares and packed in trays.

This formula should yield approximately 30 pounds of candy.

HONEY COCONUT BARS

Sugar	8 lb
Corn Syrup	12 lb
Clover Honey	2 qt
Water	1½ qt
Coconut Butter	12 oz
Vanilla Flavor	3 teaspoonfuls
Macaroon Coconut	12 lb

All of the ingredients, excluding the flavor and coconut, are placed in a large copper pan on the stove and stirred until dissolved. When the batch begins to boil it is covered and allowed to steam for a few minutes. The sides are washed down and a thermometer is placed in the batch, which is cooked to 242°F. The coconut is then stirred in gradually, and the flavor is added and mixed through thoroughly. The batch is then turned out on a greased slab between iron bars, spread out evenly to about 1½ inch thickness, and allowed to cool on the slab until hard. It is then sliced and cut into thin strips to the desired size, rolled in fine granulated sugar, and finally packed in trays.

This formula will yield approximately 40 pounds of candy.

HONEY COCONUT BONBON CENTERS

Fine Macaroon Coconut	20 lb
Sugar	10 lb
Liquid Clover Honey	3 qt
Corn Syrup	20 lb
Water	3 qt
Vanilla Flavor	6 teaspoonfuls

The coconut and honey are placed in an open-type mixing machine, which is started at low speed and kept running while the batch is prepared.

The sugar and corn syrup are placed in a copper pan on the stove and stirred until dissolved. When the batch begins to boil, it is covered and allowed to steam for a few minutes. The sides are then washed down and a thermometer is placed in the batch, which is cooked to 232°F. The batch is poured into the mixing machine on top of the coconut, and the vanilla flavor is added while the coconut and syrup are being mixed thoroughly. The batch is then transferred to the rolling machine and cut into small balls. These balls are allowed to dry and are then ready for dipping in white, yellow, green, and pink bonbon cream.

This batch will yield approximately 55 pounds of centers. (*See* BONBONS, WHITE AND YELLOW.)

HONEY COCONUT EGGS

Macaroon Coconut	10 lb
Pure Clover Honey	1 qt
MARSHMALLOW WHIP (*see*)	2 lb
Sugar	5 lb
Water	1 qt
Corn Syrup	10 lb
Vanilla Flavor	2 teaspoonfuls

The coconut is placed in an open-type mixing machine running at low speed. The honey is poured in and the marshmallow whip added. The mixer is kept in motion.

The sugar, water, and corn syrup are placed in a copper pan on the stove and stirred until dissolved. The sides are washed down, and the batch is cooked to 236°F. The syrup is poured out on top of the coconut while the machine is running, and the flavor is added and mixed through. The batch is then turned out on a greased marble and about 5 pounds are set aside and colored yellow. The white portion of the batch is allowed to cool and is spread out flat. The yellow portion is placed in the center, the white batch is wrapped around it, and the ends are closed. The batch is then rolled out in long rolls, cut into pieces, and weighed to the size desired. These are then rolled and shaped into eggs, which are allowed to cool until hard. They are then coated in a dark vanilla sweet chocolate. When set, the eggs are packed in boxes.

The batch will yield approximately 45 pounds of honey coconut eggs.

HONEY COCONUT ROYALS, CHOCOLATE-COVERED: *See* CHOCOLATE-COVERED HONEY COCONUT ROYALS.

HONEY CORDIALS

Sugar	8 lb
Clover Honey	2 lb
Water	1½ qt
Cream of Tartar	1 teaspoonful
Slow-Set Citrus Pectin	1 oz
Vanilla Flavor	1 teaspoonful
Invertase	½ teaspoonful

The pectin is mixed dry with 3 pounds of the sugar and then placed on the stove in a small pan which contains ½ quart of water and the cream of tartar. This mixture is stirred until dissolved. The balance of sugar, water, and the honey are added to the batch. Stirring is continued until all these ingredients are thoroughly dissolved. When the batch begins to boil, it is chilled with a cupful of cold water. The sides are washed down and the scum is removed. When the batch is heated to 234°F, it is removed from the stove. The flavor and invertase are stirred in rapidly and the batch is cooled for 5 minutes. It is then cast in bottle-shaped impressions molded in clean, warm, dry starch which are placed in a warm room overnight. These are then removed, brushed carefully, and coated carefully with an especially heavy, dark vanilla sweet chocolate. The cordials are allowed to set and are then wrapped in colored aluminum foil and packed in boxes.

The batch will yield approximately 20 pounds of cordials.

HONEYCOMB SPONGE, CHOCOLATE-COVERED: *See* CHOCOLATE-COVERED HONEYCOMB SPONGE.

HONEY EUCALYPTUS

The HONEY TABLET (*see*) formula is followed, but when the batch is ready to pour, 1 gram of paregoric is added and mixed through. Then the batch is finished in the same way.

This batch will yield 30 pounds of candy.

Note: These tablets may be recommended as a splendid throat pastille.

HONEY FILBERT CREAMS

Sugar	50 lb
Fresh 18% Light Cream	9 qt
Cream of Tartar	4 teaspoonfuls
Pure Liquid Clover Honey	6 lb
Invertase	2 teaspoonfuls
Roasted Filberts (chopped and cleaned)	4 lb
Vanilla Flavor	2 teaspoonfuls

The sugar, cream, and cream of tartar are placed in a large copper pan on the stove and stirred until dissolved. When the batch begins to boil, the sides are washed down and stirring is continued. The batch is cooked to a soft ball or about 248°F. The honey is stirred in rapidly and the batch is poured out on a clean, cooled, dampened cream beater and cooled rapidly for about 7 minutes or until the batch registers 110°F. The beater is then set in motion and the flavor and the nuts are added while the batch is being creamed. When the batch creams it is allowed to sweat back for 15 minutes. It is then kneaded smooth, passed through the cut-roll machine, and allowed to cool on the trays. The creams are coated in high-quality milk chocolate and packed as usual.

This batch will yield about 68 pounds of filbert creams.

Note: For approximate yield of chocolate-coated creams, *see* HAND-ROLLED CREAM CENTERS, VANILLA NUT.

HONEY FILBERT SUNDAE TOPPING

Sugar	7 lb
Clover Honey	2 lb
Corn Syrup	1 lb
Water	2½ qt
150 Grade Citrus Pectin	1 oz
Chopped Roasted Blanched Filberts	4 lb
Vanilla Flavor	1 teaspoonful
Glycerin	4 teaspoonfuls

The pectin is mixed dry with 2 pounds of the sugar. One quart of water is placed in a clean copper pan on the stove. The pectin sugar is added and stirred until completely dissolved. The remaining sugar and water are then added and again stirred until the sugar is dissolved. The honey and corn syrup are added rapidly, and the batch is cooked to 219°F. The batch is moved to a cool spot and the chopped, roasted filberts are stirred in rapidly. The flavor and glycerin are added and mixed thoroughly. The filbert sundae topping is poured into clean glass jars and closed with airtight lids. It is then ready for use.

The batch will yield approximately 10 quarts of honey filbert sundae topping.

HONEY MARZIPAN

RUNNING CREAM (*see*)	5 lb
ALMOND PASTE (*see*)	10 lb
Pure Clover Honey	5 lb
Macaroon Coconut	2 lb
6x Powdered Sugar	5 lb
Vanilla Flavor	2 teaspoonfuls
Pistachio Flavor	1 teaspoonful
Yellow Color	1 teaspoonful

All of the listed ingredients are placed in an open-type mixing machine. The machine is set in motion at low speed for about

5 minutes. Then the machine is shifted to medium speed and run for about 10 minutes. The paste is then moved to a clean marble slab and kneaded until it is smooth. The honey marzipan is formed into a round roll, about 2½ inches in diameter. It is then cut into small slices and placed on clean trays. When dry, the marzipan is placed in crystal pans for an overnight crystal. The following morning it is drained, allowed to dry, and packed in boxes.

The batch will yield aproximately 30 pounds of honey marzipan.

HONEY NOUGAT

Sugar	12 lb
Corn Syrup	8 lb
Water	2 qt
Chopped Coconut Butter	1 lb
Pure Clover Honey	4 lb
Chopped Almonds	4 lb
Chopped Pistachio Nuts	½ lb
Vanilla Flavor	3 teaspoonfuls
MARSHMALLOW WHIP (*see*)	6 lb

The sugar, corn syrup, and water are placed in a large copper pan on the stove and stirred until dissolved. When the batch begins to boil, it is covered and allowed to steam for a few minutes. The sides are washed down and a thermometer is placed in the batch, which is cooked to 262°F. The coconut butter is stirred into the batch and the honey is added and stirred in gradually. The marshmallow whip is added and beaten with a large spatula until it is well mixed. The nuts and vanilla flavor are then stirred in. The batch is then turned out on trays, which are about 2½ inches in height, lined and covered with wafer paper. This is left to set until the following day, when it is sliced down and cut in oblongs. It is then wrapped in moistureproof cellophane.

This batch will yield approximately 40 pounds of candy.

HONEY NOUGAT, CHOCOLATE-COVERED: *See* CHOCOLATE-COVERED HONEY NOUGAT.

HONEY TABLETS

Powdered Gelatin (dissolved in 2 qt of water)	2 lb
Sugar	15 lb
Corn Syrup	5 lb
Water	1½ qt
Clover Honey	5 lb
Citric Acid (dissolved in water)	2 teaspoonfuls
Pineapple Flavor	2 teaspoonfuls

The sugar, corn syrup, and water are placed in a copper pan on the stove and stirred until dissolved. When the batch begins to boil, the sides are washed down and a thermometer is placed in the batch. The honey is added gradually, and the batch is stirred constantly in order to keep it from scorching. The batch is cooked to 254°F and is then removed from the stove. The gelatin is stirred in rapidly; the batch is allowed to cool for about 15 minutes and the scum is then removed from the top. The citric acid and the flavor are then stirred in, and the batch is then cast in small impressions which are molded in clean starch. The batch is allowed to set overnight. The tablets are then removed, brushed free of starch, placed in a large, dry, clean pan and sprinkled with gum arabic solution. This is rubbed well over the tablets until they are all damp and clear looking. They are then spread out on boards lined with wax paper. When dry, they are packed as desired.

Note: Gum Arabic solution is made by dissolving 1 ounce of gum arabic in ½ quart of water.

This batch will yield aproximately 30 pounds.

HONEYDEW CHOCOLATES

Water	4 qt
Pure Clover Honey	8 lb
Sugar	22 lb
Cream of Tartar	2 teaspoonfuls
Melted Corn Syrup	2 lb
Fine-Chopped Pecans	2 lb
Invertase	1 teaspoonful

The sugar, honey, water, and cream of tartar are placed in a copper pan on the stove and stirred until dissolved. When the batch begins to boil, it is covered and allowed to steam for a few minutes. The sides are washed down and a thermometer is placed in the batch, which is cooked to 244°F. The melted corn syrup is stirred in rapidly, and the batch is poured out on a clean, cooled, dampened cream beater and cooled rapidly to 100°F. The machine is then set in motion, and the pecans and invertase are added. After the batch has creamed to a lump, it is allowed to sweat back for 15 minutes and then remelted in a steam bath over low heat to the proper running consistency. The batch is then cast in mint-wafer impressions, molded in clean dry starch, and allowed to set overnight. The creams are then removed, brushed free of starch, spread out on clean boards, and coated in a high-quality dark vanilla sweet chocolate. After they are set, the chocolates are packed in boxes.

The batch will yield approximately 60 pounds of chocolates.

HONEYDEW ICE

Water	20 qt
Sugar	20 lb
Lemon Juice	1½ qt
Yellow Color	½ teaspoonful
Honeydew Pulp	6 qt

The honeydew pulp is obtained from ripe, sweet honeydews. The seeds are carefully picked out, and the pulp is chopped fine, placed in a large mixing bowl, and set aside. The water, sugar, juice, and color are placed in the mixing can and stirred until all the sugar is dissolved. The mix is then strained into the freezer and, when the ice is about half-way frozen, the honeydew pulp is added. The ice is completed according to instructions found under VANILLA ICE CREAM NO. 1.

The batch will yield approximately 32 quarts.

HOPJES: *See* DUTCH HOPJES.

HOREHOUND DROPS NO. 1

Sugar	24 lb
Corn Syrup	6 lb
Water	3 qt
Horehound Extract	1½ oz
Burned Sugar	2 tablespoonfuls

The sugar, water, and corn syrup are placed in a copper pan on the stove and stirred until dissolved. When the batch begins to boil, it is covered and allowed to steam for a few minutes. The sides are washed down and a thermometer is placed into the batch, which is cooked to 320°F. The batch is then poured on a greased slab and colored with burned sugar to a dark shade. The edges are folded together while stirring in the horehound extract. Folding is continued until the batch is cool enough to handle, when it is taken over to the spinning table in front of the batch warmer. It is then passed through a drop-roller machine and allowed to cool on the table. The drops are broken apart and blended with a half-and-half mixture of magnesia powder and powdered sugar. They are then put through a sieve and packed in tins.

The batch will yield approximately 28½ pounds.

HOREHOUND DROPS NO. 2: *See* SAVING WASTE.

I

ICE CREAM: *See* FROZEN CONFECTIONS and *Individual Ice Cream Formulas,* e.g., CHOCOLATE ICE CREAM, VANILLA ICE CREAM.

ICE-CREAM BARS

Coconut Butter	4 oz
Corn Syrup	4 lb
Sugar	20 lb
Cream of Tartar	1 teaspoonful
Water	3 qt
Chopped Bitter Chocolate	1 lb
MARSHMALLOW WHIP (*see*)	2½ lb
Vanilla Flavor	1 teaspoonful
Strawberry Flavor	1 teaspoonful
Pink Color	

The sugar, coconut butter, cream of tartar, corn syrup, and water are placed in a copper pan on the stove and stirred until dissolved. When the batch begins to boil, it is covered and allowed to steam for a few minutes. The sides are washed down and a thermometer is placed in the batch, which is cooked to 256°F. The marshmallow whip is then stirred in and two thirds of the batch are poured out on a greased slab. The chopped chocolate is stirred in the remaining portion of the batch, and this is also poured out on a slab and kept soft. One half of the white batch is colored pink and flavored strawberry and the remaining white portion is flavored vanilla. Then all three portions are pulled separately on the hook and placed on the spinning table. The white portion is formed into a flat batch and the pink is spread out evenly on top of the white; finally the chocolate is placed on top of the pink. The batch is then formed in flat shape, spun out in bars, and cut into sections about 1½ inches in width and 3 inches in length. The candy is then wrapped as usual.

This batch will yield approximately 30 pounds of candy.

ICE-CREAM FREEZER, NEAPOLITAN, THREE-TUBE

This machine freezes 100 gallons per hour per tube. The total batch is 300 gallons. It produces three-flavor slices. (Courtesy of the Cherry Burrell Corporation, Cedar Rapids, Iowa.)

ICE-CREAM FREEZER, VOGT

This machine freezes 500 gallons of ice cream per hour at 100 percent overrun. (Courtesy of the Cherry Burrell Corporation, Cedar Rapids, Iowa.)

ICE-CREAM STORAGE TANK, RECTANGULAR

This tank holds 1000 gallons of ice cream, saves space, and is easy to clean. (Courtesy of the Cherry Burrell Corporation, Cedar Rapids, Iowa.)

ICED COCONUT DAINTIES, LEMON

Sugar	10 lb
Corn Syrup	12 lb
Water	1 qt
Macaroon Coconut	10 lb
Lemon Flavor	4 teaspoonfuls
Yellow Color	1 teaspoonful

The sugar, corn syrup, and water are placed in a clean copper pan on the stove and stirred until dissolved. When the batch begins to boil, it is covered and allowed to steam for a few minutes. The sides of the batch are washed down and a thermometer is placed in the batch, which is cooked to 230°F. About one half of the coconut is stirred into the batch while cooking and stirring constantly. The batch is cooked until it forms a soft ball when tested in cold water (*see* COLD-WATER TEST). Then the coloring and the flavor are added. The balance of the coconut is stirred in and mixed through. The batch is turned out on a greased slab between iron bars, spread out evenly with a palette knife, rolled out to about ½ inch thickness, and allowed to cool until hard. It is then sliced and cut in small oblongs. The candy is now ready for icing. For the icing formula *see* ICING FOR COVERING SOFT CANDIES. After being dipped in icing, this batch will yield about 45 pounds of candy.

ICED COCONUT DAINTIES, LOGANBERRY

Sugar	8 lb
Corn Syrup	12 lb
Water	1 qt
Loganberry Pulp	5 lb
Violet Color	1 teaspoonful
Raspberry Flavor	2 teaspoonfuls
Macaroon Coconut	12 lb

The sugar, corn syrup, and water are placed in a large copper pan on the stove and stirred until dissolved. When the batch begins to boil, it is covered and allowed to steam for a few minutes. The sides of the batch are washed down and a thermometer is placed in the batch, which is cooked to 232°F. The loganberry pulp is stirred into the batch rapidly, half of the coconut is added, and the batch is stirred and cooked until it forms a soft ball when tested in cold water. The balance of the coconut is then stirred in, and the color and flavor are added and mixed thoroughly. The batch is then turned out on a greased slab, spread out evenly, and rolled out to about ½ inch thickness. It is allowed to cool and is cut into small oblong pieces. It is then ready for icing, and is packed as usual.

This batch will yield about 50 pounds of candy after dipping in icing.

ICED COCONUT DAINTIES, ORANGE

The formula for ICED COCONUT DAINTIES, LEMON *(see)* is followed, but the batch is colored a bright orange and flavored with oil of orange. It is finished in the usual manner. After being dipped in icing, this batch will yield about 45 pounds of candy.

ICELAND ROLLIES

Batch No. 1

Sugar	9 lb
Corn Syrup	3 lb
Fresh 18% Cream	3 qt
Salt	1 teaspoonful
Butter	8 oz
Silver Sheet Gelatin (soaked in ½ cup of warm water)	½ oz
Chopped Bitter Chocolate	2 lb
FRENCH CREAM *(see)*	5 lb
Vanilla Flavor	2 teaspoonfuls

The sugar, fresh cream, corn syrup, and salt are placed in a copper pan on the stove and stirred until dissolved. The sides are washed down and the batch is stirred and cooked to 240°F. The butter is then added to the batch while it is still on the stove, and the batch is stirred until the butter is melted. The batch is then moved near the slab, and the soaked gelatin is stirred in. The chocolate is added together with the French cream and flavor. The batch is stirred until all the added ingredients are mixed thoroughly. It is then turned out on a greased marble and cooled rapidly until it is firm enough to handle. It is then formed into long rolls, about 8 inches in length and ½ inch in diameter. The rolls are now ready for dipping.

Batch No. 2

Sugar	3 lb
Corn Syrup	1 lb
Fresh Light Cream	1 qt
FRENCH CREAM (*see*)	1 lb
Red Color	1 teaspoonful
Raspberry Flavor	2 teaspoonfuls
Chopped Pecans	10 lb

The sugar, corn syrup, and cream are placed in a small copper pan on the stove and stirred until dissolved. The batch is stirred and cooked to 236°F. It is then taken near the marble slab and the color and flavor are added. The cream is stirred in rapidly. The cooled long rolls are dipped rapidly into this red sauce and rolled in the chopped pecans which are spread out on the marble. This procedure is continued until all the rolls are dipped and covered with pecans. They are allowed to set, then cut into small rollies, and packed in flat shallow pans or dishes.

This formula will yield approximately 40 pounds of candy.

ICING: *See* ICING FOR EASTER EGGS.

ICING FOR COVERING SOFT CANDIES

6x Sieved Powdered Sugar	30 lb
Instant Powdered Egg Albumen	1 lb
Water	4 qt
Cream of Tartar	2 teaspoonfuls
White Vanilla Flavor	4 teaspoonfuls

The water and powdered sugar are placed in an open beater and started at low speed. After the beater has run for about 5 minutes, the egg albumen is added and the speed is changed to medium. Whipping is continued for another 5 minutes. The cream of tartar is added and the machine is shifted to high speed. The flavor is added and whipping is continued until the batch is white and spongy. The icing is then ready for use.

For dipping, the icing is spread out on the dipping marble, and the centers are dipped in the icing the same way as centers are dipped in chocolate coating. The iced candies are deposited on wax-paper-lined dipping trays, and the trays are placed in a cool room until the candies are dry. They are then packed in the usual manner.

This formula will yield approximately 35 pounds of icing.

ICING FOR EASTER EGGS

Instant Egg Albumen	1 lb
Water	2 qt
Powdered Gum Arabic Solution	2 teaspoonfuls
Cream of Tartar	1 teaspoonful
6x Sieved Powdered Sugar	10 lb (approximately)

The water and 2 pounds of the sugar are placed in an open mixing machine. The machine is started at low speed and allowed to run for 2 minutes. The egg albumen is then added and the machine is changed to medium speed. The cream of tartar and gum arabic are added while the batch is whipped a little longer. More powdered sugar is added and the machine is shifted to high speed. Then still more powdered sugar is added as whipping is continued. The icing is whipped until it is light and spongy and stands up firmly. When a portion of the icing is dipped out in a large wooden spoon and it stands up well and does not sink, the icing is ready for decorating.

INVERTASE: *See* CREAMS AND EASTER EGGS.

ITALIAN CARAMELS, CHOCOLATE

The formula for ITALIAN CARAMELS, VANILLA is followed, adding 4 pounds of bitter chocolate, which is stirred in and melted after the batch is cooked. Four pounds of walnuts are then stirred in, the batch is poured out on a greased slab, and completed as usual.

This batch will yield approximately 44 pounds of candy.

ITALIAN CARAMELS, VANILLA

Sugar	8 lb
Corn Syrup	12 lb
Coconut Butter	1 lb
Salt	1 tablespoonful
Fresh 18% Cream	2 qt
Evaporated Milk	8 lb
BONBON CREAM (*see*)	8 lb
Vanilla Flavor	4 teaspoonfuls
Filberts	4 lb

The sugar, corn syrup, salt, coconut butter, and cream are placed in a large copper pan on the stove and stirred until dissolved. When the batch begins to boil, the evaporated milk is stirred in gradually. This stirring is continued, and the batch is cooked to a medium ball when tested in cold water. It is then removed to a stand and the bonbon cream, flavor, and filberts are stirred in and mixed thoroughly. The batch is then poured out on a greased slab between iron bars, spread out evenly with a palette knife, and allowed to cool until hard. It is then passed through a sizing machine, cut into square caramels, and wrapped in moistureproof cellophane.

This batch will yield approximately 40 pounds of candy.

ITALIAN CREAM CHOCOLATES

Sugar	40 lb
Water	8 qt
Cream of Tartar	3 teaspoonfuls
MARSHMALLOW WHIP (*see*)	10 lb
Vanilla Flavor	6 teaspoonfuls
Invertase	3 teaspoonfuls

The sugar, water, and cream of tartar are placed in a copper pan on the stove and stirred until dissolved. When the batch begins to boil, it is covered and allowed to steam for a few minutes. The sides are washed down and a thermometer is placed in the batch, which is cooked to 244°F. The batch is then poured out on a clean, cooled, dampened cream beater and sprinkled with a few drops of cold water on top. It is then cooled down rapidly to about 100°F. The marshmallow whip, flavor, and invertase are added and the beater is set in motion and kept running until the batch creams. The batch is allowed to set back for 20 minutes and the cream is remelted in a steam-bath copper pan over low heat. The cream is then cast in clean starch-molded impressions in shapes of round-top pointers. The creams are allowed to set overnight and they are then removed, brushed free of starch, and coated with dark vanilla sweet chocolate. They are allowed to set on the dipping trays and are then packed in boxes.

This formula will yield approximately 60 pounds of cream centers, 75 pounds of chocolates if machine coated, and 85 pounds if hand coated.

ITALIAN NOUGAT

Batch No. 1

Honey	7 lb
Corn Syrup	3 lb
Instant Egg Albumen	4 oz
Water	½ qt

The honey, corn syrup, egg albumen, and water are placed in a medium open beater and beaten at medium speed.

Batch No. 2

Water	½ qt
Sugar	6 lb
Corn Syrup	5 lb
Coconut Butter	8 oz
Chopped Almonds	3½ lb
Chopped Pistachio Nuts	½ lb
Vanilla Flavor	3 teaspoonfuls

The sugar, corn syrup, coconut butter and water are placed in a copper pan on the stove and stirred until dissolved. When the batch begins to boil, it is covered and allowed to steam for a few minutes. The sides are washed down and a thermometer is placed in the batch, which is cooked to 268°F. This batch No. 2 is then poured out in a slow stream and added to batch No. 1 in the beater. The resulting batch is beaten at high speed until thick, and then the speed is changed to low while the nuts and flavor are added and mixed through. The batch is then turned out on trays, lined, and covered with wafer paper. It is allowed to set overnight and is then cut and wrapped as usual.

This formula will yield approximately 30 pounds of candy.

J

JAGUAR THROAT TABLETS

Citrus Pectin, Grade 150	3 oz
Sugar	24 lb
Corn Syrup	6 lb
Water	4 qt
Powdered Jaguar Gum	8 oz
Green Color	1 teaspoonful
Oil of Peppermint	2½ teaspoonfuls

The pectin is mixed dry with 4 pounds of the sugar and placed in a clean copper pan. One quart of water is added to the mixture. The pan is placed on the stove and stirred until the sugar dissolves. The balance of the sugar and water and half of the corn syrup are added on the stove, and the batch is stirred until it dissolves. When the batch begins to boil, its sides are washed down and it is cooked to 258°F. The jaguar gum is stirred into the remaining half of the corn syrup, which has previously been melted. The gum is stirred in until it makes a paste. Then the paste is stirred into the hot batch rapidly, and the batch is removed and cooled for 10 minutes. After the color and flavor are stirred in, the batch is cast in small tablet impressions which are molded in clean dry starch and allowed to set overnight. The following morning the tablets are removed, brushed free of starch, and put in a large clean pan. They are then sprinkled with gum arabic solution and rubbed through until they are damp, clear, and bright. Then they are spread out on boards lined with wax paper and allowed to dry. When dry, they are packed as desired.

This batch should yield approximately 35 pounds of candy.

JELLED MINTS, LEMON

The formula for JELLED MINTS, LIME (*see*) is followed, and the two-batch system is continued. In batch No. 1, for the center, yellow color and lemon flavor are used. The batch is finished in the usual manner and will yield approximately 40 pounds of candy.

JELLED MINTS, LIME

Batch No. 1

Sugar	10 lb
Slow-Set Citrus Pectin	½ oz
Water	1½ qt
Green Color	1 teaspoonful
Oil of Limes	1 teaspoonful
Cream of Tartar	½ teaspoonful

Batch No. 2

Sugar	30 lb
Cream of Tartar	2 teaspoonfuls
Water	4 qt
Oil of Peppermint	3 teaspoonfuls

In making jelled mints, it is essential that both batches are started at the same time and are poured out on two separate slabs.

For batch No. 1, the pectin is mixed dry with 2 pounds of the sugar and placed with the water in a copper pan on the stove and stirred until dissolved. The balance of the sugar and the cream of tartar are added rapidly and stirred until dissolved. The sides are washed down, the batch is cooked to 264°F, and it is then poured out on a cold, greased slab. The edges are then folded together and, while cooling rapidly, the color and flavor are added and mixed through. When the batch is cold, it is pulled on the hook for 3 to 4 minutes. It is then taken over to the spinning table and kept for the center.

For batch No. 2, the sugar, water, and cream of tartar are

placed in a copper pan on the stove and stirred until dissolved. When the batch begins to boil, it is covered and allowed to steam for a few minutes. The sides are washed down and a thermometer is placed in the batch, which is cooked to 268°F. The batch is poured out on a cold, greased slab and the edges are folded together. While the batch cools rapidly, the flavor is added and mixed through. When the batch is hard enough, it is taken over to the pulling machine and pulled for about 7 minutes. It is then placed on the spinning table and formed flat. The small green batch is placed in the center and the white batch is wrapped over it. The composite batch is kept in flat shape and passed through the mint-cutting machine. It is placed on boards and allowed to set overnight in a warm room. After they cream, the mints are broken apart and packed in airtight tins.

This batch should yield approximately 40 pounds.

JELLED MINTS, ORANGE

The formula for JELLED MINTS, LIME (*see*) is followed and the two-batch system is continued. In batch No. 1 for the center, orange color and oil of orange are used. The batch is finished in the usual manner and will yield approximately 40 pounds of candy.

JELLED MINTS, RASPBERRY

The formula for JELLED MINTS, LIME (*see*) is followed and the two-batch system is continued. In batch No. 1 for the center, red color and raspberry flavor are used. Otherwise the batch is finished as usual and should yield approximately 40 pounds of candy.

Note: Jelled mints may be packed assorted or in separate flavors.

JELLY: *See* SOFT GOODS and *Individual Jelly Formulas,* e.g., ORANGE JELLY, LIME FRUIT JELLY. *See also* JELLY FRUIT CUTS, JELLY FRUIT SLICES, SPICED STRING JELLY.

JELLY BEANS, ASSORTED

Sugar	50 lb
Corn Syrup	50 lb
Water	8 gal
Best-Cooking Starch (dissolved in 10 qt of water)	10 lb
Cream of Tartar	1 oz
Acetic Acid	1 teaspoonful

The sugar, water, and corn syrup are placed in a large steam-jacketed kettle. The steam is turned on and the batch is stirred until it dissolves and begins to boil. The dissolved cornstarch is then added gradually to the cooking batch. While the batch continues cooking the cream of tartar and the acetic acid are added. The batch is stirred and cooked to an extra-heavy firm jelly or until it comes off the paddle in heavy strings or to 226°F. The batch is then poured into the depositing machine and cast into little egg impressions molded in clean warm starch. These are then set in a hot room overnight and the following day they are removed and sifted free of starch. The jelly beans are then dampened lightly with simple syrup and rolled in fine granulated sugar. They are then spread out on boards to dry and are ready for the revolving pan. Before the revolving-pan process begins, a gum arabic syrup solution is prepared as follows:

Powdered Gum Arabic (dissolved in 3 qt of water)	3 lb
Water	2 qt
Sugar	15 lb
Corn Syrup	15 lb
Citric Acid (dissolved in 12 oz of water)	4 oz

Colors: Yellow —Lemon Flavor
Orange—Orange Flavor
Green —Lime Flavor
Red —Cinnamon Flavor
Violet —Violet Flavor
Black —Licorice Anise Flavor

The sugar and water are placed in a copper pan on the stove and stirred until dissolved. The solution is allowed to boil for a minute and the corn syrup is then stirred in rapidly. The gum arabic solution is stirred in and mixed through thoroughly, and the syrup is then moved to a stand near the revolving pan.

For panning the jelly beans, the syrup is divided into six parts. One part is colored yellow, flavored lemon, and one third of the acid solution is mixed in. About 20 pounds of the jelly beans are placed in the running revolving pan and some of the yellow syrup is run on the beans until they are wet. Some powdered sugar is immediately applied to the wet beans and they are kept from sticking together by running a hand through them to separate them. This process is continued and more sugar is added. The revolving pan is allowed to run until dry, and then more yellow syrup and sugar are added and the process is continued until the beans reach the required size. When ready for polishing, they are removed and spread out on clean boards.

The same process is followed with orange syrup and added acid solution, which is applied to another 20 pounds of beans. Then another lot with green syrup and the acid solution follows. The fourth lot is done with the red syrup (no acid) and cinnamon flavor. Lot five is prepared with the violet syrup (no acid) and flavored violet. The last lot is made with the black licorice syrup and flavored anise.

When all the beans are processed for polishing, the revolving pan is prepared in the following manner: About 1 pound of hard wax is melted and poured slowly into the revolving pan while it is in motion. This is allowed to run until the wax hardens on the surface of the pan. About one third of the beans is then placed into the revolving pan and a few more small pieces of the hard wax are placed in the pan, which is allowed to run until the beans are nice and glossy. This process is continued until all the beans are polished. They are then ready to be packed as desired.

Note: The yield on this formula may vary in poundage. It will depend on the amount of sugar absorbed in the panning process.

JELLY BLACKBERRIES

The formula for JELLY STRAWBERRIES (*see*) is followed, but the batch is colored with charcoal black and flavored with blackberry. The sanding sugar is also colored with black charcoal. The batch is finished as usual and should yield approximately 65 pounds.

JELLY FRUIT CUTS, LEMON

Water	6 qt
Sugar	10 lb
Corn Syrup	10 lb
Slow-Set Citrus Pectin	4 oz
Sodium Citrate	1 oz
Citric Acid	2 oz
Yellow Color	1 teaspoonful
Oil of Lemon	3 teaspoonfuls
CREAM FONDANT (*see*)	20 lb
Yellow Color	1 teaspoonful
Oil of Lemon	3 teaspoonfuls
Diced Citron	3 lb

The citric acid and the sodium citrate are dissolved separately in warm water. The water is placed in a clean copper pan on the stove and, when boiling, the pectin is mixed dry with 3 pounds of the sugar and stirred in rapidly. When boiling again, the sodium citrate and one half of the citric acid are stirred in rapidly. Then the balance of the sugar and the corn syrup are stirred in gradually. After the batch is cooked to 228°F, the color and flavor are added while stirring in the balance of citric acid. The batch is poured out on a slab lined with plain packing paper and spread out between bars to 1/8 inch thickness.

For the second operation, the fondant is melted in a steam pan to 165°F, and the color, flavor, and the citron are stirred in and mixed thoroughly. This fondant is then poured out and spread evenly on top of the jelly.

Finally, another batch of lemon jelly is cooked following the first formula. This is poured on top of the fondant, spread out evenly and allowed to cool until it is set. The batch is then cut into small squares; they are dampened and sugared. When the squares are dry, they are placed in crystal pans and crystallized to an overnight crystal as usual (*see* CRYSTALLIZING).

The batch will yield approximately 55 pounds.

JELLY FRUIT CUTS, LIME

The formula for JELLY FRUIT CUTS, LEMON (*see*) is followed, but the jelly batch is colored a delicate green and flavored with oil of lime. The fondant for the center is also colored green and flavored lime. Three pounds of diced lime peel are stirred in, and the batch is finished in the usual manner.

This formula will yield approximately 55 pounds of candy.

JELLY FRUIT CUTS, ORANGE

The formula for JELLY FRUIT CUTS, LEMON (*see*) is followed with the following exceptions: The jelly batch is colored and flavored orange. The fondant for the center is also colored and flavored orange and 3 pounds of diced orange peel are mixed in. The batch is finished as usual and will yield approximately 55 pounds of candy.

JELLY FRUIT CUTS, STRAWBERRY

The formula for JELLY FRUIT CUTS, LEMON (*see*) is followed, but the jelly batch is colored a delicate pink and flavored strawberry. Three pounds of strawberry jam are mixed through the fondant.

This formula should yield approximately 55 pounds of candy.

JELLY FRUIT SLICES, LEMON

Sugar	22 lb
Corn Syrup	16 lb
Water	12 qt
Slow-Setting Citrus Pectin	8 oz
Sodium Citrate (dissolved in water)	2 oz
Citric Acid (dissolved in water)	4 oz
Yellow Color	2 teaspoonfuls
Oil of Lemon	4 teaspoonfuls
MARSHMALLOW WHIP (*see*)	2 lb

The water is placed in a large copper pan on the stove and heated to the boiling point. The pectin is mixed dry with 7 pounds of the sugar and added to the boiling water, while stirring constantly. When boiling again, the dissolved sodium and one third of the dissolved citric acid are added, while stirring in the balance of the sugar and the corn syrup gradually. The batch is cooked to 228°F and is then removed to a place near the slab.

Two small portions, about 5 pounds each are taken from the batch and poured into small pans; the largest portion is kept in the cooking pan. The first small portion is colored yellow, flavored with 1 teaspoonful of oil of lemon and 1 teaspoonful of citric acid. This is mixed through well, poured out rapidly on the paper-lined slab, and spread out rapidly to about $\frac{1}{16}$ inch thickness.

The marshmallow whip is now stirred into the second small portion. This is flavored with 1 teaspoonful of oil of lemon. One teaspoonful of citric acid is also stirred in. This is then poured out on top of the first layer, spread out rapidly, and covered evenly.

The remaining large portion is colored yellow and flavored with 2 teaspoonfuls of oil of lemon. Before the citric acid is added, the molds are prepared in the following manner: A small round stick, as long as the width of the starch board and about 4 inches in diameter, is used to make impressions of half-moons in the starch. A funnel with a large opening is used for casting. Now 4 teaspoonfuls of citric acid are stirred into the large batch. The batch is then run out through the funnel rapidly into the starch molds and allowed to set. The jellies are then removed, cleaned of starch, and placed on boards lined with wax paper.

The lemon-peel batch on the slab is turned upside down on the slab and sprinkled with some starch. The paper on top is dampened and removed, and sugar is sprinkled on top. After the lemon-peel batch is turned over again with the marshmallow side up, the lemon jelly bars are wrapped around with the peel, leaving the bottom as it is. The wrapped rolls are rolled in fine granulated sugar and allowed to set until the peels adhere to the bars. The jelly is then sliced into fine thin slices and rolled in sugar. After drying, the slices are placed in crystal pans (it is not advisable to place too many in a pan) and crystallized as usual (*see* CRYSTALLIZING).

The formula will yield about 55 pounds of candy.

JELLY FRUIT SLICES, LIME

The formula for JELLY FRUIT SLICES, LEMON (*see*) is followed, but the clear batch is colored a bright green. The marshmallow portion remains white and is flavored with oil of lime. The batch is finished in the usual manner and should yield 55 pounds of candy.

Note: All pectins referred to in our formulas are of the powdered type.

JELLY FRUIT SLICES, ORANGE

The formula for JELLY FRUIT SLICES, LEMON (*see*) is used, with the following exceptions: The batch is colored a bright orange. The small marshmallow portion remains white and is flavored with oil of orange. It is finished in the same manner as the lemon batch and should yield 55 pounds of candy.

JELLY RASPBERRIES

The formula for JELLY STRAWBERRIES (*see*) is followed, but the batch is colored a deep red and flavored with raspberry. A deep red color is used also for the sanding sugar. Otherwise the batch is completed in the same manner as the strawberry jellies.

This batch will yield approximately 65 pounds.

JELLY STRAWBERRIES

Agar-Agar	8 oz
Water	12 qt
Sugar	25 lb
Corn Syrup	20 lb
Citric Acid (dissolved in water)	5 oz
Strawberry Flavor	5 teaspoonfuls
Pink Color	2 teaspoonfuls
Pink Sanding Sugar	10 lb
Stemmed Strawberry Leaves	As desired

The agar-agar and water are placed in a large copper pan on the stove and stirred until dissolved. When boiling, the sugar is stirred in gradually. Stirring is continued while the corn syrup is added, and the batch is cooked to 224°F. It is then removed to a cool spot and poured out into a clean, cool, copper pan. The color and flavor are stirred in, and the batch is cooled to about 100°F. At this point, the citric acid is stirred in and the batch is cast in starch into strawberry-shaped impressions and allowed to set overnight. The jellies are then brushed free of starch, dampened, and rolled in fine granulated sugar which has been prepared and colored pink to give the appearance of a real strawberry. The strawberries are then spread out on boards and allowed to dry. The stemmed strawberry leaves are then pierced through the flat side of each strawberry, and they are then ready to pack as desired.

This batch will yield approximately 65 pounds of candy.

JERSEY DAIRY CREAM CHOCOLATES, COFFEE

The formula for JERSEY DAIRY CREAM CHOCOLATES, VANILLA (*see*) is followed, but the cream is flavored with sufficient concentrated coffee extract for a pronounced coffee flavor. Otherwise the chocolates are handled and finished as usual.

The batch will yield approximately 70 pounds of chocolates.

JERSEY DAIRY CREAM CHOCOLATES, STRAWBERRY

The formula for JERSEY DAIRY CREAM CHOCOLATES, VANILLA (*see*) is followed, but the cream is colored pink and flavored strawberry. Otherwise the creams are handled and finished in the usual manner.

The batch will yield approximately 70 pounds of chocolates.

JERSEY DAIRY CREAM CHOCOLATES, VANILLA

Sugar	15 lb
Corn Syrup	5 lb
Salt	1 teaspoonful
Fresh 20% Light Cream	7 qt
Vanilla Flavor	5 teaspoonfuls
RUNNING CREAM (*see*)	25 lb

The sugar, corn syrup, salt, and cream are placed in a large copper pan on the stove and stirred until dissolved. When the batch begins to boil, the sides are washed down, stirring process is continued, and the batch is cooked until it forms a soft ball when tested in cold water or until the thermometer registers 240°F. The running cream is stirred in gradually and the flavor is mixed in thoroughly. The cream is then cast into impressions of the desired shapes and molded in clean dry starch. The creams are allowed to set overnight and are then removed, brushed free of starch, spread out on boards lined with wax paper, coated in high-quality semisweet dark chocolate, and finished with a small loop design. After cooling and setting, they are packed in boxes.

The batch will yield approximately 70 pounds of chocolates.

JORDAN ALMONDS, ASSORTED, SMOOTH

Gum Arabic Solution

Gum Arabic	6 lb
Granulated Sugar	2 lb
Water	6 qt

The powdered gum arabic is mixed with the water. The sugar is added into a small copper pan on the stove over low heat and stirred until dissolved. After straining through a fine sieve, the solution is ready for use.

Syrup for Coating

Fine Granulated Sugar	75 lb
Powdered Cornstarch	3½ lb
Medium-Size Raw Almonds	50 lb

The almonds are placed in a revolving pan heated to about 110°F. While the pan is in motion, the almonds are sprinkled with some gum arabic solution. The pan is kept running while the almonds become somewhat wet on the surface. About 3 minutes later, a second charge of gum arabic is applied and then a third until about 1 quart of gum arabic is used up and the almonds are well covered and gummy. The powdered starch is dusted over the almonds a little at a time until they are completely covered. They are then placed on clean trays overnight and are ready for syrup coating as follows:

10 pounds of the sugar and 1½ quarts of water are placed in a small copper pan on the stove, stirred until dissolved, and cooked to 222°F. Then 1½ pounds of the gum arabic solution are added to this syrup and mixed through well. The dried, starch-dusted almonds are placed in the revolving pan, which is tempered to about 90°F. The pan is set in motion and about ½

quart of the syrup is applied to the almonds, which are allowed to roll until they are dry. Similar charges with the syrup are treated until all of this syrup is used up and the almonds end up dry and develop a light sugar dust.

Then 10 pounds of the fine granulated sugar and 1½ quarts of water are placed in a small copper pan on the stove, stirred until dissolved, and cooked to 224°F; 1½ pounds of the gum arabic solution are stirred into the syrup and the same operation in the revolving pan as with the first batch of syrup follows until all the second batch of syrup is used up.

Now 10 pounds of sugar and 1½ quarts of water are cooked to 222°F, with 1½ pounds of gum arabic solution mixed in. Again the almonds are treated as they were with the previous batch of syrup.

Again 10 pounds of the sugar and 1½ quarts of water are cooked to 224°F, mixed with 1½ pounds of the gum arabic solution, and again the almonds are treated as before with syrup. The almonds are then removed and placed on clean trays.

Finally 35 pounds of sugar and 6 quarts of water are placed in a copper pan on the stove and stirred until dissolved. When the batch begins to boil, the sides are washed down, the batch is cooked to 225°F, and 5 pounds of the gum arabic solution are stirred in. This syrup is divided into four parts. Part one is colored a deep yellow and flavored lemon; part two is colored a bright green and flavored pistachio; part three is colored a deep red and flavored cherry, and part four is colored a bright purple and flavored violet. The almonds are then divided into four parts and each part is placed in the revolving pan and treated with colored syrup. The almonds are allowed to roll until they are wet. Then some powdered sugar is applied to them, a little at a time, until they are smooth and dry. The almonds are now removed to clean boards lined with wax paper and spread out, with not too many on a board. The boards are covered with wax paper and allowed to dry overnight. The same process is followed with the other three colors, and the almonds are now ready for polishing the following day.

Before polishing the almonds, the revolving pan is spread with a mixture of melted beeswax and carnauba wax (enough to form a surface of ½ inch thickness) while the pan is in motion. About 25 pounds of the sugar-coated almonds (free of all sugar dust) are placed in the pan. The pan is then started and 2 solid cakes of the wax are placed in the pan and allowed to run until the almonds are glossy and smoothly polished. It should take 1 to 1½ hours for a good polishing job. The almonds are then carefully removed from the pan with a round sieve, placed on clean, paper-lined trays, and kept covered for protection against moisture and dust. This process is continued until all the almonds are polished. They are then packed assorted as desired.

Note: No yield can be estimated on Jordan almonds. The amount of powdered sugar to be used varies also. In general, when the syrup is applied to the almonds, enough sugar must be added to absorb all the added syrup. The amount will depend on how large almonds are desired.

K

KISSES: *See Individual Kisses Formulas,* e.g., CALIFORNIA FRUIT KISSES and CHOCOLATE COCONUT KISSES.

L

LADY CARAMELS, CHOCOLATE

The formula for LADY CARAMELS, VANILLA (*see*) is followed, with the addition of 3 pounds of cut-up bitter chocolate and 3 pounds of pecan pieces, which are stirred into the batch before pouring out on the slab. The batch is then finished as usual and should yield 36 pounds.

LADY CARAMELS, VANILLA

Sugar	12 lb
Corn Syrup	8 lb
Coconut Butter	1 lb
Salt	1 tablespoonful
Fresh 18% Light Cream	4 qt
Gelatin (soaked in a little water)	2 oz
Black Walnuts	3 lb
Vanilla Flavor	3 teaspoonfuls
Evaporated Milk	6 lb

The sugar, corn syrup, coconut butter, salt, and fresh cream are placed in a large copper pan on the stove and stirred until dissolved. When the batch begins to boil, the evaporated milk is stirred in gradually. The stirring is continued and the batch is cooked to a firm ball when tested in cold water. The batch is then removed to a stand and the soaked gelatin is stirred in and whipped with a spatula for a few minutes. The black walnuts and the flavor are then stirred in and the batch is poured out on a greased slab, spread out evenly between the iron bars, and allowed to cool until hard enough. It is then passed through the sizing machine, cut in square caramels, and wrapped in moisture-proof cellophane.

It should yield 30 pounds.

LEMON CORDIALS

The VANILLA CORDIALS (*see*) formula is followed up to the flavoring. The batch is flavored with 1½ teaspoonfuls of oil of lemon and 1 teaspoonful of citric acid solution. The cordials are then finished in the same way as the vanilla.

The batch will yield about 20 pounds of cordials.

LEMON CREAMS

The VANILLA CREAMS (*see*) formula is followed, but the cream is flavored with lemon emulsion and 4 teaspoonfuls of citric acid solution are added. The creams are finished in the usual way.

Thirty pounds of cream centers will yield about 40 pounds of chocolates when chocolate coated by enrobing machine, or 45 pounds when hand dipped.

LEMON CRIMPED RIBBON CANDY: *See* CRIMPED RIBBON CANDY, LEMON.

LEMON FILLED BUTTER CUPS: *See* FILLED BUTTER CUPS, LEMON.

LEMON FRUIT AND NUT PILLOWS: *See* FRUIT AND NUT PILLOWS, LEMON.

LEMON FRUIT CUTS: *See* FRUIT CUTS, LEMON.

LEMON FRUIT JELLY

Sugar	25 lb
Corn Syrup	20 lb
Agar-Agar	8 oz
Water	12 qt
Yellow Color	2 teaspoonfuls
Oil of Lemon	3 teaspoonfuls
Citric Acid (dissolved in a little water)	2 oz

The agar-agar and water are placed in a large copper pan on the stove and stirred constantly. When the water begins to boil, the sugar is stirred in gradually. The batch is stirred constantly and the corn syrup is added. The batch is cooked to 223°F and then poured into another pan. It is moved to a cool spot, the color and flavor are stirred in, and the jelly is cooled to a moderate lukewarm (about 90°F). Then the citric acid is added and mixed through well, and the jelly is cast in fancy impressions molded in starch. After setting overnight, the jellies are taken out, sifted free of starch, dampened with a little syrup, and rolled in fine granulated sugar. When dry, the jellies are placed in crystal pans and crystallized in the usual manner (*see* CRYSTALLIZING).

This batch should yield approximately 55 pounds of candy.

LEMON FRUIT MELLOWS: *See* FRUIT MELLOWS, LEMON.

LEMON GUMDROPS: *See* GUMDROPS, LEMON.

LEMON ICE CREAM

Large Fresh Lemons	½ dozen
Fresh 20% Cream	4 qt
Fresh 40% Cream	2 qt
Condensed Milk	2 qt
Sugar	8 lb
Lemon Emulsion	1 oz
Yellow Color	1 teaspoonful

The lemons are finely grated. The grating is mixed with the lemon emulsion in a mixing bowl and set aside. The creams, condensed milk, and sugar are placed in the mixing can and stirred with a long paddle until the sugar is dissolved. The mix is then strained into the freezer. When the cream freezes about half way, the emulsified lemon flavoring is added. Otherwise the cream is finished as VANILLA ICE CREAM NO. 1 (*see*).

The batch will yield approximately 20 quarts.

LEMON ICED COCONUT DAINTIES: *See* ICED COCONUT DAINTIES, LEMON.

LEMON JELLED MINTS: *See* JELLED MINTS, LEMON.

LEMON JELLY FRUIT CUTS: *See* JELLY FRUIT CUTS, LEMON.

LEMON JELLY FRUIT SLICES: *See* JELLY FRUIT SLICES, LEMON.

LEMON LOLLYPOPS: *See* LOLLYPOPS, LEMON.

LEMON SALT-WATER TAFFY: *See* SALT WATER TAFFY, LEMON.

LEMON SHERBET

Fresh Milk	20 qt
Sugar	20 lb
100 Grade Citrus Pectin	4 oz
Lemon Juice	2½ qt
Yellow Color	½ teaspoonful
Pure Lemon Emulsion	1 oz

The pectin is mixed dry with 5 pounds of the sugar and placed in a small pan on the stove with 4 quarts of milk. The heat is adjusted to low, and the batch is stirred and heated to the boiling point. It is then poured into the mixing can. The balance of milk and sugar, the juice, and the color are added. The mixture is stirred until all the sugar is dissolved and the flavor is then stirred in. The mix is cooled off rapidly and strained into the freezer. Freezing, storing, and ripening of the sherbet are completed as usual (*see* VANILLA ICE CREAM NO. 1).

The batch will yield approximately 30 quarts.

LEMON SOUR BALLS: *See* SOUR BALLS, LEMON.

LEMON STICK CANDY: *See* STICK CANDY, LEMON.

LEMON STICKS, CHOCOLATE-COVERED: *See* CHOCOLATE-COVERED LEMON STICKS, CREAM FILLED.

LEMON SYRUP NO. 1

Ripe, Juicy Lemons	1 dozen
Sugar	8½ lb
Water	2½ qt
Yellow Liquid Color	¼ teaspoonful
FRUIT ACID SOLUTION (*see*)	4 teaspoonfuls

The peel of the lemons is grated into a saucepan, and the juice is extracted and added. To the lemon extract 1½ pounds of sugar and 1 cup of water are added. This mixture is covered and allowed to soak overnight. The following morning, the extract is mixed with the balance of the sugar and the water. It is placed in a small copper pan on the stove and heated to the boiling point. The mix is then poured into a large percolator and allowed to percolate for 2½ hours. The color and the acid solution are then added and mixed thoroughly. The syrup is cooled and is ready for use.

The batch will yield aproximately 5 quarts of lemon syrup.

LEMON SYRUP NO. 2

SIMPLE SYRUP (*see*)	4 qt
Essence of Lemon Peel	1 oz
Pure Oil of Lemon	½ teaspoonful
FRUIT ACID SOLUTION (*see*)	4 teaspoonfuls
Liquid Yellow Color	¼ teaspoonful

All of the ingredients are placed in a clean stone jug. They are stirred and mixed thoroughly and kept covered to be used as desired.

This batch will yield approximately 4 quarts of lemon syrup.

LEMON TABLETS

Crystal "A" Sugar	18 lb
Corn Syrup	4 lb
Water	3 qt
Powdered Tartaric Acid	2 oz
Oil of Lemon	3 teaspoonfuls

The sugar, water, and corn syrup are placed in a clean copper pan on the stove and stirred until dissolved. When the batch begins to boil, it is covered and allowed to steam for a few minutes. The sides are washed down and a thermometer is placed in the batch, which is cooked to 330°F. The batch is poured on a greased slab, and the edges are folded together while stirring in the oil of lemon and tartaric acid. This folding is continued until the batch is cool enough, when it is taken over to the spinning table in front of the batch warmer. It is then passed through a tablet-roller machine and allowed to cool on the table. The tablets are broken apart and they are then put through a sieve and packed into tins.

Note: In order to keep the slab cool, the water underneath may be turned on for a few minutes. The water is then shut off so that the slab does not get too cold.

The batch will yield approximately 20½ pounds of candy.

LEMONADES

The formula for MINTY NIBBLES (*see*) can be followed with the exceptions: The center is flavored lemon and 1 teaspoonful of citric acid is added. The jacket is colored and flavored yellow and lemon. A small red stripe is used. The batch is then finished as usual and should yield 28 pounds of candy.

LICORICE CARAMELS: *See* SAVING WASTE.

LICORICE FRUIT CUTS: *See* FRUIT CUTS, LICORICE.

LICORICE JELLIES: *See* SAVING WASTE.

LICORICE SALT-WATER TAFFY: *See* SAVING WASTE.

LICORICE STICK CANDY: *See* STICK CANDY, LICORICE.

LICORICE SQUARES

The same process is followed as for the CINNAMON SQUARES (*see*), but black color is used for coloring and oil of aniseed for flavor.

The batch will yield approximately 22½ pounds.

LIME BARLEY SUGAR CUTS: *See* BARLEY SUGAR CUTS, LIME.

LIME CHOCOLATE STRAWS: *See* CHOCOLATE STRAWS, LIME.

LIME CRIMPED RIBBON CANDY: *See* CRIMPED RIBBON CANDY, LIME.

LIME FRUIT AND NUT PILLOWS: *See* FRUIT AND NUT PILLOWS, LIME.

LIME FRUIT CUTS: *See* FRUIT CUTS, LIME.

LIME FRUIT JELLIES

The LEMON FRUIT JELLIES (*see*) formula is followed, but the batch is colored a delicate green and flavored with oil of lime. It is finished as usual and should yield approximately 55 pounds of candy.

LIME FRUIT MELLOWS: *See* FRUIT MELLOWS, LIME.

LIME GUMDROPS: *See* GUMDROPS, LIME.

LIME ICE

Water	20 qt
Sugar	20 lb
Lemon Juice	2½ qt
Pure Lime Emulsion	2 oz
Green Color	1 teaspoonful

The ingredients are placed in the mixing can and stirred until all the sugar is dissolved and the mixture registers 20 degrees on the syrup gage. The mix is then strained into the freezer and frozen as in VANILLA ICE CREAM NO. 1 (*see*).

The batch will yield approximately 30 quarts.

LIME JELLED MINTS: *See* JELLED MINTS, LIME.

LIME JELLY FRUIT CUTS: *See* JELLY FRUIT CUTS, LIME.

LIME JELLY FRUIT SLICES: *See* JELLY FRUIT SLICES, LIME.

LIME LOLLIPOPS: *See* LOLLIPOPS, LIME.

LIME SALT-WATER TAFFY: *See* SALT-WATER TAFFY, LIME.

LIME SOUR BALLS: *See* SOUR BALLS, LIME.

LIME STICK CANDY: *See* STICK CANDY, LIME.

LIME SYRUP

SIMPLE SYRUP (*see*)	4 qt
Pure Oil of Limes	4 teaspoonfuls
FRUIT-ACID SOLUTION (*see*)	2 oz
Liquid Green Color	½ teaspoonful

The ingredients are placed in a stone crock, stirred, and mixed thoroughly. The syrup is then ready for use.

The batch will yield approximately 4 quarts of lime syrup.

LIME TABLETS

The LEMON TABLETS (*see*) formula is followed, but the batch is colored a delicate green color and flavored with oil of limes. It is then finished in the usual manner.

The batch will yield approximately 20½ pounds of candy.

LIME WALNUT PILLOWS: *See* WALNUT PILLOWS, LIME.

LIQUID CHERRIES

FRENCH CREAM (*see*)	15 lb
Small Drained Maraschino Cherries	2-qt jar
Cherry Flavor	2 teaspoonfuls

The cherries are placed in a fine screen sieve and allowed to drain for 1 hour. The French cream is placed in a double-boiler bonbon-dipping pot over low heat and stirred until melted. It is then flavored with the cherry extract and adjusted to a fluid dipping consistency. The cherries are then dipped one at a time and deposited on trays lined with wax paper and allowed to cool for about an hour. They are then coated in a high-quality milk chocolate and finished with a tight curl mark. After they have set, they are cupped and packed in boxes.

The batch will yield the exact weight of all the ingredients used in its making.

LIQUID CHERRIES, CAST

RUNNING CREAM (*see*)	20 lb
Small Drained Maraschino Cherries	2 qt
Cherry Flavor	2 teaspoonfuls
Invertase	1 teaspoonful

The cherries are placed on a fine-screen sieve and allowed to drain for about an hour. The running cream is remelted on a steam bath pan over low heat to the right temperature and consistency. The flavor and invertase are stirred in. About 1 dozen starch boards are levelled and molded with a cherry mold. The cream is then cast lightly into the molds, forming a small bottom in each mold. The cherries are then placed in the cream-bottomed molds and filled up with the melted cream. After setting, they are carefully removed from the starch, brushed clean, and dipped in dark vanilla sweet chocolate. After they have set and are cool, the cherries are cupped and packed in boxes.

The batch will yield approximately 40 pounds of liquid cherries.

LIQUID PINEAPPLES

Drained Pineapple Cubes	1 can
FRENCH CREAM (*see*)	15 lb
Pineapple Flavor	2 teaspoonfuls

The pineapple cubes are placed on a fine-screen sieve and allowed to drain for about 2 hours. The French cream is remelted in a double-boiler bonbon-dipping pot over low heat to the right temperature and consistency. The flavor is then stirred in and the pineapple is dipped as usual and deposited on dipping trays. The pineapple is allowed to set until cool; it is then coated in dark vanilla sweet chocolate and finished with a long narrow ring mark. The chocolates are allowed to cool on the trays and they are then cupped and packed in boxes.

The batch will yield the exact weight of all the ingredients used in its making.

LIQUID STRAWBERRIES

Best Processed Drained Strawberries	1 can
FRENCH CREAM (*see*)	15 lb
Strawberry Flavor	2 teaspoonfuls

The strawberries are drained for about 2½ hours. The cream is remelted as usual, and the flavor is stirred in. Then the melted cream is adjusted for dipping and the berries are dipped and covered entirely with the cream. After standing until cool, they are dipped in a high-quality dark vanilla chocolate. They are allowed to cool on the trays and are then packed in bonbon cups and boxed as usual.

The yield of this batch will total the weight of all the ingredients used in its making.

LOG CABIN PECAN ROLL

Batch No. 1

Sugar	6 lb
Corn Syrup	4 lb
Water	1 qt
Chopped Coconut Butter	8 oz
MARSHMALLOW WHIP (*see*)	3 lb
Chopped Cherries	1 lb
Chopped Pineapple	1 lb
Chopped Raw Almonds	1 lb
Chopped Pistachio Nuts	8 oz
Vanilla Flavor	3 teaspoonfuls

The sugar, corn syrup, and water are placed in a copper pan on the stove and stirred until dissolved. When the batch begins to boil, it is covered and allowed to steam for a few minutes. The sides are washed down and a thermometer is placed in the batch, which is cooked to 254°F. The batch is removed from the heat, the coconut butter is added and the batch is allowed to cool for about 10 minutes. The marshmallow whip is stirred in gradually. The flavor is added and the fruits and nuts are stirred in gradually. The batch is whipped with the paddle until it is thick, and is then turned out on a greased slab. It is formed into rolls about 8 inches long and 1½ inches in diameter. The rolls are now ready for caramel dipping.

Batch No. 2

Sugar	4 lb
Corn Syrup	3 lb
Salt	1 teaspoonful
Coconut Butter	4 oz
Fresh 18% Cream	1 qt
Evaporated Milk	4 lb
Vanilla Flavor	2 teaspoonfuls
Pecan Pieces	20 lb

The sugar, corn syrup, salt, coconut butter, and cream are placed in a copper pan on the stove and stirred until dissolved. The sides are washed down, stirring is continued, and the evaporated milk is added gradually. The batch is cooked until it forms a soft ball when tested in cold water (*see* COLD-WATER TEST). The batch is moved near the slab, and the vanilla is stirred in. The pecan pieces are spread out on the slab, which is lined with iron bars. The nougat rolls are dipped in the caramel. Two dipping forks are used. They are then rolled in the pecans and set aside to cool. Then they are sliced and are ready for counter sale. To keep the candies fresh it is best to stock them in the original rolls and slice small quantities.

This formula should yield approximately 50 pounds of candy.

LOG NUT ROLL

Batch No. 1

Sugar	6 lb
Corn Syrup	4 lb
Salt	1 teaspoonful
Coconut Butter	4 oz
Fresh 18% Cream	2 qt
Fresh Butter	1 lb
Vanilla Flavor	2 teaspoonfuls
MARSHMALLOW WHIP (*see*)	2 lb

The sugar, corn syrup, salt, coconut butter, and cream are placed in a copper pan on the stove and stirred until dissolved,

The sides are washed down, the stirring is continued, and the butter is added. The batch is cooked until it forms a soft ball when tested in cold water (*see* COLD-WATER TEST). The batch is then moved near the slab and cooled for about 10 minutes. The flavor and marshmallow whip are stirred in and whipped with the paddle until thick. The batch is then turned out on the greased slab and formed into long rolls in preparation for caramel dipping.

Batch No. 2

Sugar	4 lb
Corn Syrup	6 lb
Salt	1 teaspoonful
Coconut Butter	8 oz
Evaporated Milk	10 lb
Vanilla Flavor	2 teaspoonfuls
English Walnut Pieces	25 lb

The sugar, corn syrup, salt, coconut butter, and half of the milk are placed in a copper pan on the stove and stirred until dissolved. The sides are washed down and the balance of the milk is stirred in gradually. The batch is cooked until it forms a soft ball when tested in cold water (*see* COLD-WATER TEST). The batch is moved near the slab and the flavor is stirred in. The rolls are dipped in the caramel and rolled in the walnuts spread out on the slab. The rolls are then set aside to cool and they are sliced and stocked as usual.

This formula should yield approximately 60 pounds of candy.

LOGANBERRY ICE

Water	20 qt
Sugar	20 lb
Lemon Juice	2½ qt
Pink Color	½ teaspoonful
Loganberry Pulp	5 qt

The water, sugar, lemon juice, and color are placed in the mixing can and stirred until all the sugar is dissolved and the mixture registers 20 degrees on the syrup gage. The mix is then strained into the freezer, and when the ice is about halfway frozen, the loganberry pulp is added. The ice is completed as VANILLA ICE CREAM NO. 1 (*see*).

The batch will yield approximately 32 quarts.

LOGANBERRY ICED COCONUT DAINTIES: *See* ICED COCONUT DAINTIES, LOGANBERRY.

LOLLIPOPS, CARAMEL

Sugar	20 lb
Corn Syrup	5 lb
Water	3 qt
Butter	8 oz
Salt	1 tablespoonful
Vanilla Flavor	3 teaspoonfuls
Burned Sugar	2 tablespoonfuls

The sugar, water, and corn syrup are placed in a copper pan on the stove and stirred until dissolved. When the batch begins to boil, it is covered and allowed to steam for a few minutes. The sides are washed down and a thermometer is placed in the batch, which is cooked to 330°F. The butter and salt are stirred in, and the batch is allowed to boil again for about a minute. The batch is then poured out on a greased slab and colored with burned sugar to medium dark. The edges are folded together while stirring in the vanilla flavor. Folding and kneading are continued until the batch is cool enough. Then it is transferred to the lollipop machine in front of the batch warmer. It is then shaped round and passed through the machine. It is allowed to cool on the table on the other end of the machine and wrapped ready for packing in boxes.

The batch will give about 24 pounds.

LOLLIPOPS, LEMON

Sugar	40 lb
Corn Syrup	10 lb
Water	6 qt
Yellow Color	2 teaspoonfuls
Powdered Citric Acid	2 oz
Flavor	5 teaspoonfuls
Lollipop Sticks	

The sugar, water, and corn syrup are placed in a large copper pan on the stove and stirred until dissolved. When the batch begins to boil, it is covered and allowed to steam for a few minutes. The sides are washed down and a thermometer is placed in the batch, which is cooked to 330°F. The batch is poured out on a greased slab and colored a bright yellow. The edges are folded together while stirring in the citric acid and the oil of lemon. Folding and kneading are continued until the batch is cool enough. Then it is transferred to the lollipop machine in front of the batch warmer. The batch is shaped round and passed through the lollipop machine. It is allowed to cool on the table on the other end of the machine. The pops are then wrapped in waxed paper or moistureproof cellophane by hand or wrapping machine. They are then packed in boxes.

This batch will give about 37½ pounds.

LOLLIPOPS, LIME

The formula for LOLLIPOPS, LEMON (*see*) is followed, but a bright-green color is used with oil of limes for flavor. The batch is finished as usual and will yield approximately 37½ pounds.

LOLLIPOPS, MARATHON TYPE

Sugar	24 lb
Corn Syrup	6 lb
Water	3½ qt
Anise Flavor	3 teaspoonfuls
Hard Candy Cherry Drops	1 lb
Licorice Sticks	2 lb

The sugar, water, and corn syrup are placed in a copper pan and stirred until dissolved. When the batch begins to boil, it is allowed to steam for a few minutes. The sides are washed down and a thermometer is placed in the batch, which is cooked to 300°F. The batch is poured on a greased slab, and the edges are folded together while the oil of anise is stirred in. Folding is continued until the batch is cool enough. It is then transferred to the hook and pulled to a good fluffy white; then it is taken over to the spinning table in front of the batch warmer and kept very soft. It is then spun and cut with the shears into small pieces about 1 ounce each. A licorice stick is placed on each piece, which is flattened out, and a cherry is then placed in the center of each sucker. After cooling on the table, the lollipops are wrapped and packed in boxes.

The batch will yield approximately 31½ pounds.

LOLLIPOPS, ORANGE

The formula for LOLLIPOPS, LEMON *(see)* is used, but the batch is colored and flavored orange. It is finished as usual.

The batch will yield approximately 37½ pounds.

LOLLIPOPS, RASPBERRY

The formula for LOLLIPOPS, LEMON is used, but the batch is colored a bright red and raspberry flavor is added. It is finished as usual and will yield approximately 37½ pounds.

LOLLIPOPS, TOASTED-COCONUT

Sugar	20 lb
Corn Syrup	5 lb
Water	3 qt
Fine Desiccated Coconut	4 lb
Vanilla Flavor	3 teaspoonfuls

The sugar, water, and corn syrup are placed in a large pan on the stove and stirred until dissolved. When the batch begins to boil, it is covered and allowed to steam for a few minutes. The sides are washed down and a thermometer is placed in the batch, which is cooked to 300°F. The heat is then reduced and the coconut is stirred in. The stirring is continued until the batch is toasted to a light brown color. It is then poured out on a greased slab and the edges are folded together, while stirring in the vanilla flavor. Folding and kneading are continued until the batch is cool enough, when it is transferred to the lollipop machine in front of the batch warmer. It is then shaped round and passed through the machine. The lollipops are allowed to cool on the table at the other end of the machine and are then wrapped ready to pack in boxes.

Note: The best grade of sticks is used for the continuous lollipop machine so that the machine runs smoothly without interruption.

The batch will yield approximately 27½ pounds.

LONDON CARAMELS, CHOCOLATE

The formula for LONDON CARAMELS, VANILLA (*see*) is followed, but the vanilla flavor and 4 pounds of cut-up bitter chocolate are stirred into the batch before it is poured. It is finished in the usual manner and should yield approximately 40 pounds of caramels.

LONDON CARAMELS, CHOCOLATE NUT

The formula for LONDON CARAMELS, VANILLA is followed, but 4 pounds of pecan pieces are stirred into the batch before it is poured. It is finished in the usual manner and should yield approximately 44 pounds of caramels.

LONDON CARAMELS, PISTACHIO

The formula for LONDON CARAMELS, VANILLA (*see*) is followed, but before pouring the batch, green color and pistachio flavor are stirred in. The batch is completed as usual and should yield approximately 40 pounds.

LONDON CARAMELS, RASPBERRY

The formula for LONDON CARAMELS, VANILLA is followed, but before pouring, red color and raspberry flavor are stirred in. The batch is finished in the usual manner and should yield approximately 40 pounds.

LONDON CARAMELS, VANILLA

Sugar	12 lb
Corn Syrup	13 lb
Coconut Butter	1 lb
Salt	1 tablespoonful
Light Fresh 18% Cream	6 qt
Evaporated Milk	10 lb
Fresh Dairy Butter	2 lb
Vanilla Flavor	4 teaspoonfuls

The sugar, corn syrup, coconut butter, salt, and cream are placed in a large copper pan on the stove and stirred until dissolved. When the batch begins to boil, the evaporated milk is stirred in gradually. Stirring is continued. When the batch becomes thick the butter is stirred in and cooking is continued until the batch forms a medium firm ball when tested in cold water (*see* COLD-WATER TEST). The vanilla flavor is then stirred in, and the batch is poured out on a greased slab lined with iron bars. It is spread out evenly and allowed to cool until hard. It is then sized down, cut into square caramels, and wrapped in moisture-proof cellophane.

This batch will yield approximately 40 pounds.

LONDON CARAMELS, VANILLA NUT

The formula for LONDON CARAMELS, VANILLA is followed, with the addition of 4 pounds of black walnuts, which are stirred into the batch before pouring. The batch is finished as usual and should yield approximately 44 pounds of caramels.

LORRAINE CHOCOLATES, BITTER ALMOND

The formula for LORRAINE CHOCOLATES, WALNUT (*see*) is used, but the center mixture is flavored with 1 teaspoonful of oil of bitter almond and chopped roasted almonds are kneaded into the mixture on the slab. The centers are cut into small oblong shapes and coated with milk chocolate. The tops are sprinkled with chopped pistachio nuts. Otherwise they are finished in the usual manner.

The batch will yield approximately 35 pounds of chocolates.

LORRAINE CHOCOLATES, COCONUT

The LORRAINE CHOCOLATES, WALNUT (*see*) formula is followed, but the mixture is flavored with Roman punch flavor. No nuts are used. The centers are rolled out, and cut into small balls, which are coated in milk chocolate and rolled in fine macaroon coconut. The balls are cooled and packed in the usual manner.

The batch will yield approximately 35 pounds of chocolates.

LORRAINE CHOCOLATES, FILBERT

The formula for LORRAINE CHOCOLATES, WALNUT (*see*) is followed, but 1½ pounds of chopped roasted filberts are kneaded into the mixture on the slab instead of walnuts. The centers are dipped in milk chocolate and rolled in chocolate sprinkles. Otherwise the chocolates are finished in the usual manner.

The batch will yield about 35 pounds of chocolates.

LORRAINE CHOCOLATES, WALNUT

Chopped Walnuts	1½ lb
Melted Dark Sweet Chocolate	12 lb
Melted FRENCH CREAM (*see*)	3 lb
Liquid Pure Honey	2½ lb
Fresh 20% Light Cream	2 qt
Vanillin	¼ teaspoonful

The melted French cream is stirred gradually into the melted chocolate. The fresh cream and the honey are then added, and the mixture is stirred continuously until it is whipped. It is then turned out on a clean marble and allowed to cool while the vanillin is mixed through. The chopped walnuts are kneaded in and the batch is rolled out and cut into small balls, which are dipped in high quality milk chocolate and rolled in fine chopped walnuts. After cooling they are packed in boxes.

The batch will yield approximately 35 pounds of chocolates.

M

MALTED MILK TAFFY

Sugar	12 lb
Corn Syrup	8 lb
Salt	1 teaspoonful
Water	1½ qt
Coconut Butter	4 oz
Butter	1 lb
Malted-Milk Extract	1 tablespoonful

The sugar, corn syrup, water, coconut butter, and salt are placed in a large copper pan on the stove and stirred until dissolved. When the batch begins to boil, it is covered and allowed to steam for a few minutes. The sides are washed down, and a thermometer is placed in the batch, which is cooked to 280°F. The butter is then stirred into the batch, mixed through thoroughly, and the batch is then poured out on a greased slab. The edges are folded together, while stirring in the malted milk extract. The batch is cooled rapidly and, when cool enough, it is taken over to the pulling machine and pulled for about 6 minutes. It is then transferred to the spinning table in front of the batch warmer, where it is formed into a round batch. It is then cut and wrapped.

This batch will yield approximately 20 pounds of candy.

MAPLE COCONUT

Sugar	8 lb
Water	1 qt
Corn Syrup	12 lb
BONBON CREAM (*see*)	5 lb
Maple Flavor	2 tablespoonfuls
Macaroon Coconut	10 lb

The sugar, corn syrup, and water are placed in a copper pan on the stove and stirred until dissolved. When the batch begins to boil, it is covered and allowed to steam for a few minutes. The sides are washed down and a thermometer is placed in the batch, which is cooked to 244°F. The bonbon cream is stirred in and, when it is melted, the coconut is stirred in gradually. The flavor is added and mixed thoroughly. The batch is then turned out on a greased slab, stretched out evenly, and allowed to cool. It is then cut into large squares and packed in trays.

This formula should yield approximately 40 pounds of candy.

MAPLE COCONUT BONBONS: *See* BONBONS, MAPLE COCONUT.

MAPLE COCONUT SQUARES, CHOCOLATE-COVERED: *See* CHOCOLATE-COVERED MAPLE COCONUT SQUARES.

MAPLE-CREAM COCONUT SQUARES: *See* CREAM COCONUT SQUARES, MAPLE.

MAPLE CREAMS, CAST

The VANILLA CREAMS, CAST (*see*) formula is followed, but the cream is flavored with 1 teaspoonful of vanilla and 4 teaspoonfuls of maple flavor. The creams are finished as usual. Thirty pounds of cream centers will yield about 40 pounds of chocolates when chocolate coated by enrobing machine, or 45 pounds when hand dipped.

MAPLE ICE CREAM

Fresh 20% Cream	16 qt
Condensed Milk	4 qt
100 Grade Citrus Pectin	4 oz
Granulated Sugar	8 lb
Maple Sugar	4 lb
Caramel Color	2 oz
Vanilla Flavor	1 oz
Maple Flavor	1½ oz

The pectin is mixed dry with 2 pounds of the white sugar and placed in a double boiler on the stove. Two quarts of the cream are added, and the mixture is stirred until it begins to boil. Then it is poured into a large ice-cream can. The balance of the cream and the white sugar, the condensed milk, and the maple sugar are added into this can and stirred with a large paddle until the sugar is dissolved. The caramel color and the flavors are stirred in. The mix is strained, cooled as usual, and poured into the freezer. The same freezing process is followed as for VANILLA ICE CREAM NO. 1 (*see*).

The batch will yield approximately 40 quarts.

MAPLE MARSHMALLOW SQUARES: *See* MARSHMALLOW SQUARES, MAPLE.

MAPLE PECAN BONBONS: *See* BONBONS, MAPLE PECAN.

MAPLE PECAN CHOCOLATES, OLD-FASHIONED

Crystal "A" Sugar	30 lb
Maple Sugar	10 lb
Water	8 qt
Cream of Tartar	3 teaspoonfuls
MARSHMALLOW WHIP (*see*)	3 lb
Vanilla Flavor	1 teaspoonful
Maple Flavor	3 teaspoonfuls
Invertase	2 teaspoonfuls
Pecan Pieces	5 lb

The sugars, water, and cream of tartar are placed in a copper pan on the stove and stirred until dissolved. When the batch begins to boil, it is covered and allowed to steam for a few minutes. The sides are washed down and a thermometer is placed in the batch, which is cooked to 243°F. The batch is poured out on a cooled, dampened cream beater and cooled rapidly to 100°F. The marshmallow whip, flavors, and invertase are added, the beater is set in motion and left in operation until the batch creams. The batch is then allowed to sweat back for 20 minutes and it is then remelted as usual (*see* CREAM CENTERS, CAST). The starch boards are imprinted with round flat-top molds and the pecan pieces are deposited in these impressions. The cream is then cast in the impressions on top of the pecans and allowed to set overnight. The creams are then removed, brushed free of starch, and coated with a dark vanilla sweet chocolate. After they set on the dipping trays, the maple pecans are packed in boxes.

This formula should yield about 55 pounds of centers and 75 pounds of machine coated centers.

MAPLE PECAN CREAM CENTERS, HAND-ROLLED

The formula for VANILLA-NUT CREAMS, HAND-ROLLED is followed until the batch is cooled on the cream beater. The marshmallow whip and invertase are then added and the batch is flavored with 2 teaspoonfuls of vanilla and 4 teaspoonfuls of maple flavor. The beater is set in motion and, while it is running, 5 pounds of pecan pieces are added. The creaming of the batch is then finished and the creams are completed in the usual way.

Note: The average ratio for using chocolate coating for hand-rolled centers is approximately 20 percent for enrober coating and 30 percent for hand coating.

MAPLE PECAN PATTIES

Crystal "A" Sugar	5 lb
Maple Sugar	3 lb
Water	1 qt
Cream of Tartar	1 teaspoonful
FRENCH CREAM (*see*)	6 lb
Pecan Pieces	4 lb
Maple Flavor	2 teaspoonfuls
Glycerin	2 teaspoonfuls

The sugars, water, and cream of tartar are placed in a clean copper pan on the stove and stirred until dissolved. When the batch begins to boil, it is covered and allowed to steam for a few minutes. The sides are washed down and a thermometer is placed in the batch, which is cooked to 240°F. The batch is then removed from the stove and the French cream, flavor, and glycerin are all added to the batch and mixed thoroughly. The pecans are then stirred in and the batch is spooned into fritter rings on a wax paper-lined slab with a large spoon. This process is continued until the entire batch has been spooned into patties. The patties are allowed to cool until firm enough to handle. They are then wrapped and packed as usual.

This formula should yield approximately 20 pounds of patties.

MAPLE PECAN PRALINES

Crystal "A" Sugar	7 lb
Maple Sugar	3 lb
Cream of Tartar	1 teaspoonful
Water	2 qt
Salt	2 teaspoonfuls
Maple Flavor	4 teaspoonfuls
Small Halves of Pecans	2½ lb

The sugars, cream of tartar, water, and salt are placed in a copper pan on the stove and stirred until dissolved. When the batch begins to boil, it is covered and allowed to steam for a few minutes. The sides are washed down and a thermometer is placed in the batch, which is cooked to 242°F. The batch is then moved near the table and allowed to cool for about 5 minutes. The flavor is added and the batch is stirred around the sides of the pan until it appears to be cloudy. The pecans are then added and mixed through. The pecan pralines are then spooned out with a soup ladle and placed on heavy sheets of waxed paper which have been spread out on the table. The pralines are formed into round cakes, which are allowed to cool until hard. They are then wrapped and packed carefully.

This formula should yield approximately 15 pounds of candy.

MAPLE WALNUT BOARDWALK FUDGE: *See* BOARDWALK FUDGE, MAPLE WALNUT.

MAPLE WALNUT BUTTER TOFFEE

Maple Sugar	4 lb
Crystal "A" Sugar	12 lb
Corn Syrup	4 lb
Salt	2 teaspoonfuls
Water	2 qt
Ground Walnuts	4 lb
Butter	2 lb
Maple Flavor	2 teaspoonfuls
Bicarbonate of Soda	2 teaspoonfuls

The sugars, corn syrup, salt, and water are placed in a copper pan on the stove and stirred until dissolved. When the batch begins to boil, it is covered and allowed to steam for a few minutes. The sides are washed down and a thermometer is placed in the batch, which is cooked to 300°F. The butter is then stirred in and the walnuts are added while the batch is stirred continuously. The cooking is continued until the batch forms a hard crisp ball when tested in cold water (*see* COLD-WATER TEST). The bicarbonate of soda is then stirred in rapidly and the batch is poured out on a warm, greased slab. It is spread out evenly to a thin sheet and cut with an adjustable cutter into 1½ inch squares. These are allowed to cool on the boards and are then coated with a high-quality milk chocolate. The tops are sprinkled with a generous amount of ground walnuts. After setting, the toffee is packed in boxes.

The batch will yield approximately 45 pounds of toffee.

MAPLE WALNUT CREAMS

Crystal "A" Sugar	25 lb
Maple Sugar	15 lb
Water	7 qt
Cream of Tartar	2 teaspoonfuls
Melted Corn Syrup	4 lb
Vanilla Flavor	1 teaspoonful
Maple Flavor	2 teaspoonfuls
Chopped Walnuts	5 lb

The sugars, water, and cream of tartar are placed in a copper pan on the stove and stirred until dissolved. When the batch begins to boil, it is covered and allowed to steam for a few minutes. The sides are then washed down and a thermometer is placed in the batch, which is cooked to 254°F. The melted corn syrup is stirred in rapidly, and the batch is poured out on a cooled, dampened cream beater and allowed to cool for about 8 minutes. The nuts and flavors are added, and the beater is set in motion and allowed to run until the batch creams to a lump. It is allowed to sweat back for 20 minutes and is then transferred to the rolling table, which is dusted with flour. The batch is then kneaded, rolled out, and cut into small rolls. After these rolls are cool and dry, they are placed in crystal pans, which are only half filled. They are then crystallized in cool syrup, using the overnight process (*see* CRYSTALLIZING). The candies are then drained, dried, and packed as usual.

This batch will yield approximately 60 pounds.

MAPLE WALNUT EGGS, HAND-ROLLED

Crystal "A" Sugar	30 lb
Maple Sugar	10 lb
Water	8 qt
Cream of Tartar	2 teaspoonfuls
Melted Corn Syrup	5 lb
Chopped English Walnuts	5 lb
MARSHMALLOW WHIP (*see*)	3 lb
Vanilla Flavor	1 teaspoonful
Maple Flavor	4 teaspoonfuls
Invertase	2 teaspoonfuls

The sugars, water, and cream of tartar are placed in a clean copper pan on the stove and stirred until dissolved. When the batch begins to boil, it is covered and allowed to steam for a few minutes. The sides are washed down and a thermometer is placed in the batch, which is cooked to 256°F. The melted corn syrup is stirred in rapidly, and the batch is poured out on a clean, cooled, dampened cream beater and allowed to cool for 7 minutes. The flavors, invertase, and marshmallow whip are added, and the beater is set in motion. The chopped walnuts are added, and creaming is continued until the batch forms a lump. The batch is then kneaded soft and is cut in the cut-roll machine to the desired size. The eggs are allowed to cool on boards and are then coated in a high-quality milk chocolate. They are finished with the mark "M" on top. After the finished eggs have cooled on trays, they are packed in boxes.

The batch will yield approximately 75 pounds of maple walnut eggs.

MAPLE WALNUT PARFAIT

Fresh 20% Cream	16 qt
Instant Powdered Egg Yolks	1 lb

Granulated Sugar	10 lb
Crushed Maple Sugar	5 lb
Maple Flavor	1½ oz
Chopped Walnut Pieces	3½ lb

The maple sugar and 4 quarts of the cream are placed in a copper pan on the stove and stirred and heated to the boiling point. This mixture is then poured into the mixing can. The granulated sugar is mixed with the yolks and then placed into the can. The balance of the cream is poured in and the mixture is stirred until all the sugar is dissolved. The mix is then cooled off and strained into the freezer as usual. When the cream freezes about halfway, the flavor and nuts are added. To complete the parfait the directions under VANILLA ICE CREAM NO. 1 (*see*) are followed.

The batch will yield approximately 35 quarts.

MAPLE WALNUT SEAFOAM FUDGE: *See* SEAFOAM FUDGE, MAPLE WALNUT.

MAPLE WALNUT SEAFOAM KISSES: *See* SEAFOAM KISSES, MAPLE WALNUT.

MAPLE WALNUT SUNDAE TOPPING

Crystal "A" Sugar	6 lb
Maple Sugar	3 lb
Corn Syrup	1 lb
Water	2½ qt
150 Grade Citrus Pectin	1 oz
Chopped English Walnuts	5 lb
Maple Flavor	4 teaspoonfuls
Glycerin	4 teaspoonfuls

The pectin is mixed dry with 2 pounds of the white sugar. One quart of water is placed in a small copper pan on the stove over low heat. The pectin sugar is added and stirred until dissolved. The balance of the water and white sugar and the maple sugar are added to the batch and again stirred until the sugar is completely dissolved. The corn syrup is stirred in rapidly, and the batch is cooked to 219°F. It is then removed to a cool spot and the chopped nuts are stirred in rapidly. The flavor and glycerin are added and mixed thoroughly. This sundae topping is then poured into clean jars with airtight lids and may be used as desired.

The batch will yield approximately 10 quarts of maple walnut sundae topping.

MARATHON LOLLIPOPS: *See* LOLLIPOPS, MARATHON TYPE.

MARSHMALLOW BANANAS

Batch 1

Gelatin (soaked in 10 qt of water)	3 lb
6x Powdered Sugar	30 lb
Corn Syrup	10 lb

Batch 2

Sugar	30 lb
Corn Syrup	10 lb
Water	4 qt
Banana Flavor	1½ oz
Yellow Color	2 teaspoonfuls

The soaked gelatin and the powdered sugar are placed in the beater and started at medium speed. They are beaten until stiff. Then the 10 pounds of corn syrup are added and the beating is continued.

In the meantime, batch 2 is cooked to 240°F. This syrup is then poured in a fine stream into the beater, which is now turned on to high speed. The color and flavor are added and the batch is beaten until it weighs about 3 pounds and 4 ounces to the gallon. It is then moved into the depositor, run out in banana impressions molded in dry starch, and allowed to set overnight. The bananas are then placed in a hot room of about 100°F for another day. They are then removed, sifted, and packed in boxes.

This formula should yield approximately 75 pounds of candy.

MARSHMALLOW BEATER

This beater has a stainless-steel water jacket. It whips the marshmallow continuously until it is ready for the depositor. (Courtesy of the Savage Brothers Co., Chicago, Illinois.)

MARSHMALLOW BELMONTS, CHOCOLATE-COVERED: *See* CHOCOLATE-COVERED MARSHMALLOW BELMONTS.

MARSHMALLOW CREAM

Sugar	30 lb
Corn Syrup	20 lb
Water	12 qt
Best Cooking Starch (dissolved in 8 qt of water)	6 lb
Instant Egg Albumen	1 lb
Water	2 qt
Pure Vanilla Extract	1½ oz

The sugar, corn syrup, and water are placed in a large steam-jacketed kettle and stirred until dissolved. When the batch begins to boil, the soaked cooking starch is stirred in gradually. The batch is cooked to a firm jelly, or until it drops off the paddle in long strings, or to 224°F. The batch is then poured out into the beater and started on a lower speed for about 5 minutes. The albumen and the 2 quarts of water are then added, and the beater is shifted to high speed. Beating is continued until the batch is white and light. The vanilla extract is then added and beaten for another 2 minutes. The marshmallow is then placed in dry, clean cans and covered with airtight lids. This marshmallow is used for the fountain and, before using, it is diluted with SIMPLE SYRUP (*see*) to the right consistency.

This batch should yield approximately 65 pounds of cream.

MARSHMALLOW FIGURES AND SHAPES

Sugar	40 lb
Slow-Set Pactin	4 oz
Corn Syrup	30 lb
Water	8 qt
Cream of Tartar	3 teaspoonfuls
Powdered Gelatin (dissolved in 8 qt of water)	2 lb
Powdered Jaguar Gum	1 lb
Pure Vanilla Flavor	2 oz

The pectin is mixed dry with 5 pounds of sugar, placed in a large copper pan on the stove with 2 quarts of the water and stirred until dissolved. The cream of tartar with the balance of the sugar and water are placed into the kettle and stirred until dissolved. The sides are washed down and the batch is cooked to 234°F and removed from the stove. The jaguar gum is stirred into the corn syrup, and this mixture is warmed up and added to the pectinized sugar batch. The soaked gelatin is placed into the beater, and the syrup and sugar batches are poured into the beater. The beater is started at slow speed for 5 minutes, and then shifted to high speed. Beating of the batch is continued until the marshmallow is white and light, weighing about 1 pound to the quart. The vanilla flavor is then stirred in and the batch is placed into a warm depositor. The marshmallow is cast in fancy figures and shapes, molded in clean, dry starch, and placed in a hot room at about 100°F overnight. The marshmallows are then removed, brushed off, and packed as usual.

This batch should yield approximately 75 pounds of marshmallow.

MARSHMALLOW PEANUTS

Sugar	40 lb
Corn Syrup	20 lb
Water	2 qt
Gelatin (soaked in 6 qt of water)	1½ lb
CREAM FONDANT (*see*)	10 lb
Scarlet Red Color	3 teaspoonfuls
Vanilla Flavor	1½ oz

The sugar, corn syrup, water, and gelatin are placed in the beater and run at medium speed for about 10 minutes. The fondant is added and the beater is then shifted to high speed. The color and flavor are added, and beating is continued until the marshmallow weighs about 3½ pounds to the gallon. The batch is then moved to the depositor and run in peanut impressions molded in dry starch. The candies are allowed to set overnight and then placed in a hot room of 100°F for another day. The peanuts are then removed, sifted, and packed.

This batch should yield approximately 75 pounds of candy.

MARSHMALLOW SPREAD

Sugar	35 lb
Corn Syrup	40 lb
Water	16 qt
Best Cooking Starch (dissolved in 8 qt of water)	8 lb
Instant Egg Albumen (dissolved in 2 qt of water)	1 lb
Pure Vanilla Extract	2 oz

The sugar, corn syrup, and water are placed in a large steam copper kettle and stirred until dissolved. When the batch begins to boil, the dissolved cooking starch is stirred in gradually. Stirring is continued and the batch is cooked to a firm jelly, or until it runs off the paddle in long strings, or to 224°F. The heat is then shut off and the batch is poured into the beater, which is started at low speed for about 10 minutes. The 2 quarts of water and the albumen are then added rapidly, and the beater is shifted to high speed. The marshmallow is beaten until it is white and light. The vanilla flavor is then added and stirred in when the beater is shut off. The marshmallow is poured into cans and covered with airtight lids. It is then ready for use as a spread.

This batch should yield approximately 90 pounds of spread.

MARSHMALLOW SQUARES, MAPLE

The formula for MARSHMALLOW SQUARES, VANILLA (*see*) is followed, but the batch is flavored with 2 ounces of maple flavor and otherwise completed the same way as the vanilla batch.

This batch should yield approximately 65 pounds of squares.

MARSHMALLOW SQUARES, MINT

The formula for MARSHMALLOW SQUARES, VANILLA (*see*) is followed, but the batch is colored green and flavored with 3/4 ounce triple-distilled oil of peppermint. It is then completed in the usual manner and should yield approximately 65 pounds of squares.

MARSHMALLOW SQUARES, STRAWBERRY

The formula for MARSHMALLOW SQUARES, VANILLA (*see*) is followed, but the batch is colored pink and flavored with strawberry. It is then completed in the usual manner and should yield approximately 65 pounds of squares.

MARSHMALLOW SQUARES, VANILLA

Gelatin (dissolved in 2 qt of water)	2 1/4 lb
Sugar	24 lb
Invert Sugar	6 lb
Water	3 qt
Cream of Tartar	2 teaspoonfuls
Corn Syrup	30 lb
Vanilla Flavor	1 1/2 oz

The dissolved gelatin is placed in the marshmallow beater. The sugar, water, invert sugar, and cream of tartar are placed in a large copper pan on the stove and stirred until dissolved. When the batch begins to boil, the sides are washed down and the batch is cooked to 236°F. The corn syrup is then stirred in and the batch is poured slowly into the beater on top of the gelatin. The beater is run at low speed for about 5 minutes, and then shifted to high speed. Beating continues until the batch is white and light; then the flavor is added and allowed to mix in. The batch is then poured out on a slab lined with heavy wrapping paper and spread out to about 3/4 inch thickness. After the batch is set, the top is dusted with starch mixed with powdered sugar. The batch is then turned upside down, and the paper is soaked lightly and then removed. The marshmallow is then cut with an adjustable cutter into small squares and placed on boards dusted with starch. The marshmallows are now ready for coating in chocolate or rolling in toasted coconut.

This batch should yield approximately 65 pounds of squares.

MARSHMALLOW WHIP

Water	3 qt
Sugar	50 lb
Corn Syrup	50 lb
Powdered Gelatin (soaked in 1/2 pint of water)	8 oz
Instant Egg Albumen	1 1/2 lb
Water	2 1/2 qt

The water, sugar, and corn syrup are placed in a large copper pan on the stove and stirred until dissolved. When the batch begins to boil, it is covered and allowed to steam for a few minutes. The sides are washed down and a thermometer is placed in the batch, which is cooked to 242°F. The soaked gelatin is placed in the beater and the batch is poured into the beater, which is run at slow speed for about 5 minutes. The 2 1/2 quarts of water and the albumen are added gradually. The beater is shifted to full speed, and the marshmallow is beaten until it is white and spongy. The marshmallow is then poured into cans and kept well covered. It is now ready for use in cream centers, fudge, nougats, etc.

This batch should yield about 110 pounds of marshmallow.

Note: Instant egg albumen is a finely powdered egg albumen. It is recommended for use because it eliminates soaking. However, ordinary egg albumen should also give good results if soaked the day before use. Two quarts of water are sufficient to soak 1 pound of albumen.

MARSHMALLOWS

There are two kinds of marshmallows: the egg albumen and the gelatin types. The egg-albumen marshmallow is used mostly for dipping in the manufacture of chocolates and nut goodies. Although some gelatin marshmallow is likewise dipped in chocolate, most of it is consumed as is.

Marshmallow contains a considerable amount of moisture which is held in suspension by gelatin or albumen. Therefore, the starch-cast marshmallow must set rapidly, i.e., overnight. If the setting is unduly prolonged, i.e., if the hot room is not hot enough, some of the moisture may be released and crusting will result. A crusty marshmallow is one that has picked up sticky, inedible lumps of starch. The starch used for cast marshmallows must be reasonably dry. It should not contain more than 7 percent moisture in dry weather, and should contain only 5 percent in sticky, humid weather. An expert candymaker can gage the percentage of moisture in starch by picking up a handful and feeling it.

When the marshmallow is cast, the starch should be cold, i.e., at natural room temperature. Coolness hastens the setting. Before the batches are poured out into the depositor, their temperature should be checked and regulated so that the best results will be obtained. The depositor temperatures for running mashmallow vary from 120° to 140°F. The heavier the body of the marshmallow, the hotter the depositor should be.

Since the marshmallow department is usually housed jointly with the jelly and gumdrop department, as a rule, these products share a hot room. The depositing machine and starch buck which are used in making jellies and gumdrops are also used in making marshmallows. The most important machine in the marshmallow department is that solo performer, the marshmallow beater. This machine should be up to date in every respect. It must do an effective beating job in a limited time. Some machines will beat a 100-pound batch in about 10 minutes; this is good time, but the process should never take longer than 15 to 20 minutes. Marshmallow beaters should be checked at least once a year and, if any defects show up, they should be repaired or replaced with new equipment. The whipping paddles are the most important part of the beater. A well-whipped marshmallow will have a good body and will eat well, whereas a marshmallow that is whipped slowly and with insufficient thoroughness will turn out flat and tough to the teeth.

In this encyclopedia there are some formulas for so-called "slab gelatin" marshmallows. These marshmallows require no hot room and no starch casting. They are cut, usually into squares, with a dampened adjustable cutter. They are easy to make, but the formulas should be followed with considerable care. It is possible, but not desirable, to use a weaker gelatin when making starch-cast marshmallows, but when making "slab gelatin" marshmallows the "200 bloom test" gelatin is necessary. This gelatin produces an excellent marshmallow suitable for both coating in chocolate and toasting in coconut. While it is impossible to make an egg albumen slab marshmallow because it will not set well, it is possible to combine egg albumen with gelatin and make a good slab marshmallow. However, only an expert candymaker can handle this tedious process. That is why a formula for this combination slab marshmallow is not included. The ingenious candymaker can probably develop his own formula.

For the various marshmallow formulas, see the *Individual Names,* such as, MARSHMALLOW WHIP, SNOWFLAKE MARSHMALLOWS, etc.

MARSHMALLOWS, CHOCOLATE-COVERED: *See* CHOCOLATE-COVERED MARSHMALLOWS.

MARZIPAN ALMOND GOODIES

FONDANT MARZIPAN (*see*)	15 lb
Green Color	½ teaspoonful
Oil of Bitter Almonds	1½ teaspoonfuls
Chopped Roasted Almonds	3 lb

The marzipan is colored a delicate green color. The flavor is mixed in and the marzipan is kneaded to a smooth paste. It is then rolled out and cut into small oblongs, which are coated halfway (bottoms only) in milk chocolate. The bottoms are then dipped in the chopped almonds, and the goodies are deposited on dipping trays. They are allowed to set and are then packed in boxes.

The batch will yield approximately 20 pounds of marzipan almond goodies.

MARZIPAN BARS

FONDANT MARZIPAN (*see*)	20 lb
ALMOND PASTE (*see*)	10 lb
Colors	As desired
Flavors	To suit

Ten pounds of the fondant marzipan are colored green and flavored with 1 teaspoonful of pistachio flavor. The other 10 pounds of the fondant marzipan are colored red and flavored with 1 teaspoonful of the raspberry flavor. The almond paste is kneaded until it is smooth, and rolled out in a large sheet on the marble. The green marzipan is kneaded smooth, rolled out in a flat sheet, and placed on top of the almond paste. The red marzipan is then kneaded, rolled out, and placed on top of the green. The three-color batch is then folded and rolled jelly roll fashion. Then it is cut into bars about 1½ inches in diameter and 4 inches long. The bars are coated with fine milk chocolate and allowed to set. They are then wrapped and packed in the usual manner.

The batch will yield approximately 40 pounds of marzipan bars.

MARZIPAN CANDIES

Marzipan, a confection with an almond base, is usually sold in high-class retail shops. Translated literally, the word means almond confection. Marzipan has been made in Europe for centuries. These candies are usually shaped like tiny fruits, flowers, and vegetables and colored accordingly. After some practice they can be shaped with the fingers or simple kitchen tools. The confectioner who has some artistic background can use it to produce marzipan fruits that resemble natural fruits.

In order to match the natural fruit and vegetable colors, marzipan candies are coated with melted fondants, melted chocolate, glacéed syrup, colored sugar, and tinted coconut after they are shaped. The melted fondant and glacéed syrup are the means by which various colors are applied to the marzipan. The colors are mixed in the liquids and then applied on the marzipan. Chocolate is used to give a black or brown color. Tinted coconut and colored sugar are used where a blend of two or three colors is wanted on one piece.

For faster and more artificial work, marzipan molds may be obtained at any supply house. Before they are used the molds are usually cleaned, greased slightly with pure olive oil, and wiped with a soft cloth.

Almonds are the most important ingredient in marzipan. The confectioner should select carefully the brand he uses. Although California and Italian almonds may be used with good results, Spanish Valencia almonds are preferred for top-quality work.

The almond paste and marzipan fondant formulas call for bitter almond oil as an ingredient. Although the amount to be used is listed in the ingredients, the confectioner should test this flavor on his own to determine more exactly the proper amount. Bitter almond oil is made in various strengths by almost every flavor house. Even if the flavor house specifies the amount of flavor to be used for every 100 pounds of candy, it is safer for the confectioner to make his own test and eliminate guesswork. If not enough flavor is used, the natural almond quality will be flat. If excessive flavor is used, the taste will be more bitter than that of bitter almonds.

An effective method for testing the strength of bitter almond oil: The confectioner takes 1 pound of neutral cream fondant and mixes in the bitter almond oil, measuring by grams. The confectioner keeps tasting the fondant until he gets the taste he wants. When that happens, he multiplies the number of grams for larger batches.

It is advisable to make marzipan in small quantities and carry little stock since the shelf life of these confections cannot be assured. An exception is almond paste. Some manufacturers produce it as a specialty and pack it in airtight tins. In this way the moisture is retained for a long time and the almond paste remains edible.

For confectioners who command a good price for their products, there are two ways of packing marzipan so that its shelf life will be prolonged. Each piece may be wrapped in aluminum foil. The marzipan wrapped in this fashion may be packed in regular boxes. Or the marzipan may be packed unwrapped in fancy airtight tins that are wrapped in moistureproof cellophane.

For the various marzipan formulas, see the *Individual Names,* such as, ALMOND PASTE, MARZIPAN DATES, MARZIPAN BARS, etc.

MARZIPAN CHERRY GOODIES

FONDANT MARZIPAN (*see*)	15 lb
Red Color	½ teaspoonful
Cherry Flavor	1 teaspoonful
Chocolate Rice	5 lb

The marzipan is colored red and flavored with cherry. It is kneaded until it is smooth, and the cherry marzipan is cut and rolled into small balls. The balls are coated in dark sweet chocolate halfway (bottoms only). The bottoms are dipped in chocolate rice and the goodies are deposited on dipping trays. They are allowed to set and are then packed in boxes.

The batch will yield approximately 20 pounds of marzipan cherry goodies.

MARZIPAN DATES

FONDANT MARZIPAN (*see*)	10 lb
Pitted Soft Juicy Dates	10 lb
Orange Color	½ teaspoonful
Oil of Orange	1 teaspoonful

The marzipan fondant is kneaded until it is smooth. The color and flavor are added and mixed thoroughly. The batch is then rolled out and cut into small oblongs. The dates are stuffed with the marzipan and placed on clean trays lined with wax paper. The stuffed dates are allowed to dry and are then placed in crystal pans and crystallized (*see* CRYSTALLIZING) in an overnight crystal. The usual crystallizing procedure is followed.

The batch will yield approximately 25 pounds of marzipan dates.

MARZIPAN PECAN GOODIES

FONDANT MARZIPAN (*see*)	15 lb
Pecan Halves	4 lb
Green Color	1 teaspoonful
Pistachio Flavor	1½ teaspoonfuls

The marzipan is colored green and flavored pistachio. It is then kneaded until smooth, rolled out, and cut into small pieces. Each piece is put between two pecan halves and the pecans are placed on clean trays. This process is continued until all of the green marzipan is used up. The goodies are then placed in crystal pans and crystallized (*see* CRYSTALLIZING) in an overnight crystal.

The batch will yield approximately 23 pounds of marzipan pecan goodies.

MARZIPAN PINEAPPLE

Glacéed Pineapple Slices	6 lb
Fondant Marzipan	12 lb
Yellow Color	1 teaspoonful
Pineapple Flavor	1 teaspoonful

The marzipan is colored yellow and flavored pineapple. It is kneaded until it is smooth and rolled out in a sheet about ¼ inch thick. The pineapple slices are covered with marzipan on both sides; each slice is cut into eight pieces and placed on clean trays lined with wax paper. After the marzipan pineapple is dry, it is transferred into crystal pans for an overnight crysal (*see* CRYSTALLIZING). The following morning, the marzipan pineapple is drained, allowed to dry, and packed in boxes.

The batch will yield approximately 22 pounds of marzipan pineapple.

MARZIPAN PISTACHIO GOODIES

FONDANT MARZIPAN (*see*)	15 lb
Green Color	½ teaspoonful
Pistachio Flavor	1½ teaspoonfuls
Chopped Pistachio Nuts	2 lb

The marzipan is colored green and flavored pistachio. It is kneaded until it is smooth, rolled out, and cut into small balls. The balls are then coated halfway (bottoms only) with dark vanilla chocolate. The bottoms are dipped in the chopped pistachio nuts, and the goodies are deposited on dipping trays. They are allowed to set as usual and then packed in boxes.

The batch will yield approximately 20 pounds of marzipan pistachio goodies.

MARZIPAN PRUNES

FONDANT MARZIPAN (*see*)	15 lb
Pitted Oregon Prunes	10 lb
Vanilla Flavor	1 teaspoonful
Pink Color	½ teaspoonful

The marzipan fondant is kneaded on the marble. The color and flavor are added and the marzipan is kneaded again until it is smooth. The color is then adjusted to a delicate pink. The mass is then rolled out and cut into oblongs. The prunes are stuffed with these oblongs. They are then spread out and allowed to dry on trays. The prunes are then placed in crystal pans and crystallized in an overnight crystal (*see* CRYSTALLIZING).

The batch will yield approximately 30 pounds of marzipan prunes.

MARZIPAN SAUSAGE

FONDANT MARZIPAN (*see*)	20 lb
ALMOND PASTE (*see*)	5 lb
Colors	As needed
Flavors	To suit

The fondant marzipan is divided into two parts. Part one is colored yellow and flavored with 1 teaspoonful of Jamaica rum flavor. Part two is colored red and flavored with 1 teaspoonful of cherry flavor. The red marzipan is kneaded and rolled out flat. The yellow is kneaded, rolled out, and placed on top of the red. The almond paste is colored orange; then it is kneaded, rolled out, and placed on top of the yellow. The three-color batch (orange on yellow on red) is then rolled like a jelly roll. Then it is rolled to increase its length and doubled up. Rolling out and doubling up are continued until the batch has been doubled up six times. It is finally rolled out to 2 inches in diameter and cut into 10-inch-long bars. These are coated in milk chocolate and allowed to set. Then they are cut into thin slices and wrapped in tin foil. The slices are packed in boxes.

The batch will yield approximately 32 pounds of marzipan sausage.

MARZIPAN SNOW CAPS

FONDANT MARZIPAN (*see*)	15 lb
Vanilla Flavor	1 teaspoonful
Macaroon Coconut	10 lb

The marzipan is flavored vanilla and kneaded until it is smooth. It is then cut and rolled into small balls. The balls are coated with dark sweet chocolate, rolled in the macaroon coconut, and spread out on wax-paper-lined trays. They are allowed to set and are then packed in boxes.

This batch will yield approximately 25 pounds.

MARZIPAN STRAWBERRIES

FONDANT MARZIPAN (*see*)	15 lb
Pink Color	1 teaspoonful
Pure Strawberry Extract	1½ teaspoonfuls

The marzipan is colored pink and flavored strawberry. It is kneaded until smooth; then it is rolled out and cut into medium-size pieces, which are shaped into strawberries by hand. The strawberries are allowed to dry; then they are placed in crystal pans and crystallized in an overnight crystal (*see* CRYSTALLIZING). The following morning, they are drained and allowed to dry. Green paper hulls are stuck in each berry. They are then cupped and packed in one-layer boxes.

The batch will yield approximately 18 pounds of marzipan strawberries.

MARZIPAN SWEET POTATOES

FONDANT MARZIPAN (*see*)	12 lb
Dutch Cocoa Powder	1 lb
Vanilla Flavor	1½ teaspoonfuls

The marzipan is flavored vanilla, kneaded until smooth, and then rolled out and cut into small pieces. The cocoa powder is spread out on a clean board, and the marzipan pieces are rolled with damp hands and covered with the cocoa powder. The potatoes are then placed on a clean board lined with wax paper. When thoroughly dry, they are packed in boxes.

The batch will yield approximately 13 pounds of marzipan sweet potatoes.

MARZIPAN WALNUT GOODIES

FONDANT MARZIPAN (*see*)	15 lb
Walnut Halves	5 lb
Pink Color	1 teaspoonful
Strawberry Flavor	1½ teaspoonfuls

The marzipan is colored pink and flavored strawberry. It is kneaded until smooth; then it is rolled out and cut into small pieces. Each piece is placed between two walnut halves and formed into marzipan walnuts. This process is continued until all the pink marzipan is used up. The marzipan walnuts are then placed in crystal pans and crystallized in an overnight crystal (*see* CRYSTALLIZING). They are then drained, dried, and packed as usual.

The batch will yield approximately 23 pounds of marzipan walnut goodies.

MELLOWS: *See individual Mellows Formulas,* e.g., FRUIT MELLOWS, SPEARMINT MELLOWS.

MELTING CREAM: *See* BANQUET MINT WAFERS.

MENTHOL DROPS

Sugar	25 lb
Water	3½ qt
Cream of Tartar	2½ teaspoonfuls
Oil of Menthol	3 teaspoonfuls
Green Coloring	

The sugar, water, and cream of tartar are placed in a clean copper pan on the stove and stirred until dissolved. When the batch begins to boil, it is covered and allowed to steam for a few minutes. The sides are washed down and a thermometer is placed in the batch, which is cooked to 338°F. The batch is poured on a greased slab, and the edges are folded together while stirring in the green color and the oil of menthol. Folding and kneading are continued until the batch is cool enough, when it is taken over to the spinning table in front of the batch warmer. It is then passed through a drop-roller machine and allowed to cool on the table. The drops are broken apart, put through a sieve, and packed into tins.

The batch will yield approximately 23½ pounds.

Note: A very delicate green color is desirable.

MEXICAN KISSES

Maple Sugar	4 lb
Crystal "A" Sugar	3 lb
Water	1 qt
Cream of Tartar	1 teaspoonful
Maple Flavor	2 teaspoonfuls
FRENCH CREAM (*see*)	6 lb
Pecan Pieces	4 lb
Glycerin	2 teaspoonfuls

The sugars, water, and cream of tartar are placed in a small copper pan on the stove and stirred until dissolved. When the batch begins to boil, it is covered and allowed to steam for a few minutes. The sides are washed down and a thermometer is placed in the batch, which is cooked to 238°F. The batch is then removed from the heat, and the French cream and flavor are added and stirred until the cream is melted. Then the nuts are stirred in and the glycerin is added and mixed thoroughly. The batch is then spooned out into small kisses and deposited on wax-paper-lined trays. The kisses are allowed to cool until they can be taken from the paper clean. They are then packed in boxes.

This formula should yield approximately 18 pounds of candy.

MINT CLEAR DROPS

Crystal "A" Sugar	20 lb
Water	3½ qt
Cream of Tartar	2 teaspoonfuls
Triple-Distilled Oil of Mint	2 teaspoonfuls

The sugar, water, and cream of tartar are placed in a copper pan on the stove and stirred until dissolved. When the batch begins to boil, it is covered and allowed to steam for a few minutes. The sides are washed down and a thermometer is placed in the batch, which is cooked to 328°F. The batch is poured out on a greased slab and cooled for a moment. The edges are folded together while stirring in the oil of mint. Folding and kneading are continued until the batch is cool enough. Then it is taken over to the spinning table in front of the batch warmer. It is then passed through a drop-roller machine and allowed to cool. The drops are broken apart, put through a sieve and then packed into tins.

The batch will give about 19 pounds of candy.

MINT CUTTING MACHINE: *See* AFTER-DINNER MINTS.

MINT MARSHMALLOW SQUARES: *See* MARSHMALLOW SQUARES, MINT.

MINT SALT-WATER TAFFY: *See* SALT-WATER TAFFY, MINT.

MINT SAUCE

Sugar	7 lb
Corn Syrup	3 lb
Water	3 qt
150 Grade Citrus Pectin	1½ oz
Liquid Green Color	½ teaspoonful
Oil of Peppermint	2 teaspoonfuls

The pectin is mixed dry with 2 pounds of the sugar. One quart of the water is placed in a copper pan on the stove and the pectin sugar is added. The batch is stirred until the sugar is dissolved. Then the remaining sugar and water are added and the batch is again stirred until the sugar is dissolved. The corn syrup is added and the batch is cooked to 218°F. It is removed to a cool spot and the color and flavor are mixed in thoroughly. The sauce is then cooled and poured into a clean stone crock, which is covered. It is then ready for use.

The batch will yield approximately 8 quarts of mint sauce.

MINT SAUCE ICE CREAM

Mint Sauce

Water	1½ qt
100 Grade Citrus Pectin	2 oz
Sugar	2½ lb
Corn Syrup	2 lb
Green Color	¼ teaspoonful
Oil of Peppermint	½ oz

Ice Cream Mix

Fresh 20% Cream	4 qt
Fresh 40% Cream	2 qt
Condensed Milk	2 qt
Sugar	8 lb

The pectin is mixed dry with 1 pound of sugar and placed in a small pan on the stove. Then 1½ quarts of water are added and the mixture is stirred until dissolved. The balance of the sugar and corn syrup are added and the batch is stirred and cooked to 220°F. The sauce is then transferred to a clean pan and the green color and mint flavor are stirred in. It is then set aside to cool.

For the ice cream mix the light cream, condensed milk, and the heavy cream are poured into the mixing can. The sugar is added and the mixture is stirred until the sugar is dissolved. It is then strained into the freezer and, when the cream freezes about halfway, the cooled mint sauce is added. The ice cream is then finished as under VANILLA ICE CREAM NO. 1 (*see*).

The batch will yield approximately 20 quarts.

MINT SCRAP: *See* SAVING WASTE.

MINT STICKS, CHOCOLATE-COVERED: *See* CHOCOLATE-COVERED MINT STICKS.

MINTS: *See* AFTER-DINNER MINTS and *Individual Mint Formulas,* e.g., CREAM MINTS, JELLED MINTS.

MINTY NIBBLES

Center

Soft Coconut Butter	4 lb
Sieved 6x Powdered Sugar	6 lb
Triple-Distilled Oil of Peppermint	2 teaspoonfuls

The coconut butter is melted over low heat and thickened with the powdered sugar to a thick paste. The mint is added and mixed in thoroughly. This is used for the center and kept very warm until ready for use.

Jacket

Sugar	20 lb
Water	3½ qt
Cream of Tartar	2 teaspoonfuls
Red Color	As desired
Oil of Peppermint	2 teaspoonfuls

The sugar, water, and cream of tartar are placed in a copper pan on the stove and stirred until dissolved. When the batch begins to boil, it is covered and allowed to steam for a few minutes. The sides are washed down and a thermometer is placed in the batch, which is cooked to 330°F. The batch is poured out on a cooled, greased slab between iron bars. The edges are folded together while the mint flavor is stirred in. A small portion of the batch (about 8 ounces) is colored red for a stripe and is kept soft near the batch warmer. The batch on the slab is folded together and cooled rapidly until cool enough to handle. It is then transferred to the pulling machine and pulled for about 2 minutes. It is then taken to the spinning table in front of the batch warmer, where it is kneaded with gloved hands until smooth. One third of the batch is taken off and kept soft near the stripe. The balance of the batch is stretched out flat and the warm center is placed in the middle of the flat batch. The flat batch is then folded over and the edges are closed to seal the center. The batch is then stretched out and doubled up. This process is continued until the batch is doubled up four times. The folded batch is kept short and square. The remaining portion of the batch (the one third which was taken off) is stretched out large enough to hold the folded batch. The folded batch is placed in the center of the remaining flat portion, which is folded over so that it encloses the folded center batch. The batch is kept round, and the red portion for the stripe is spun out into eight fine, even stripes. These are used to stripe the batch all around. The batch is formed in spinning shape and is spun out into fine sticks. It is then passed through a lump-cutting machine and allowed to cool on the table. The candies are then packed in tins or jars.

This formula should yield approximately 28 pounds of candy.

MIXER FOR REMELTING CREAM, MODEL D-4

This machine has a steam-jacketed kettle. It mixes and remelts cream fondant until it is ready for casting in starch. (Courtesy of the Savage Brothers Co., Chicago, Illinois.)

MOGUL, MODEL M-100

This is a three-in-one unit. It deposits liquid soft candies into the starch impressions; it brushes the starch off the candies; then it refills the starch boards and prints the impressions in the leveled starch boards. (Courtesy of National Equipment Corporation, New York, New York.)

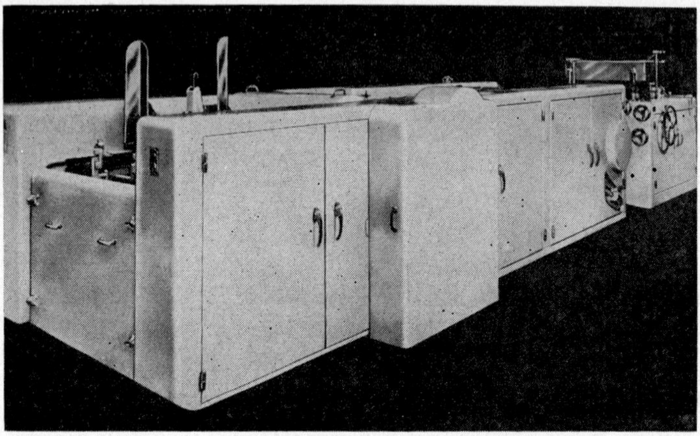

MOLASSES COCONUT BONBON CENTERS

Macaroon Coconut	20 lb
Molasses	3 qt
Brown Sugar	4 lb
Crystal "A" Sugar	6 lb
Corn Syrup	20 lb
Water	3 qt
Vanilla Flavor	4 teaspoonfuls

The coconut and molasses are placed in the mixing machine, which is set at medium speed. The sugars, corn syrup, and water are placed in a copper pan on the stove and stirred until dissolved. The sides are then washed down and the batch is cooked to 233°F. It is then poured on top of the coconut into the mixing machine. The flavor is added and mixing is continued for about 3 minutes. The batch is transferred to a slab, cooled for a few minutes, and it is then kneaded, placed in the rolling machine, and cut into small balls. These balls are allowed to cool on boards, and they are then ready for dipping in bonbon cream (*see* BONBONS, WHITE AND YELLOW) for maple and chocolate colors.

The batch should yield approximately 60 pounds of centers.

MOLASSES COCONUT CARAMELS, CHOCOLATE-COVERED: *See* CHOCOLATE-COVERED MOLASSES COCONUT CARAMELS.

MOLASSES COCONUT HAYSTACKS

Water	½ qt
Sugar	4 lb
Corn Syrup	6 lb
Molasses	1 qt
Long-Thread Coconut	6 lb
Vanilla Flavor	2 teaspoonfuls

The sugar, water, corn syrup, and molasses are placed in a copper pan on the stove and stirred until dissolved. The sides are washed down and the batch is cooked to 236°F. The flavor and the coconut are stirred in rapidly and mixed thoroughly. The batch is then turned out on the slab and allowed to cool for about 5 minutes. The coconut batch is formed into small stacks with hands dampened in water. These stacks are placed in lines over the entire slab and are allowed to cool until dry. They are then packed in single-layer flat boxes.

This formula should yield approximately 18 pounds of candy.

MOLASSES COCONUT ROYALS, CHOCOLATE-COVERED: *See* CHOCOLATE-COVERED MOLASSES COCONUT ROYALS.

MOLASSES COCONUT SQUARES

Sugar	8 lb
Corn Syrup	12 lb
Molasses	4 qt
Coconut Butter	12 oz
Water	1 qt
Vanilla Flavor	3 teaspoonfuls
Long-Thread Coconut	10 lb

The sugar, corn syrup, molasses, coconut butter, and water are placed in a large copper pan on the stove and stirred until dissolved. When the batch begins to boil, the sides are washed down and a thermometer is placed in the batch, which is cooked to 230°F. Then the thermometer is removed and stirring begins. Some of the coconut is gradually added and the batch is stirred and cooked on a low flame for about 7 minutes or until half of the coconut is used up. The flame is then turned off and the balance of the coconut is added and mixed thoroughly. The flavor is then stirred in and the batch is poured out on a greased slab between iron bars and spread out evenly with a palette knife. The batch is allowed to cool on the slab until it is firm enough to handle. It is then cut into squares and packed on flat pans lined with wax paper.

This formula should yield approximately 40 pounds of candy.

MOLASSES COCONUT TAFFY

Sugar	8 lb
Water	1 qt
Corn Syrup	8 lb
Molasses	2 qt
Coconut Butter	8 oz
Vanilla Flavor	2 teaspoonfuls
Long-Thread Coconut	6 lb

The sugar, water, corn syrup, molasses, and coconut butter are placed in a copper pan on the stove and stirred until dissolved. When the batch begins to boil, the sides are washed down and a thermometer is placed in the batch, which is cooked to 242°F. The coconut and flavor are then stirred in thoroughly. The batch is poured out on a greased slab between iron bars and spread out with a palette knife to approximately 1 inch thickness. The batch is allowed to cool, and then sliced and cut into squares. It is wrapped in moistureproof cellophane.

This batch will yield approximately 25 pounds of candy.

MOLASSES KISSES

Sugar	6 lb
Corn Syrup	12 lb
Open-Kettle Molasses	2 qt
Coconut Butter	8 oz

Water	1 qt
Salt	1 tablespoonful
Butter	8 oz
MARSHMALLOW WHIP (*see*)	1 lb
Vanilla Flavor	3 teaspoonfuls

The sugar, water, corn syrup, molasses, salt, and coconut butter are placed in a large copper pan on the stove and stirred until dissolved. When the batch begins to boil, the sides are washed down and the butter is added. Stirring is continued and the batch is cooked until it forms a medium ball when tested in cold water (*see* COLD-WATER TEST). The marshmallow whip and vanilla flavor are then stirred in and the batch is poured out on a cold, greased slab between iron bars. The edges are folded together, and the batch is cooled rapidly. When hard enough, the batch is taken to the pulling machine and pulled for about 7 minutes. It is then taken to the spinning table or wrapping machine, cut, and wrapped as usual.

This batch will yield approximately 25 pounds of candy.

MOLASSES MELLOW BARS

Sugar	8 lb
Corn Syrup	12 lb
Molasses	2 qt
Water	1 qt
Salt	1 tablespoonful
Coconut Butter	8 oz
Butter	1 lb
MARSHMALLOW WHIP (*see*)	2 lb
Vanilla Flavor	3 teaspoonfuls

The sugar, corn syrup, molasses, water, salt, and coconut butter are placed in a copper pan on the stove and stirred until dissolved. When the batch begins to boil, the sides are washed down, the butter is added, and stirring and cooking are continued until the batch forms a medium ball when tested in cold water. The marshmallow whip and vanilla flavor are stirred in and the batch is poured out on a cold, greased slab between iron bars. The edges are folded together and the batch is cooled rapidly. When hard enough, it is taken to the pulling machine and pulled for about 7 minutes. It is then transferred to the spinning table, formed into a flat batch, and spun out in bars about 2 inches in width. It is cut about 3 inches long, allowed to cool on the table, and wrapped as usual.

This batch will yield approximately 25 pounds of candy.

MOLASSES MELLOWS

Sugar	8 lb
Corn Syrup	12 lb
Molasses	2 qt
Coconut Butter	8 oz
Salt	2 teaspoonfuls
Evaporated Milk	6 lb
Butter	1 lb
Chopped Black Walnuts	3 lb
Vanilla Flavor	2 teaspoonfuls
MARSHMALLOW WHIP (*see*)	2 lb

The sugar, corn syrup, molasses, coconut butter, salt, and milk are placed in a copper pan on the stove and stirred until dissolved. The sides are washed down and the batch is stirred and cooked to 258°F. The butter and marshmallow whip are stirred in and mixed thoroughly. The batch is then poured out on a cooled, greased slab and the flavor is stirred in while the edges are folded together. The batch is cooled rapidly and then pulled on the machine or hook for 7 minutes. It is then taken over to the slab again and the chopped nuts are kneaded and mixed in. The batch is then rolled and spread out evenly to ¾ inch thickness. It is then sliced and cut into oblongs and wrapped in moisture-proof cellophane.

This formula should yield approximately 35 pounds of candy.

MOLASSES NOUGAT

Sugar	8 lb
Corn Syrup	12 lb
Molasses	2 qt
Water	1 qt
Salt	1 teaspoonful
Coconut Butter	8 oz
Chopped Walnuts	3 lb
MARSHMALLOW WHIP (*see*)	5 lb
Vanilla Flavor	3 teaspoonfuls

The sugar, corn syrup, molasses, water, and salt are placed in a large copper pan on the stove and stirred until dissolved. When the batch begins to boil, it is covered and allowed to steam for a few minutes. The sides are washed down and a thermometer is placed in the batch, which is cooked to 256°F. The coconut butter is stirred in while the marshmallow whip is added, stirred, and beaten with a large spatula. Beating is continued until the batch is thick. The chopped walnuts and the vanilla flavor are then added and mixed through. The batch is then turned out on trays which are lined and covered with wafer paper. It is allowed to set overnight, and is cut and wrapped as usual.

This formula will yield approximately 30 pounds.

MOLASSES PEANUT PUFFS

Center

Sugar	4 lb
Corn Syrup	4 lb
Water	½ qt
Roasted Chopped Peanuts	5 lb

The water, sugar, and corn syrup are placed in a copper pan on the stove and stirred until dissolved. When the batch begins to boil, the sides are washed down and a thermometer is placed in the batch, which is cooked to 230°F. The peanuts are then stirred in and mixed thoroughly. This center should be kept warm until ready for use.

Jacket

Sugar	20 lb
Water	3½ qt
Cream of Tartar	2 teaspoonfuls
Molasses	½ qt
Salt	1 tablespoonful
Vanilla	½ oz

The sugar, water, and cream of tartar are placed in a copper pan on the stove and stirred until dissolved. When the batch begins to boil, it is covered and allowed to steam for a few minutes. The sides are washed down and a thermometer is placed in the batch, which is cooked to 340°F. The molasses and salt are then stirred in and the batch is poured on a greased slab. The edges are folded together while stirring in the extract of vanilla. Folding and kneading are continued until the batch is cool enough to handle. The batch is then partly pulled on the hook and twisted until it shows a gloss and bright golden color. It is then taken over to the spinning table in front of the batch warmer, kneaded, and formed into a flat batch. The peanut batch is then placed in the center of the jacket and wrapped around with the end closed up. The batch is then spun out and cut on a puff-cutting machine. When cool enough, it is packed in tins.

The batch will yield approximately 32 pounds.

MOLASSES PEANUT TAFFY

Sugar	8 lb
Corn Syrup	12 lb
Molasses	4 qt
Water	1 qt
Salt	2 tablespoonfuls
Jumbo Roasted and Blanched Peanuts	25 lb

The peanuts are kept warm until ready for use. All other ingredients are placed in a large copper pan on the stove and stirred until dissolved. When the batch begins to boil, it is covered and allowed to steam for a few minutes. The sides are washed down and a thermometer is placed in the batch, which is cooked to 250°F. The thermometer is removed and some peanuts are gradually stirred into the batch. Continuously stirring and cooking on a slow flame, this process is continued until two thirds of the peanuts have been added and the batch is cooked to a hard ball. Cooking is then stopped and the remaining peanuts are mixed in. The batch is then poured on a greased slab and spread out to about ½ inch thickness. While cooling, it is cut into small sheets. When cool enough, it is packed in waxed-paper-lined boxes with waxed paper between each sheet.

This batch will yield approximately 50 pounds of candy.

MOLASSES PEPPERMINT KISSES

Sugar	8 lb
Corn Syrup	12 lb
Molasses	2 qt
Water	1 qt
Coconut Butter	12 oz
Salt	1 tablespoonful
Oil of Peppermint	2 teaspoonfuls

The sugar, corn syrup, molasses, water, salt, and coconut butter are placed in a large copper pan on the stove and stirred until dissolved. When the batch begins to boil, the sides are washed down and stirring process is continued. The batch is cooked until it forms a medium ball when tested in cold water (*see* COLD-WATER TEST). It is then poured out a cold, greased slab between iron bars. The edges are folded together while the oil of peppermint is stirred in. The batch is cooled rapidly and, when hard enough, it is taken to the pulling machine and pulled for about 7 minutes. It is then taken to the spinning table or wrapping machine and cut and wrapped as usual.

This batch will yield approximately 20 pounds of candy.

MOLASSES PEPPERMINTS

Sugar	20 lb
Water	3½ qt
Cream of Tartar	2 teaspoonfuls
Molasses	½ qt
Oil of Peppermint	2 teaspoonfuls

The sugar, water, and cream of tartar are placed in a clean copper pan on the stove and stirred until dissolved. When the batch begins to boil, it is covered and allowed to steam for a few minutes. The sides are washed down and a thermometer is placed in the batch, which is cooked to 340°F. While the molasses is stirred in, the batch is poured on a greased slab. The edges are folded together while stirring in the oil of peppermint. Folding and kneading are continued until the batch is cool enough to handle. A small piece (about 1 pound) is taken off for the stripe and is kept warm near the batch warmer. The main batch is pulled on the hook to a light consistency. It is then taken over to the spinning table in front of the batch warmer, kneaded, and formed into a flat batch. The clear lump is spun out in very fine stripes and drawn all around the flat batch on both sides. It is then passed through a drop-roller machine and allowed to cool on the table. The drops are broken apart, put through a sieve, and packed in tins.

The batch will yield approximately 19½ pounds.

MOLASSES PEPPERMINT SALT-WATER TAFFY: *See* SAVING WASTE.

MOLASSES PLANTATIONS: *See* SAVING WASTE.

MOLASSES WALNUT TAFFY

Sugar	8 lb
Corn Syrup	12 lb
Molasses	2 qt
Coconut Butter	8 oz
Water	1 qt
Salt	1 tablespoonful
MARSHMALLOW WHIP (*see*)	1 lb
Vanilla Flavor	2 teaspoonfuls
Chopped English Walnuts	2½ lb

The sugar, corn syrup, molasses, coconut butter, salt, and water are placed in a copper pan on the stove and stirred until dissolved. When the batch begins to boil, the sides are washed down. Stirring and cooking are continued until the batch forms a medium ball when tested in cold water (*see* COLD-WATER TEST). Then the marshmallow whip is added and mixed through. The flavor and chopped walnuts are then stirred in. The batch is poured out on a cold, greased slab between iron bars and the edges are folded together. The batch is cooled rapidly and, when cool enough, it is taken to the pulling machine and pulled for 5 minutes. It is then transferred to the spinning table, where it is cut and wrapped.

This batch will yield approximately 25 pounds of candy.

N

NEAPOLITAN COCONUT SQUARES: *See* COCONUT SQUARES, NEAPOLITAN TYPE.

NEW ENGLAND PEANUT BRITTLE: *See* PEANUT BRITTLE, NEW ENGLAND TYPE.

NORMANDY CHOCOLATES, BITTER ALMOND

The formula for NORMANDY CHOCOLATES, CREAM NUT (*see*) is followed, but the mixture is colored pale green and flavored with 2 teaspoonfuls of oil of bitter almonds and 1½ pounds of chopped roasted almonds are kneaded into the mixture. The batch is then rolled out, cut into small oblongs, and dipped in fine milk chocolate. Each piece is topped with chopped pistachio nuts. After setting, they are packed in boxes.

The batch will yield approximately 38 pounds of chocolates.

NORMANDY CHOCOLATES, CREAM NUT

Cocoa Butter	10 lb
Dry Milk Powder	12 oz
Sifted 6x Powdered Sugar	12 lb
Vanillin	¼ teaspoonful
Chopped Brazil Nuts	2 lb

The cocoa butter is melted and cooled off. The powdered sugar and milk powder are mixed in and stirred continuously until the batch is well whipped. It is then turned out on the slab and the vanillin and Brazil nuts are added and mixed in thoroughly. The batch is rolled out and cut into small balls, which are coated with a high-quality milk chocolate. After setting, they are packed in boxes.

The batch will yield approximately 38 pounds of chocolates.

NORMANDY CHOCOLATES, FILBERT

The formula for NORMANDY CHOCOLATES, WALNUT (*see*) is followed, but 1½ pounds of chopped roasted filberts and ½ pound of pure honey are kneaded into the mixture on the slab. The batch is finished in the usual manner.

The batch will yield approximately 38 pounds of chocolates.

NORMANDY CHOCOLATES, PECAN

The formula for NORMANDY CHOCOLATES, WALNUT (*see*) is followed all the way until the mixture is turned out on the slab. Then 1½ pounds of chopped pecans are kneaded into the batch, which is finished in the usual manner.

The batch will yield approximately 38 pounds of chocolates.

NORMANDY CHOCOLATES, WALNUT

Sifted 6x Powdered Sugar	12 lb
Cocoa Butter	10 lb
Cocoa Powder	2½ lb
Vanillin	¼ teaspoonful
Chopped Walnuts	1½ lb

The cocoa butter is melted in a steam-bath pan over low heat and allowed to cool off by constant stirring. The sifted powdered sugar is added gradually and stirring is continued until the mixture looks foamy. The powdered cocoa and vanillin are added and mixed in thoroughly. The batch is then turned out on the slab; the chopped walnuts are added and mixed through. The batch is then kneaded, rolled out, and cut and rolled into small balls, which are coated in a high-quality dark vanilla sweet chocolate. After setting, they are packed in boxes.

The batch will yield approximately 38 pounds of chocolates.

NOUGAT, TURKISH STYLE: *See* TURKISH NOUGAT.

NOUGATINES, CHOCOLATE-COVERED: *See* CHOCOLATE-COVERED NOUGATINES.

NOUGATS: *See* CHEWING CANDIES.

NUT ROLL: *See* LOG NUT ROLL.

NUT SQUARES, ALMOND

Use the NUT SQUARES, WALNUT (*see*) formula, substituting light roasted almonds for walnuts.

The batch will yield approximately 15 pounds of candy.

NUT SQUARES, BRAZIL

Use the NUT SQUARES, WALNUT (*see*) formula, substituting 5 pounds of chopped Brazil nuts for walnuts.

The formula will yield approximately 15 pounds.

NUT SQUARES, WALNUT

Sugar	7 lb
Corn Syrup	3 lb
Water	1 qt
Salt	1 tablespoonful
English Walnut Pieces	5 lb

The sugar, corn syrup, water, and salt are placed in a copper pan on the stove and stirred until dissolved. When the batch begins to boil, it is covered and allowed to steam for a few minutes. The sides are washed down and a thermometer is placed in the batch, which is cooked to 290°F. The nuts are stirred in gradually and cooked for 2 or 3 minutes. The batch is then poured out on a greased slab, the edges are folded together, and the mass is rolled out between iron bars about 1½ inches high. After cooling for about 5 minutes, the batch is sliced down with a sharp knife and cut into squares. When cool enough, the candy is packed in tins.

The formula will yield approximately 15 pounds of candy.

NUT TOPPINGS: *See* FOUNTAIN SYRUPS and *Individual Nut Toppings,* e.g., CHOCOLATE ALMOND CRUNCH, MAPLE WALNUT SUNDAE TOPPING.

NUTS AND FRUITS: *See* GLAZED NUTS AND FRUITS.

O

OPEN-KETTLE MOLASSES: *See* CHEWING CANDIES and COCONUT CANDIES.

OPERA CHOCOLATES, BLACK-WALNUT

Sugar	30 lb
Invert Sugar	4 lb
Salt	2 teaspoonfuls
Fresh 20% Light Cream	8 qt
Vanilla Flavor	2 teaspoonfuls
Finely Chopped Black Walnuts	3½ lb
Melted Corn Syrup	2 lb
Invertase	2 teaspoonfuls

The sugar, invert sugar, salt, and cream are placed in a copper pan on the stove and stirred until dissolved. When the batch begins to boil, the sides are washed down. Stirring process is continued and the batch is cooked to a soft ball or about 244°F. The melted corn syrup is stirred in rapidly and the batch is poured out on a clean, cooled, dampened cream beater. The top is sprinkled with a few drops of cold water and the batch is cooled rapidly to about 100°F. The flavor and invertase are added and the beater is set in motion. The nuts are added while the batch is creaming. After it has creamed to a lump, the batch is allowed to sweat back for 15 minutes and it is then kneaded soft, put through a cutting machine, and cut into small oblongs. These are allowed to cool on the trays and are then coated with a dark vanilla sweet chocolate. They are topped with some chocolate sprinkles and allowed to set. The chocolates are then packed in the usual manner.

The batch will yield approximately 65 pounds of chocolates.

OPERA CHOCOLATES, PEACH

The formula for OPERA CHOCOLATES, BLACK-WALNUT (*see*) is followed, but when the batch is poured out on a clean, dampened cream beater, peach flavor and invertase are added. The machine is set in motion and 3 pounds of pure peach pulp are added to the batch while it is in the creaming process. It is then completed in the same way as the black-walnut batch, with the exception that the centers are rolled round and no chocolate sprinkles are used for topping.

The batch will yield approximately 65 pounds of chocolates.

OPERA CHOCOLATES, PINEAPPLE

The formula for OPERA CHOCOLATES, BLACK-WALNUT (*see*) is followed up to the point where the batch is poured out on the cream beater. It is cooled rapidly to 100°F and 2 teaspoonfuls of pineapple flavor and the invertase are added. The machine is then set in motion and 3 pounds of pure pineapple pulp are added to the batch while it is still in the creaming process. The batch is then finished in the usual manner.

It will yield about 65 pounds.

OPERA CREAMS, CRYSTALLIZED

Sugar	40 lb
Corn Syrup	5 lb
Fresh 18% Light Cream	8 qt
Melted Corn Syrup	5 lb
Salt	2 teaspoonfuls
Vanilla Flavor	4 teaspoonfuls
Maple Flavor	3 teaspoonfuls
Rum-Coffee Flavor	3 teaspoonfuls
Melted Bitter Chocolate	1½ lb
Invertase	2 teaspoonfuls

The sugar, raw corn syrup, cream, and salt are placed in a large copper pan on the stove and stirred until dissolved. When the batch begins to boil, the sides are washed down. Stirring is continued and the batch is cooked to a soft ball or about 246°F. The melted corn syrup is stirred in rapidly and the batch is then poured out on a dampened, cooled cream beater and allowed to cool for about 8 minutes. The vanilla flavor and the invertase are then added and the machine is set in motion and kept running until the batch creams to a lump. It is then allowed to sweat back for about 20 minutes. The cream is then remelted to lukewarm in a steam-bath copper pan; it is stirred constantly while melting. The batch is divided into four parts as follows: Part 1 remains as is; part 2 is flavored maple; part 3 is flavored rum-coffee, and part 4 is flavored with the melted bitter chocolate. The remelted cream is then cast in fancy-shaped impressions, molded in clean dry starch, and allowed to set overnight. The creams are then removed, brushed free of starch, and placed in crystal pans, which are only half filled. They are crystallized as usual by the overnight process. They are then drained, dried, and packed as usual (*see* CRYSTALLIZING).

The batch should yield approximately 65 pounds.

ORANGE BONBONS: *See* BONBONS, ORANGE.

ORANGE CORDIALS

The VANILLA CORDIALS (*see*) formula is followed, but the batch is flavored with 1½ teaspoonfuls of oil of orange and 1 teaspoonful of citric acid solution. The batch is then finished in the usual manner.

It will yield approximately 20 pounds of cordials.

ORANGE CREAMED BARS

The formula for RASPBERRY CREAMED BARS (*see*) is followed, but the batch is colored orange and flavored with oil of orange. It is then finished in the usual manner and should yield approximately 16 pounds of candy.

ORANGE CREAMS, HAND-ROLLED

Sugar	50 lb
Water	10 qt
Cream of Tartar	2 teaspoonfuls
Melted Corn Syrup	5 lb
MARSHMALLOW WHIP (*see*)	3 lb
Invertase	2 teaspoonfuls
Citric Acid Solution	3 teaspoonfuls
Orange Emulsion	4 teaspoonfuls
Orange Pulp	3 lb

The sugar, water, and cream of tartar are placed in a large copper pan on the stove and stirred until dissolved. When the batch begins to boil, it is covered and allowed to steam for a few minutes. The sides are then washed down and a thermometer is placed in the batch, which is cooked to 250°F. The batch is poured on a cooled, dampened cream beater and cooled rapidly for about 6 minutes. The marshmallow whip and invertase are then added and the beater is set in motion. While it is running, the citric acid and orange pulp and flavor are added. When the batch is creamed, it is allowed to sweat back for 15 minutes. The batch is then kneaded smooth and passed through the cut-roll machine. When cool, the centers are coated in dark semisweet vanilla or high-quality milk chocolate.

This formula should yield approximately 68 pounds of cream centers.

Note: For approximate yield of chocolate-covered centers, *see* HAND-ROLLED CREAM CENTERS, VANILLA NUT.

ORANGE FRUIT AND NUT PILLOWS: *See* FRUIT AND NUT PILLOWS, ORANGE.

ORANGE FRUIT CUTS: *See* FRUIT CUTS, ORANGE.

ORANGE FRUIT JELLIES

The LEMON FRUIT JELLIES (*see*) formula is followed, but the batch is colored a bright orange and flavored with oil of orange. It is finished in the usual manner and should yield approximately 55 pounds of candy.

ORANGE FRUIT MELLOWS: *See* FRUIT MELLOWS, ORANGE.

ORANGE GUMDROPS: *See* GUMDROPS, ORANGE.

ORANGE ICE

Water	20 qt
Sugar	20 lb
Lemon Juice	2½ qt
Orange Emulsion Flavor	2½ oz
Orange Color	2 teaspoonfuls

The ingredients are placed in the mixing can and stirred until all the sugar is dissolved and the mixture registers 20 degrees on the syrup gage. The mix is then strained into the freezer and completed in the same manner as VANILLA ICE CREAM NO. 1.

The batch will yield approximately 30 quarts.

ORANGE ICED COCONUT DAINTIES: *See* ICED COCONUT DAINTIES, ORANGE.

ORANGE JELLED MINTS: *See* JELLED MINTS, ORANGE.

ORANGE JELLY

Sugar	50 lb
Corn Syrup	30 lb
Water	40 qt
Cooking Starch (soaked in 10 qt of water)	8 lb
Cream of Tartar	1 oz
Acetic Acid	1 teaspoonful
Oil of Orange	2½ oz
Orange Color	½ oz

The sugar, corn syrup, and water are placed in a large steam cooking kettle, the steam is turned on, and the batch is stirred until it dissolves and begins boiling. The dissolved starch is then added to the cooking batch gradually. When the batch boils again, the cream of tartar and acetic acid are added, and the batch is stirred and cooked until it drops off the palette knife in long strings or about 224°F. The steam is then shut off and the batch is colored a bright orange. After the flavor is stirred in, the batch is poured out on the depositor and cast in warm starch. The starch boards are set in the hot room overnight. They are then sifted free from starch, dampened with simple syrup, and the jelly is then rolled in fine granulated sugar and allowed to dry. It is then placed in pans and crystallized as usual (*see* CRYSTALLIZING). This batch should yield 105 pounds of candy.

ORANGE JELLY FRUIT CUTS: *See* JELLY FRUIT CUTS, ORANGE.

ORANGE JELLY FRUIT SLICES: *See* JELLY FRUIT SLICES, ORANGE.

ORANGE LOLLIPOPS: *See* LOLLIPOPS, ORANGE.

ORANGE SALT-WATER TAFFY: *See* SALT-WATER TAFFY, ORANGE.

ORANGE SHERBET

Fresh Milk	20 qt
Sugar	20 lb
100 Grade Citrus Pectin	4 oz
Lemon Juice	1½ qt
Orange Color	1 teaspoonful
Pure Orange Emulsion	2½ oz

The pectin is mixed dry with 5 pounds of the sugar and placed in a copper pan on the stove over low heat. Four quarts of the milk are added rapidly and the batch is stirred and heated to the boiling point. It is then poured into the mixing can. The balance of milk and sugar, the juice, color, and flavor are all added into the mixing can. The mixture is then stirred until all the sugar is dissolved. The mix is cooled rapidly and strained into the freezer. Freezing, storing, and ripening of the sherbet are completed as usual (*see* VANILLA ICE CREAM NO. 1).

The batch will yield approximately 30 quarts.

ORANGE SOUR BALLS: *See* SOUR BALLS, ORANGE.

ORANGE STICKS, CHOCOLATE-COVERED: *See* CHOCOLATE-COVERED ORANGE STICKS, CREAM-FILLED.

ORANGE SYRUP

Thick-Skinned Oranges	10
Lemons	2
Sugar	8½ lb
Water	5½ pints
Liquid Orange Color	½ teaspoonful
Fruit Acid Solution	4 teaspoonfuls

The peel from the lemons and oranges is grated into a saucepan and 1½ pounds of sugar and 1 cup of water are mixed in. This orange extract is covered and soaked overnight. The following morning, the balance of sugar and water and the orange extract are heated and then percolated for 2½ hours. The color and acid solution are mixed in. The syrup is cooled and is then ready for use.

The batch will yield approximately 6 quarts of orange syrup.

ORANGE TABLETS

The formula for LEMON TABLETS (*see*) is followed, using oil of orange for flavoring and coloring the batch a bright orange color.

The batch will yield approximately 20½ pounds of candy.

ORANGEADES

The formula for MINTY NIBBLES (*see*) is followed, but the center is flavored orange, and 1 teaspoonful of citric acid is added. The jacket batch is colored and flavored orange. A small red stripe is used. Otherwise, the batch is finished as usual and should yield approximately 28 pounds of candy.

ORIENTAL TOFFEE

Sugar	10 lb
Corn Syrup	10 lb
Coconut Butter	8 oz
Evaporated Milk	2 lb
Fresh 18% Cream	2 qt
Butter	1 lb
Chopped Dates	2 lb
Roasted Chopped Filberts	1 lb
Vanilla Flavor	2 teaspoonfuls
Salt	1 teaspoonful

The sugar, corn syrup, salt, coconut butter, and cream are placed in a copper pan on the stove and stirred until dissolved. When the batch begins to boil, the evaporated milk is added gradually while the batch is stirred constantly. When the batch cooks thick, the butter is added and stirring continued. Cooking is continued until the batch forms a medium ball when tested in cold water (*see* COLD-WATER TEST). The chopped filberts, dates, and vanilla flavor are stirred in and mixed thoroughly. The batch is then poured out on a greased, cold slab and the edges are folded together. The batch is cooled rapidly and when firm enough to handle, it is cut to the desired size and wrapped.

This batch will yield approximately 30 pounds of candy.

OVERNIGHT CRYSTAL: *See* CRYSTALLIZING.

P

PAN CARAMELS, CHOCOLATE

Sugar	10 lb
Coconut Butter	1 lb
Corn Syrup	14 lb
Salt	1 tablespoonful
Evaporated Milk	20 lb
Caramel Paste	5 lb
Vanilla Flavor	2 teaspoonfuls
Cut-Up Bitter Chocolate	4 lb

The sugar, corn syrup, salt, coconut butter, and half of the milk are placed in a large copper pan on the stove and stirred until dissolved. Stirring is continued while the batch is cooking; then the caramel paste and balance of milk are stirred in gradually. The batch is cooked until it forms a medium ball when tested in cold water (*see* COLD-WATER TEST) and it is then removed to a stand. The chocolate and flavor are stirred in and the batch is then poured out on a greased slab between iron bars. It is spread out evenly with a palette knife and allowed to cool until hard. It is then passed through the sizing machine, cut in square caramels, put into bonbon cups, and finally packed in caramel trays lined with wax paper. This should yield approximately 45 pounds of candy.

PAN CARAMELS, CHOCOLATE MARSHMALLOW

The formula for PAN CARAMELS, CHOCOLATE (*see*) is followed. When the caramel is set, it is sized down to very thin sheets, which are dampened slightly. Some marshmallow sheets are brushed free from starch, dampened, placed on half of the caramel sheets, and covered with the remaining half. The caramels are then cut and packed as usual.

This batch should yield approximately 45 pounds of candy.

PAN CARAMELS, VANILLA

Sugar	10 lb
Corn Syrup	12 lb
Coconut Butter	1½ lb
Salt	1 tablespoonful
Evaporated Milk	20 lb
Caramel Paste	5 lb
Vanilla Flavor	4 teaspoonfuls

The sugar, corn syrup, salt, coconut butter, and half of the evaporated milk are placed in a large copper pan on the stove and stirred until dissolved. Stirring is continued while the batch is cooking and the caramel paste and balance of evaporated milk are stirred in gradually. The batch is cooked to a medium ball when tested in cold water (*see* COLD-WATER TEST) and is then removed to a stand. The flavor is stirred in and the batch is poured out on a greased slab between iron bars, spread out evenly with a palette knife, and allowed to cool until set. It is then passed through the sizing machine, cut in square caramels, put into bonbon cups, and finally packed in caramel trays lined with wax paper. This batch will yield approximately 40 pounds.

PAN CARAMELS, VANILLA MARSHMALLOW

The formula for PAN CARAMELS, VANILLA (*see*) is followed, but 3 pounds of dipping marshmallows are brushed free from starch and stirred into the batch before pouring. Stirring is continued until some of the marshmallows begin to melt. The batch is finished as usual.

This should yield approximately 40 pounds of candy.

PANNED CHOCOLATES, POLISHING: *See* CHOCOLATE ALMONDS, PAN-COATED.

PANNED CHOCOLATES, POLISHING: *See* CHOCOLATE ALMONDS, PAN-COATED.

PANNED GOODS

Approximately 90 percent of all candy factories in the United States do not have a panned-goods department in their plant. Some plants, however, make panned goods exclusively. Production in these plants is often based on excessive speed which cannot go hand in hand with quality. For example, jelly beans for Easter are often made at low cost and are inedible. The center piece in the Jordan almond may be of good eating quality, but the consumer must taste 8 or 9 parts of sugar before he gets to it. The same situation holds for creamed filberts. Many hard-candy and licorice centers are panned and sold to vending-machine distributors. These candies are often made by a cheap and unsatisfactory method.

In contrast to this situation, about 10 percent of the candy manufacturers in the United States find it practical to install a panned-goods department in their plant and specialize in quality. Their small jelly and coconut eggs for Easter are panned with a small amount of sugar on top. Tinted and polished in attractive pastel colors, they bring a price that is three to four times higher than the average and repeat frequently.

The formulas that are found in this encyclopedia are meant for those who run a small or medium panned-goods department in their factory and are interested in quality. Also, manufacturers who operate retail chain stores may find that instead of buying and reselling cheaply made panned candies that are out of line with their general quality products, they can set up a panned-goods department and make their own quality panned candies.

For the various panned-goods formulas, *see* the *individual names,* such as, JELLY BEANS, SUGAR-COATED PEANUTS, CREAMED ALMONDS, etc.

PANNED PASTEL COCONUT EGGS: *See* PASTEL COCONUT EGGS, PANNED.

PANNING PROCESS FOR CHOCOLATE-COATED PANNED GOODS: *See* CHOCOLATE ALMONDS, PAN-COATED.

PANNING PROCESS FOR SUGAR-COATED PANNED GOODS: *See* ALMOND DAINTIES AND SILVER DRAGEES.

PENUCHI KISSES

Crystal "A" Sugar	6 lb
Brown Sugar	2 lb
Corn Syrup	2 lb
Fresh 18% Cream	3 qt
Salt	2 teaspoonfuls
Slow-Set Citrus Pectin	2 oz
Maple Flavor	2 teaspoonfuls
FRENCH CREAM (*see*)	6 lb
Chopped English Walnuts	4 lb
Glycerin	2 teaspoonfuls

The pectin is mixed dry with the brown sugar and it is placed in a copper pan on the stove with 1 quart of the cream and stirred until dissolved. The white sugar, corn syrup, salt, and the balance of the cream are added to the batch. It is stirred and cooked to a soft ball when tested in cold water (*see* COLD-WATER TEST). The batch is then removed from the heat and the French cream is stirred in rapidly. The flavor and glycerin are added and the batch is again stirred as the nuts are added and mixed through. The batch is then spooned out into small kisses and deposited on dipping trays. The kisses are allowed to cool until dry and they are then packed in boxes as usual.

This formula should yield approximately 24 pounds of candy.

PARFAITS: *See* FROZEN CONFECTIONS and *Individual Parfait Formulas,* e.g., VANILLA PARFAIT and MAPLE-WALNUT PARFAIT.

PASTE, CHOCOLATE-COVERED: *See* CHOCOLATE-COVERED PASTE.

PASTES: *See* SOFT GOODS and *Individual Paste Formulas,* e.g., ANGORA PASTE and TURKISH RAHAT LACOUM PASTE.

PASTE TESTS: *See* TURKISH RAHAT LACOUM PASTE, LEMON.

PASTEL COCONUT EGGS, PANNED

Centers

Sugar	12 lb
Water	2 qt
Corn Syrup	20 lb
Fine Macaroon Coconut	15 lb
Vanilla Flavor	1 oz

The sugar, water, and corn syrup are placed in a copper pan on the stove and stirred until dissolved. When the batch begins to boil, it is covered and allowed to steam for a few minutes. The sides are washed down and a thermometer is placed in the batch, which is cooked to 228°F. The macaroon coconut and vanilla flavor are then stirred in and mixed through. The batch is turned out on a cold greased slab and cooled off rapidly. It is then put through a hand-roll machine, formed into little eggs, and allowed to cool off until it is hard enough for the revolving pan.

Panning

Gum Arabic (dissolved in 4 qt of water)	4 lb
Sugar	20 lb
Water	3 qt
Corn Syrup	20 lb
Colors and Flavors	To suit

The gum arabic is heated and strained through a sieve. The sugar and water are placed in a copper pan on the stove and stirred until dissolved. They are allowed to boil and removed to a stand near the revolving pan. The corn syrup and the gum arabic solution are stirred into the sugar syrup and mixed thoroughly.

The syrup is then divided into five parts and placed in five small copper pans. The colors (all deep) and flavors are prepared in each pan as follows: yellow is flavored lemon; orange is flavored orange; green is flavored pistachio; pink is flavored strawberry; and purple is flavored violet.

About one fifth of the eggs is placed in the revolving pan,

which is set in motion. The eggs are sprinkled with some of the colored syrup and then covered with some fine powdered sugar. This is continued until the eggs are dry. When all of the syrup is used up and the eggs are finished dry but not roughly, they are removed and spread on clean trays. They are allowed to dry overnight and are then ready for polishing. The same process is used for all five colors.

POLISHING

Polishing is carried out in the usual manner. The eggs are placed in the polishing pan, with small pieces of hard beeswax and carnauba wax. While the pan is revolving the wax melts and covers the surface of the pan. The eggs are polished for 1 to 1½ hours while the pan is revolving in steady motion. When they look glossy and well polished, they are removed, spread out on clean trays lined with wax paper, and kept covered until dry. They are then packed in boxes.

Note: No yield can be estimated on the coconut eggs, since the weight varies depending on the amount of powdered sugar absorbing the syrup in the panning process.

PATTIES: *See Individual Patties Formulas,* e.g., ALMOND CRISP PATTIES and CHOCOLATE COCONUT PATTIES.

PEACH ICE CREAM

Fresh 20% Cream	4 qt
Fresh 40% Cream	2 qt
Condensed Milk	2 qt
Sugar	8 lb
Preserved Peach Cubes	6 lb
100 Grade Citrus Pectin	2 oz

The pectin is mixed dry with 2 pounds of sugar and placed in a double boiler on the stove together with 1 quart of the light cream. This mixture is stirred until dissolved and heated to the boiling point. It is then poured into the mixing can and the balance of the sugar and the light cream, the condensed milk, and the heavy cream are added and stirred with a large paddle until the sugar is dissolved. This mix is then strained into the freezer and, when the cream freezes about halfway, the peaches are added. Otherwise the same freezing process is followed as for VANILLA ICE CREAM NO. 1 (*see*).

The batch will yield approximately 25 quarts.

PEACH OPERA CHOCOLATES: *See* OPERA CHOCOLATES, PEACH.

PEACH SHERBET

Fresh Milk	20 qt
Sugar	20 lb
100 Grade Citrus Pectin	4 oz
Lemon Juice	1½ qt
Pink Color	¼ teaspoonful
Pure Peach Extract	½ oz
Peach Pulp	6 qt

The pectin is mixed dry with 5 pounds of the sugar and placed in a small copper pan on the stove over low heat. Rapidly, 4 quarts of milk are added and the mixture is stirred and heated to the boiling point. It is then poured into the mixing can. The balance of the sugar and milk, the juice, color, and flavor are then placed into the can and the mix is stirred until all the sugar is dissolved. The mix is cooled rapidly and strained into the freezer as usual. Freezing is started, and when the sherbet is about halfway frozen, the peach pulp is added and the freezing is finished as usual. Storing and ripening of the sherbet are carried out in the usual manner (*See* VANILLA ICE CREAM NO. 1).

The batch will yield approximately 36 quarts.

PEACH STONES

Crystal "A" Sugar	20 lb
Water	3½ qt
Cream of Tartar	2 teaspoonfuls
Peach Flavor	3 teaspoonfuls
Powdered Tartaric Acid	2 oz

The sugar, water, and cream of tartar are placed in a clean copper pan on the stove and stirred until dissolved. When the batch begins to boil, it is covered and allowed to steam for a few minutes. The sides are washed down and a thermometer is placed in the batch, which is cooked to 330°F. The batch is then poured on a greased slab and colored bright red. The edges are folded together while stirring in the flavor and acid. Stirring is continued until the batch is cool enough; then it is taken over to the spinning table in front of the batch warmer. It is then passed through a peach-stone roller machine and allowed to cool on the table. The candy is then broken apart, put through a sieve, and packed into tins.

The batch will yield approximately 19 pounds of candy.

PEANUT BARS, SPANISH TYPE

Sugar	10 lb
Salt	1 tablespoonful
Spanish No. 1 Peanuts	10 lb

The sugar and salt are placed in a large copper pan on the stove on a low flame and stirred constantly until dissolved. The peanuts are then added and stirring is continued until the peanuts are well roasted. The batch is then poured out on a greased slab between iron bars. The edges are folded together, and the batch is spread out evenly to about ½ inch thickness. After cooling for a few minutes and being loosened underneath with a large knife, the batch is sliced and cut in bars about 3 inches long and 1 inch wide. After cooling, the bars are wrapped in moisture-proof cellophane.

This batch will yield approximately 19 pounds.

PEANUT BRITTLE, NEW ENGLAND TYPE

Sugar	10 lb
Corn Syrup	5 lb
Water	1½ qt
Peanuts, Spanish No. 1	7 lb
Salt	1 tablespoonful
Bicarbonate of Soda	2 tablespoonfuls

The sugar, water, corn syrup and salt are placed in a large clean copper pan on the stove and stirred until dissolved. When the batch begins to boil, the peanuts are added and allowed to boil until the batch becomes thicker. The batch is allowed to cook with constant stirring until the peanuts crack, or when some of the candy dipped in cold water breaks brittle when tested (*see* COLD-WATER TEST). The batch is then removed from the stove and set on a stand. The bicarbonate of soda is stirred in rapidly. The batch is then poured on a greased slab and spread out with a palette knife quite thin. It is loosened underneath with a large knife and turned on the other side. The batch is then stretched out very thin and, after cooling on the slab, it is packed in tins or in boxes.

This batch will yield approximately 20 pounds of candy.

PEANUT BUTTER EGGS

Sugar	30 lb
Water	7 qt
Cream of Tartar	2 teaspoonfuls
Melted Corn Syrup	5 lb
Vanilla Flavor	4 teaspoonfuls
Invertase	2 teaspoonfuls
MARSHMALLOW WHIP (*see*)	3 lb
Cold Stiff Peanut Butter	8 lb

The sugar, water, and cream of tartar are placed in a clean copper pan on the stove and stirred until dissolved. When the batch begins to boil, it is covered and allowed to steam for a few minutes. The sides are washed down and a thermometer is placed in the batch, which is cooked to 254°F. The melted corn syrup is stirred in rapidly, and the batch is poured out on a clean, cooled, dampened cream beater and cooled rapidly for about 8 minutes. The marshmallow whip, flavors, and invertase are added and the machine is set in motion. Creaming is continued until the batch forms a lump. It is then kneaded soft and placed in the cream tank above the cut-roll machine, leaving a space open in center from top to bottom. The peanut butter is kneaded smooth and placed in the center space. It is arranged so it stays in the center and is enveloped by the vanilla cream. The cream is then put through the cutting machine and cut as usual. After the eggs have cooled on boards, they are coated with dark sweet vanilla chocolate. When set, they are packed in boxes.

The batch will yield approximately 75 pounds of peanut-butter eggs.

Note: When the hand method is used for forming this batch, the vanilla cream is spread out on the table, the peanut butter is placed in center of the flat cream and both ends are wrapped over the peanut butter and closed all around. The batch is rolled out in long rolls about 2 inches in diameter, then cut, weighed, and shaped into eggs of the desired size.

PEANUT BUTTER KISSES, CHOCOLATE: *See* CHOCOLATE PEANUT BUTTER KISSES.

PEANUT BUTTER SNAPS

Sugar	20 lb
Water	3½ qt
Cream of Tartar	2 level teaspoonfuls
Molasses	½ qt
Peanut Butter	10 lb

The sugar, water, and cream of tartar are placed in a clean copper pan on the stove and stirred until dissolved. When the batch begins to boil, it is covered and allowed to steam for a few minutes. The sides are washed down and a thermometer is placed into the batch, which is cooked at 320°F. The molasses is then stirred in and cooking finished. The batch is poured on a greased slab. In the meantime, the peanut butter is being heated to almost hot. The edges of the batch are folded together and folding continued until cool enough. Then the batch is pulled on the hook to a golden color, taken to the spinning table in front of the batch warmer, and kneaded to a smooth surface. The batch is divided into two parts. One half is kept soft for the jacket. The other half is formed flat and the warm peanut butter is placed in center, folding over and closing up all around. It is then pulled out and folded together and doubled up four times. The jacket part is flattened out large enough to hold the center, which is wrapped in and covered all around. The batch is spun and cut on the cutting machine. When cool enough, it is packed in tins.

The batch will yield approximately 29 pounds.

PEANUT BUTTER STICKS, CREAM FILLED

Peanut Butter	10 lb
Sugar	20 lb
Cream of Tartar	2 teaspoonfuls
Water	3½ qt
Molasses	½ qt

The peanut butter is stirred and heated until almost hot. It is kept warm until it is needed.

The sugar, water, and cream of tartar are placed in a copper pan on the stove and stirred until dissolved. When the batch begins to boil, it is covered and allowed to steam for a few minutes. The sides are washed down and a thermometer is placed in the batch. The molasses is added gradually and the batch is cooked to 320°F. The batch is then poured out on a greased slab and the edges are folded together. From this point on the center spinning and cutting are handled as in making CRUNCHY MINT STICKS (*see*). After they are coated, the sticks are rolled in finely ground roasted peanuts.

The batch will yield approximately 55 pounds.

PEANUT CHEWS

Coconut Butter	2½ lb
Sugar	20 lb
Corn Syrup	30 lb
Molasses	3 qt
Salt	2 tablespoonfuls
Roasted and Blanched Virginia No. 1 Peanuts	40 lb
Evaporated Milk	20 lb
Vanilla Flavor	4 teaspoonfuls

The sugar, corn syrup, coconut butter, salt, and one half of the milk are placed in a large copper pan on the stove and stirred until dissolved. When the batch begins to boil, the molasses is stirred in gradually. Stirring is continued and the balance of the milk is stirred in. Most of the peanuts are stirred in and the batch is cooked until it forms a medium ball when tested in cold water (*see* COLD-WATER TEST). The balance of the peanuts and the flavor are stirred in and mixed through. The batch is poured out on a cold, greased slab between iron bars and spread out evenly to about 1/4 inch thickness. When cool and hard enough, it is sliced and cut into small oblongs. These are coated in dark sweet vanilla chocolate and allowed to cool until set. The chews are packed in stock boxes.

The batch will yield approximately 140 pounds of chocolate-covered peanut chews.

PEANUT CHIPS, CHOCOLATE-COVERED: *See* CHOCOLATE-COVERED PEANUT CHEWS.

PEANUT CLUSTERS, CHOCOLATE: *See* CHOCOLATE PEANUT CLUSTERS.

PEANUT CRISP, VIRGINIA TYPE

Sugar	12 lb
Corn Syrup	8 lb
Water	1 1/2 qt
Salt	1 1/2 tablespoonfuls
Roasted and Blanched Virginia Peanuts	12 lb

The sugar, water, corn syrup, and salt are placed in a copper pan on the stove and stirred until dissolved. When the batch begins to boil, it is covered and allowed to steam for a few minutes. The sides are washed down and a thermometer is placed in the batch, which is cooked to 290°F. Then it is removed from the stove, placed on a stand, and the peanuts are stirred in rapidly. The batch is then poured out on a greased slab and spread out thin with a palette knife. It is then turned on the opposite side, stretched out as thinly as possible, and allowed to cool on the slab. The candy is then packed in tins.

This batch will yield approximately 30 pounds of candy.

PEANUT CRUNCH, CHOCOLATE: *See* CHOCOLATE PEANUT CRUNCH.

PEANUT CRUNCH, NEW ORLEANS TYPE

Sugar	12 lb
Corn Syrup	8 lb
Water	1½ qt
Molasses	1 qt
Salt	1 tablespoonful
Bicarbonate of Soda	2 tablespoonfuls
Roasted and Blanched Virginia Peanuts	15 lb

The sugar, water, corn syrup, and salt are placed in a copper pan on the stove and stirred until dissolved. When the batch begins to boil, it is covered and allowed to steam for a few minutes. The sides are washed down and a thermometer is placed in the batch, which is cooked to 260°F. The molasses is mixed in and all the peanuts are stirred in gradually. The batch is stirred and cooked until the candy breaks brittle when tested in cold water (*see* COLD-WATER TEST). It is then removed from the stove, placed on a stand, and the bicarbonate of soda is stirred in rapidly. The batch is then poured out on a greased slab and spread out thinly with a palette knife. Stretching is continued until the candy is quite thin. It is then allowed to cool on the slab and packed in tins.

This batch will yield approximately 35 pounds of candy.

PEANUT NOUGAT, CHOCOLATE-COVERED: *See* CHOCOLATE-COVERED PEANUT NOUGAT.

PEANUT PATTIES

Sugar	6 lb
Corn Syrup	4 lb
Salt	3 teaspoonfuls
Water	1 qt
Spanish Peanuts	12 lb
Butter	1 lb

The sugar, corn syrup, water, and salt are placed in a copper pan on the stove and stirred until dissolved. When the batch begins to boil, it is covered and allowed to steam for a few minutes. The sides are washed down and the peanuts are added gradually. While the batch is stirred continuously for another 5 minutes, the butter is added and stirring continued until the peanuts are well roasted or until the batch is hard and crisp when tested in cold water (*see* COLD-WATER TEST). The batch is then turned out on a warm, greased slab and the edges are folded together until the batch is partly cool. It is then kept folded together in a warm spot and a few small pieces, about the size of small balls, are pulled from the batch and placed in round, greased fritter tin rings on the slab. These are flattened out with a hand scraper to fill up the rings. This process is continued, working rapidly, until the entire batch is molded into peanut patties. After the patties are cool, they are wrapped and packed as usual.

This formula should yield approximately 21 pounds of peanut patties.

PEANUT SQUARES

Sugar	12 lb
Corn Syrup	8 lb
Molasses	2 qt
Water	2 qt
Salt	2 tablespoonfuls
Butter	1 lb
Roasted and Blanched Virginia Peanuts	24 lb

The peanuts are kept warm until ready for use. The sugar, water, corn syrup, molasses, and salt are placed in a large copper pan on the stove and stirred until dissolved. When the batch begins to boil, it is covered and allowed to steam for a few minutes. The sides are washed down and a thermometer is placed in the batch, which is cooked to 260°F. The thermometer is removed and some peanuts are stirred in gradually. Still stirring and cooking, the butter is added and stirring continued until over one half of the peanuts have been added and the candy breaks crisp when tested in cold water (*see* COLD-WATER TEST). The cooking is then stopped and the remaining peanuts are stirred in while the batch is still on the hot stove. It is then poured out on a greased slab between iron bars and the edges are folded together. While cooling, the batch is spread out evenly to about 1¼ inch thickness. It is then sliced and cut into squares. After cooling, the candy is packed in square tins lined with wax paper between layers.

This batch will yield approximately 50 pounds of candy.

PEANUTS, SALTED: *See* SALTED PEANUTS.

PEANUTS, SUGAR-COATED: *See* SUGAR-COATED PEANUTS.

PECAN BALLS

SNOWFLAKE MARSHMALLOW (*see*)	10 lb
Sugar	4 lb
Corn Syrup	6 lb
Coconut Butter	8 oz
Salt	2 teaspoonfuls
Evaporated Milk	6 lb
Fresh Butter	1 lb
Vanilla Flavor	2 teaspoonfuls
Pecan Pieces	25 lb

The marshmallows are sieved, brushed free of starch, and placed on boards ready for dipping. The sugar, corn syrup, coconut butter, salt, and half of the milk are placed in a copper pan on the stove and stirred until dissolved. The sides are washed down and the batch is stirred constantly as the balance of the milk is stirred in gradually. The butter is added and stirred through. The batch is cooked until it forms a soft ball when tested in cold water. It is then moved near the slab, where the pecans are spread out. The vanilla flavor is then stirred into the batch. The marshmallows are now dipped rapidly into the caramel batch, using a dipping fork. They are then rolled in the pecans and placed on boards lined with heavy wax paper. After they have cooled and hardened, the pecan balls are packed in boxes.

This formula should yield approximately 50 pounds of candy.

PECAN CREAM CUTS

Sugar	30 lb
Water	6 qt
Cream of Tartar	2 teaspoonfuls
Melted Corn Syrup	5 lb
Pecan Pieces	4 lb
Vanilla Flavor	3 teaspoonfuls

The sugar, water, and cream of tartar are placed in a copper pan on the stove and stirred until dissolved. When the batch begins to boil, it is covered and allowed to steam for a few minutes. The sides are washed down and a thermometer is placed in the batch, which is cooked to 250°F. The melted corn syrup is stirred in rapidly and the batch is brought to a boil. It is then poured out on a cooled, dampened cream beater and allowed to cool for about 6 minutes. The pecans and vanilla flavor are added and the machine is then set in motion. The batch is turned to a creamy lump and then allowed to sweat back for about 30 minutes; then it is transferred to the work table which is dusted with flour. The batch is then kneaded, rolled out, and cut into small oblongs. When cool enough, the cuts are placed in crystal pans and crystallized with an overnight crystal (*see* CRYSTALLIZING). They are than drained as usual, allowed to dry, and packed in boxes.

The batch should yield approximately 50 lb.

PECAN GOODIES: *See* MARZIPAN PECAN GOODIES.

PECAN-PASTE BONBON CENTERS

Sugar	25 lb
Water	5 qt
Cream of Tartar	2 teaspoonfuls
Melted Corn Syrup	3 lb
Ground Pecans	4 lb
Vanilla Flavor	2 teaspoonfuls
Invertase	1 teaspoonful
Glycerin	1 teaspoonful to every 5 lb of bonbon cream

The sugar, water, and cream of tartar are placed in a copper pan on the stove and stirred until dissolved. When the batch begins to boil, it is covered and allowed to steam for a few minutes. The sides are then washed down and a thermometer is placed in the batch, which is cooked to 248°F. The melted corn syrup is stirred in rapidly, the batch is poured out on a cooled, dampened cream beating machine, and allowed to cool for 7 minutes. The nuts, flavor, and invertase are then added and the beater is set in motion and kept running until the batch creams to a lump. The batch is then allowed to sweat back for 15 minutes and is then transferred to the working table, where it is kneaded, rolled, and cut into small oblongs, which are ready to be used as centers for violet and maple bonbons.

This batch will yield approximately 40 pounds of centers.

PECAN-PASTE BONBONS: *See* BONBONS, PECAN PASTE.

PECAN ROLL: *See* LOG-CABIN PECAN ROLL.

PECANS, SALTED: *See* SALTED PECANS.

PECTIN: *See* FROZEN CONFECTIONS.

PEPPERMINT BARLEY-SUGAR CUTS: *See* BARLEY-SUGAR CUTS, PEPPERMINT.

PEPPERMINT CRIMPED RIBBON CANDY: *See* CRIMPED RIBBON CANDY, PEPPERMINT.

PEPPERMINT SOUR BALLS: *See* SOUR BALLS, PEPPERMINT.

PEPPERMINT STICK CANDY: *See* STICK CANDY, PEPPERMINT.

PEPPERMINT YELLOW JACK: *See* YELLOW JACK, PEPPERMINT.

PIGNOLIA CLUSTERS, CHOCOLATE: *See* CHOCOLATE PIGNOLIA CLUSTERS.

PIGNOLIA NUTS, SALTED: *See* SALTED PIGNOLIA NUTS.

PINEAPPLE: *See* GLAZED PINEAPPLE.

PINEAPPLE, CHOCOLATE-COVERED: *See* CHOCOLATE-COVERED PINEAPPLE.

PINEAPPLE COCONUT DAINTIES

Sugar	12 lb
Corn Syrup	8 lb
Water	2 qt
Pineapple Pulp	5 lb
Macaroon Coconut	12 lb
Yellow Color	1 teaspoonful
Pineapple Flavor	3 teaspoonfuls

The sugar, corn syrup, and water are placed in a copper pan on the stove and stirred until dissolved. When the batch begins to boil, it is covered and allowed to steam for a few minutes. The sides are washed down and a thermometer is placed in the batch, which is cooked to 240°F. The pineapple pulp is stirred in and cooked about 2 minutes and then the coconut is mixed in. The yellow color and pineapple flavor are added and mixed thoroughly. The batch is then turned out on a greased slab between iron bars and spread out evenly with a palette knife to about 1-inch thickness. It is allowed to cool until hard; then it is sliced and cut into squares, rolled in fine granulated sugar, and packed in trays.

This formula should yield approximately 45 pounds of candy.

PINEAPPLE-CREAM CHOCOLATES

Sugar	30 lb
Water	7 qt
Cream of Tartar	2 teaspoonfuls
Melted Corn Syrup	3 lb
Pineapple Pulp	3 lb
Pineapple Flavor	3 teaspoonfuls
Powdered Citric Acid	1 teaspoonful
Invertase	2 teaspoonfuls
MARSHMALLOW WHIP (*see*)	3 lb

The sugar, water, and cream of tartar are placed in a copper pan on the stove and stirred until dissolved. When the batch begins to boil, it is covered and allowed to steam for a few minutes. The sides are washed down and a thermometer is placed in the batch, which is cooked to 244°F. The melted corn syrup is stirred in rapidly. The batch is poured out on a cooled, dampened cream beater and the top is sprinkled with a few drops of cold water. The batch is cooled rapidly to 100°F. The pineapple fruit and flavor, citric acid, invertase, and marshmallow whip are all added to the batch, which is creamed. Creaming is continued until the batch forms a lump. It is then allowed to sweat back for 15 minutes and kneaded and passed through the cut-roll machine. The small balls formed are spread out on boards and, when cool enough, they are dipped in dark vanilla sweet coating. After setting, they are packed in the usual manner.

This formula should yield approximately 48 pounds of centers or 65 pounds of chocolates when hand coated.

PINEAPPLE-CREAM CUTS

Sugar	30 lb
Water	6 qt
Cream of Tartar	2 teaspoonfuls
Melted Corn Syrup	5 lb
Pineapple Pulp	5 lb
Liquid Yellow Color	1 teaspoonful
Pineapple Flavor	2 teaspoonfuls
Powdered Citric Acid	2 oz

The sugar, water, and cream of tartar are placed in a clean copper pan on the stove and stirred until dissolved. When the batch begins to boil, it is covered and allowed to steam for a few minutes. The sides are washed down and a thermometer is placed in the batch, which is cooked to 254°F. The melted corn syrup is then stirred in rapidly. The pineapple pulp is spread out on the cooled cream beater and the batch is poured out on the same cream beater and allowed to cool for about 5 minutes. The color, flavor, and citric acid are then added and the machine is set in motion and run until the batch turns into a creamy lump. It is then allowed to sweat back for about 20 minutes; then it is transferred to the work table and dusted with some flour. The batch is then kneaded, rolled out, and cut into small oblongs. The cuts are placed on boards lined with wax paper and allowed to set until hard enough to handle. They are then placed in crystal pans and crystallized by the overnight process (*see* CRYSTALLIZING). They are then drained and dried as usual and packed in boxes.

This batch will yield approximately 50 pounds.

PINEAPPLE CREAMS, HAND-ROLLED

The formula for ORANGE CREAMS, HAND-ROLLED (*see*) is followed, but the batch is flavored with 3 teaspoonfuls of pineapple flavor and 3 pounds of pineapple pulp. It is then completed in the usual manner. Seventeen pounds of centers should yield approximately 25 pounds of chocolates after dipping.

PINEAPPLE EGGS, HAND-ROLLED

Sugar	35 lb
Water	7 qt
Cream of Tartar	2 teaspoonfuls
Melted Corn Syrup	5 lb
Pineapple Pulp	3 lb
Pineapple Flavor	3 teaspoonfuls
Citric Acid Solution	2 teaspoonfuls
Invertase	1 teaspoonful
MARSHMALLOW WHIP (*see*)	3 lb
Yellow Color	1 teaspoonful

The sugar, water, and cream of tartar are placed in a clean copper pan on the stove and stirred until dissolved. When the batch begins to boil, it is covered and allowed to steam for a few minutes. The sides are washed down and a thermometer is placed in the batch, which is cooked to 254°F. The melted corn syrup is stirred in rapidly, and the batch is poured out on a clean, cooled, dampened cream beater and cooled rapidly for about 7 minutes. The pineapple pulp, flavor, citric acid, and invertase are added and the beater is set in motion. The marshmallow whip and yellow color are added and creaming is continued until the batch forms a lump. It is then transferred to the work table, which has been dusted with flour, and kneaded until soft. The cream is then put through the cut-roll machine and is cut and rolled to the desired size. The eggs are spread out on boards lined with wax paper and allowed to set until cool. They are then coated in semisweet dark chocolate, and finished with the mark "P" on top. The finished eggs are allowed to set on dipping trays and are then packed in boxes.

The batch will yield approximately 70 pounds of pineapple eggs.

PINEAPPLE FRAPPÉ

Sugar	20 lb
Corn Syrup	20 lb
Water	12 qt
Slow-Set Citrus Pectin	8 oz
Sodium Citrate (dissolved in warm water)	2 oz
Citric Acid (dissolved in warm water)	4 oz
Yellow Color	2 teaspoonfuls
Pineapple Flavor	4 teaspoonfuls
MARSHMALLOW WHIP (*see*)	5 lb

The water is placed in a large copper pan on the stove and heated to the boiling point. The pectin is mixed dry with 5 pounds of the sugar and stirred in rapidly. Stirring is continued until the mixture boils again. Now the dissolved sodium citrate and one third of the dissolved citric acid are added to the batch, stirring constantly. The balance of the sugar and the corn syrup are stirred in gradually and the batch is cooked to 228°F. It is then divided into three parts. The first part is colored yellow, flavored with 1 teaspoonful of pineapple flavor, and one third of the remaining citric acid is stirred in. This portion of the batch is poured out rapidly on a slab lined with packing paper and spread out between iron bars to about ⅛-inch thickness.

The marshmallow whip is stirred into the second portion of the batch, which is colored straw yellow and flavored with 2 teaspoonfuls of pineapple flavor and half of the remaining citric acid. This is all mixed through well and poured out rapidly on top of the first layer on the slab and spread out evenly to cover the entire layer.

The third part of the batch is colored yellow, flavored with 1 teaspoonful of pineapple flavor, and the remaining citric acid is stirred in rapidly. This portion is then poured out on top of the second layer on the slab, spread out evenly, and allowed to set. It is then sprinkled with sugar on top and the whole batch is turned upside down. The paper is dampened with a wet brush, so that it can be removed from the candy easily. The batch is then cut in small squares or oblongs, dampened, and sugared. After the candy is allowed to dry, it is placed in crystal pans and crystallized as usual (*see* CRYSTALLIZING).

This batch should yield approximately 55 pounds of candy.

PINEAPPLE ICE CREAM

Fresh 20% Cream	4 qt
Fresh 40% Cream	2 qt
Condensed Milk	2 qt
Sugar	8 lb
100 Grade Citrus Pectin	2 oz
Pineapple Pulp	6 lb

The pectin is mixed dry with 2 pounds of the sugar, and placed in a double boiler on the stove together with 1 quart of the light cream. This mixture is stirred until dissolved and heated to the boiling point. It is then poured into the mixing can and the balance of the light cream and sugar, the condensed milk, and the heavy cream are added. The mix is stirred until the sugar is dissolved and is then strained into the freezer. When the cream freezes about halfway, the pineapple pulp is added. Otherwise the same freezing process is followed as for VANILLA ICE CREAM NO. 1 (*see*).

The batch will yield approximately 25 quarts.

PINEAPPLE JELLY

Canned Crushed Pineapple	10 lb
Sugar	8 lb
Agar-Agar	4 oz
Water	8 qt
Sugar	12 lb
Corn Syrup	2 lb
Yellow Color	1 teaspoonful
Pineapple Flavor	3 teaspoonfuls

The crushed pineapple and sugar are placed in a copper pan on the stove, stirred constantly, and cooked to a firm jelly or about 222°F. This is then set aside.

Now the agar-agar and water are placed in a copper pan on the stove and stirred constantly. When the batch begins to boil, the sugar is stirred in gradually. Stirring is continued while the corn syrup is added. After the batch is cooked to 223°F, it is moved to a stand where the pineapple fruit, color, and flavor are added and mixed thoroughly. The jelly is then cast in starch, run into varishaped impressions which are printed on starch boards, and allowed to set overnight. The starch is then removed from the candies, and they are dampened and sugared. When dry, they are placed in crystal pans and crystallized in the usual manner (*see* CRYSTALLIZING).

This batch should yield approximately 45 pounds of candy.

PINEAPPLE OPERA CHOCOLATES: *See* OPERA CHOCOLATES, PINEAPPLE.

PINEAPPLE SEA-FOAM FUDGE: *See* SEA-FOAM FUDGE, PINEAPPLE.

PINEAPPLE SEA-FOAM KISSES: *See* SEA-FOAM KISSES, PINEAPPLE.

PINEAPPLE SHERBET

Fresh Milk	20 qt
Sugar	20 lb
100 Grade Citrus Pectin	4 oz
Lemon Juice	1½ qt
Yellow Color	½ teaspoonful
Pure Pineapple Extract	½ oz
Pineapple Pulp	5 qt

The pectin is mixed dry with 5 pounds of the sugar and placed in a copper pan on the stove over low heat. Four quarts of the milk are added rapidly, and the mixture is stirred and heated to the boiling point. It is then poured into the mixing can. The balance of the milk and sugar, the juice, color, and flavor are added and the mixture is stirred until all the sugar is dissolved. The mix is cooled rapidly and strained into the freezer. The freezing process is then started and, when the sherbet is about halfway frozen, the pineapple pulp is added, and the freezing is completed as usual. The storing and ripening of the sherbet is followed through as usual (*see* VANILLA ICE CREAM NO. 1).

The batch will yield approximately 35 quarts.

PINK CANDY CURLS: *See* CANDY CURLS, PINK.

PISTACHIO BOARDWALK FUDGE: *See* BOARDWALK FUDGE, PISTACHIO.

PISTACHIO CANDY CURLS: *See* CANDY CURLS, PISTACHIO.

PISTACHIO CREAMS

The CAST CREAM CENTERS, VANILLA (*see*) formula is followed, but the cream is colored a pale green and flavored with 4 teaspoonfuls of pistachio emulsion. Pistachio creams are usually coated with a good grade of milk chocolate and topped with green-colored coconut.

Thirty pounds of cream centers will yield about 40 pounds of chocolate when chocolate coated by enrobing machine, or 45 pounds when hand dipped.

PISTACHIO FILLED BUTTER CUPS: *See* FILLED BUTTER CUPS, PISTACHIO.

PISTACHIO GOODIES: *See* MARZIPAN PISTACHIO GOODIES.

PISTACHIO ICE CREAM

Fresh 20% Cream	4 qt
Fresh 40% Cream	2 qt
Condensed Milk	2 qt
Sugar	8 lb
Chopped Pistachio Nuts	1½ lb
Butter	12 oz
Pistachio Emulsion	1½ oz
Green Color	½ teaspoonful

The pistachio nuts are toasted in the butter; then they are salted and set aside. The light and heavy cream and condensed milk are poured into the mixing can, the sugar is added, and the mixture is stirred until the sugar is dissolved. The flavor and color are mixed in and the mix is strained into the freezer as usual. When the cream freezes about halfway, the toasted pistachio nuts are added. Otherwise the directions under VANILLA ICE CREAM NO. 1 (*see*) are followed.

The batch will yield approximately 20 quarts.

PISTACHIO LONDON CARAMELS: *See* LONDON CARAMELS, PISTACHIO.

PISTACHIO MARSHMALLOWS, GELATIN TYPE

The formula for FANCY MARSHMALLOWS, VANILLA (*see*) is followed, but the batch is colored a bright green and flavored with pistachio flavor while still in the beater. The batch is then completed as usual and should yield approximately 125 pounds.

PISTACHIO NUTS, SALTED

No. 1 Fancy Shelled Pistachio Nuts	5 lb
GUM ARABIC SOLUTION (*see*)	1 tablespoonful
Salt	1 tablespoonful

The pistachio nuts are covered with the gum arabic solution and salted in the same way as walnuts. They are then placed in a large flat pan and left in a moderately hot oven until the salted pistachio nuts are dry and toasted. Since the pistachio is a smaller nut, toasting should only take a short time (from 40 minutes to 1 hour). In order to assure more uniform toasting, the nuts should be stirred every 15 minutes.

The batch will yield approximately 5 pounds of salted pistachio nuts.

PISTACHIO SEA-FOAM FUDGE: *See* SEA-FOAM FUDGE, PISTACHIO.

PISTACHIO TOFFEE

Sugar	10 lb
Corn Syrup	10 lb
Coconut Butter	8 oz
Salt	1 teaspoonful
18% Fresh Cream	3 qt
Butter	1 lb
Chopped Pecans	1½ lb
Pistachio Flavor	3 teaspoonfuls
Green Color	

The sugar, corn syrup, coconut butter, salt, and fresh cream are placed in a copper pan on the stove and stirred until dissolved. When the batch begins to boil, the butter is added. The batch is stirred constantly and cooked until it forms a medium ball when tested in cold water (*see* COLD-WATER TEST). The batch is then colored green, the nuts and flavor are stirred in, and it is then poured out on a greased, cold slab between iron bars. The edges are folded together and the batch is cooled rapidly until it is firm enough to handle. It is then cut to the desired size and wrapped.

This batch will yield approximately 25 pounds of candy.

PLANTATIONS, CHOCOLATE-COVERED: *See* CHOCOLATE-COVERED PLANTATIONS.

POLISHING ALMONDS: *See* JORDAN ALMONDS and ALMOND DAINTIES.

POPCORN COCONUT BRITTLE

Sugar	12 lb
Corn Syrup	6 lb
Water	1½ qt
Sliced Coconut	2 lb
Fresh Popcorn	2 lb
Salt	1 tablespoonful

The sugar, water, corn syrup, and salt are placed in a copper pan on the stove and stirred until dissolved. When the batch begins to boil, it is allowed to steam for a few minutes, the sides are washed down, and a thermometer is placed in the batch, which is cooked to 260°F. The thermometer is removed and cooking and stirring are continued while adding the sliced coconut. Cooking and stirring are continued until the coconut is toasted to a light-brown color and the candy breaks crisp when tested in cold water. The pop corn is then stirred in and the batch is poured out on a greased slab and spread out thinly with a palette knife. Stretching is continued until the batch is quite thin. After cooling on the slab, is is packed into tins.

This batch will yield approximately 22 pounds of candy.

POPCORN CRISPETTES

Sugar	12 lb
Corn Syrup	6 lb
Water	1½ qt
Molasses	1 qt
Coconut Butter	12 oz
Fresh Popcorn	4 lb
Salt	1 tablespoonful

All the ingredients, excluding the popcorn, are placed in a large copper pan on the stove and stirred until dissolved. When the batch begins to boil, it is covered and allowed to steam for a few minutes. The sides are washed down and a thermometer is placed in the batch, which is cooked to 290°F. The cooking is then stopped and all the popcorn is stirred in gradually. The batch is poured out on a greased slab between iron bars and spread out evenly to about 1½-inch thickness. While it cools for a few minutes, the bottom is loosened with a large knife. It is then sliced down and cut into bars to the size desired and wrapped in moistureproof cellophane.

This batch will yield approximately 25 pounds.

POPCORN, LOOSE, IN BULK

Sugar	7 lb
Corn Syrup	3 lb
Water	½ qt
Honey	½ qt
Salt	2 teaspoonfuls
Butter	8 oz
Popcorn	3 lb

The sugar, corn syrup, water, and salt are placed in a large copper pan on the stove and stirred until dissolved. When the batch begins to boil, it is covered and allowed to steam for a few minutes. The sides are washed down and a thermometer is placed in the batch, which is cooked to 300°F. The butter and honey are stirred in and cooking is continued for 2 more minutes. Then the cooking is stopped and all the popcorn is stirred in gradually; the batch is poured out on a greased slab. It is separated, working rapidly with gloved hands, so that the popcorn is all loosened and separated with only a thin shell of candy remaining. It is then allowed to cool and packed in airtight tins.

This batch will yield approximately 15 pounds.

PRALINES: *See* MAPLE PECAN PRALINES.

PROCESSOR FOR ICE CREAM, CONE-BOTTOM, ROUND

This machine is used for mixing, heating, and cooling heavy-bodied ice cream. The fifteen-degree cone bottom assures complete unloading of the product. (Courtesy of the Cherry Burrell Corporation, Cedar Rapids, Iowa.)

PUFFED-WHEAT BRITTLE

Sugar	12 lb
Corn Syrup	6 lb
Water	1½ qt
Salt	1 tablespoonful
Butter	1 lb
Puffed Wheat	3–4 lb

The sugar, water, corn syrup, and salt are placed in a copper pan on the stove and stirred until dissolved. When the batch begins to boil, it is covered and allowed to steam for a few minutes. The sides are washed down and a thermometer is placed in the batch, which is cooked to 290°F. The butter is then stirred in and the batch is cooked for about 3 minutes; then the puffed wheat is stirred in until the batch is quite thick. It is then poured out on a greased slab and spread out as thinly as possible with a palette knife. After cooling on the slab, it is packed in tins.

This batch will yield approximately 20 pounds.

R

RAHAT LACOUM PASTE: *See* TURKISH RAHAT LACOUM PASTE.

RAISIN CLUSTERS, CHOCOLATE: *See* CHOCOLATE RAISIN CLUSTERS.

RAISIN VARIETY CARAMELS: *See* VARIETY CARAMELS, RAISIN.

RASPBERRY COCONUT SQUARES: *See* COCONUT SQUARES, RASPBERRY.

RASPBERRY CREAMED BARS

Sugar	16 lb
Water	2 qt
Cream of Tartar	1½ teaspoonfuls
Raspberry Flavor	3 teaspoonfuls
Powdered Citric Acid	1 teaspoonful
Red Color	1 teaspoonful
Glycerin	1 teaspoonful

The sugar, water, and cream of tartar are placed in a copper pan on the stove and stirred until dissolved. When the batch begins to boil, it is covered and allowed to steam for a few minutes. The sides are washed down and a thermometer is placed in the batch, which is cooked to 274°F. The batch is poured out on a cooled greased slab between bars, and the color, flavor, glycerin, and acid are added. The batch is folded together and cooled rapidly until firm. It is then placed on the pulling machine and pulled for about 7 minutes. It is then transferred to the spinning table, dusted with powdered sugar, formed into a flat batch, and spun out into bars. These are placed in a warm room overnight on wax-paper-lined boards. When they are creamed, they are broken apart and packed in cans.

This formula should yield approximately 16 pounds of candy.

RASPBERRY CREAMS

The formula for CAST CREAM CENTERS, VANILLA (*see*) is used, but the batch is flavored with 2 teaspoonfuls of raspberry flavor, 2 tablespoonfuls of concentrated raspberry pulp, and 4 teaspoonfuls of citric acid solution. The creams are then finished as usual.

Thirty pounds of cream centers will yield about 40 pounds of chocolates when chocolate coated by enrobing machine, or 45 pounds when hand dipped.

RASPBERRY FRAPPÉ

The formula for PINEAPPLE FRAPPE (*see*) is followed, but all three portions of the batch are colored red and flavored raspberry. The batch is then finished in the usual manner and should yield approximately 55 pounds of candy.

RASPBERRY FRUIT CUTS: *See* FRUIT CUTS, RASPBERRY.

RASPBERRY FRUIT MELLOWS: *See* FRUIT MELLOWS, RASPBERRY.

RASPBERRY GUMDROPS: *See* GUMDROPS, RASPBERRY.

RASPBERRY ICE

Water	20 qt
Sugar	20 lb
Lemon Juice	2½ qt
Red Color	1 teaspoonful
Raspberry Pulp	5 qt

The water, sugar, lemon juice, and red color are placed in the mixing can and stirred until the sugar is dissolved and the mixture registers 20 degrees on the syrup gage. The mix is then strained into the freezer. When the ice is about halfway frozen, the raspberry pulp is added. The ice is then completed in the usual manner (*see* VANILLA ICE CREAM NO. 1).

The batch will yield approximately 32 quarts.

RASPBERRY JELLED MINTS: *See* JELLED MINTS, RASPBERRY.

RASPBERRY JELLY

Jap Gelatin (agar-agar)	4 oz
Water	8 qt
Sugar	12 lb
Corn Syrup	3 lb
Raspberry Flavor	2 teaspoonfuls
Red Color	1 teaspoonful
Canned Raspberries	10 lb
Sugar	10 lb

The raspberry fruit and the 10 pounds of sugar are placed in a copper pan on the stove and stirred constantly while cooking. This fruit batch is cooked to a firm jelly (about 222°F) and set aside. Then the gelatin and water are placed in a copper pan on the stove and stirred constantly. When boiling, the 12 pounds of sugar are stirred in gradually. Stirring is continued and the corn syrup is added. The batch is then cooked to 223°F and removed to a stand. The raspberry fruit is stirred in, and the flavor and color are added and mixed thoroughly. The jelly is then cast in starch, run into fancy shapes, and allowed to set overnight. Then the raspberry jellies are sifted free of starch, dampened with a little syrup, rolled in fine granulated sugar, and allowed to dry. They are placed in crystal pans and crystallized (*see* CRYSTALLIZING) in the usual manner.

This batch should yield approximately 50 pounds of candy.

RASPBERRY LOLLIPOPS: *See* LOLLIPOPS, RASPBERRY.

RASPBERRY LONDON CARAMELS: *See* LONDON CARAMELS, RASPBERRY.

RASPBERRY SYRUP

SIMPLE SYRUP (*see*)	4 qt
Pure Raspberry Jam	½ qt

Pure Raspberry Extract	2 teaspoonfuls
FRUIT ACID SOLUTION (*see*)	2 oz
Liquid Red Color	1 teaspoonful

All ingredients are placed in a clean stone crock, stirred, and mixed thoroughly. The syrup is then ready for use.

The batch will yield approximately 4½ quarts of raspberry syrup.

REVOLVING PAN

When using a revolving pan for the first time, it will be necessary to rough the inner surface. This is done by the following method. Five pounds of sugar, 5 pounds of corn syrup, and ½ quart of water are cooked to 260°F. While the pan is in motion, the syrup is slowly poured into it with a small dipper. The confectioner should begin at the back of the pan and continue to draw the dipper toward the front while pouring. When the syrup hardens in the pan, it will give a useful roughness that will prevent the candy centers from sliding in the pan and will make them roll.

Building up a sugar surface on the inside of the revolving pan is necessary when grossing work on panned candies is started. But as the revolving pans continue to run for two to three weeks, excessive sugar will accumulate on the inner surface. This makes the revolving pans unwieldy and tends to lower their speed. The undesirable coating may then be removed by applying steam through the coils of the pan. The steam will soften and loosen up the sugar and make it easy to remove. The revolving pan should be cleaned in this manner after being used continuously for fifteen full days.

When revolving pans are used constantly, the gears are more likely to be affected than any other part of the machine. They should, therefore, be carefully checked. When the gears are worn out, perfect revolving is impossible and the panned candies will turn out with too much scrap. The gears should be adjusted, repaired, and replaced as necessary. The steam coils should also be closely watched. They may start leaking and when this happens, the pan cannot work properly.

REVOLVING PAN WITH STEAM COILS

This machine is used for grossing and polishing panned goods. (Courtesy of the Groen Manufacturing Co., Chicago, Illinois.)

ROASTED COCONUT BRITTLE: *See* COCONUT BRITTLE, ROASTED.

ROASTED NUTS: *See* SALTED NUTS.

ROCK CANDY

Crystal "A" Sugar	60 lb
Water	12 qt
Cream of Tartar	½ teaspoonful
Acetic Acid	½ teaspoonful

In making the original old-fashioned rock candy, old-fashioned, primitive equipment usually gives the best results. About fifteen or twenty round tin cans about 15 inches long and 10 inches in diameter are pierced 2 to 3 inches apart. A long heavy needle, like the one used for mending burlap sacks, is used for the piercing. The confectioner begins piercing each can about 2 inches from the bottom and continues piercing at regular 2-inch intervals until he reaches 2 inches from the top. The needle is threaded with white, heavy string, which is strung across the can through each pair of holes. The string is knotted at each hole so that it holds firm and does not slip back. Some brown wrapping paper is pasted around the outer side of the can to prevent the syrup from leaking out. The cans are placed in a hot room of 120°F on two parallel logs which are some distance apart and long enough to hold the entire row of cans.

The sugar, water, and cream of tartar are placed in a large clean copper pan on the stove and stirred until dissolved. When the batch begins to boil, it is chilled off with a cupful of cold water. The scum is removed and the sides are washed down. The acetic acid is added and the batch is cooked to 240°F. The syrup batch is then moved into the hot room, poured slowly into the cans, and allowed to set there undisturbed for 40 hours with the room temperature kept at 120°F. When the rock candy has formed a heavy crystal, the room is cooled off and the heavy crust on top of the cans is broken loose. The syrup around the heavy crust is drained into a receptacle by placing the cans upside down and allowing them to drain for 2 hours. The paper on the outer sides of the cans is dampened and removed. The knotted strings are cut loose at the ends, the cans are tapped gently, and the rock candy which has clustered around the strings will come out in one piece. The bars of rock candy are carefully broken apart and packed in cans or jars.

The solid waste which has collected around the top, bottom, and sides of the cans is placed in a clean copper pan. The drained syrup is poured into this pan together with 1½ quarts of water. The pan is placed on the stove over a small flame and stirred constantly until the contents dissolve. The pan is then removed to a cold spot and allowed to cool. The syrup is then bottled, closed with airtight caps, and labelled as rock candy syrup.

This batch will yield approximately 45 pounds of rock candy and about 8 quarts of rock-candy syrup.

ROLLIES: *See Individual Rollies Formulas,* e.g., SEQUOIA ROLLIES, DAISY ROLLIES.

ROMAN PUNCH ROLLIES

Sugar	8 lb
Corn Syrup	4 lb
Fresh 18% Cream	3 qt
Salt	1 teaspoonful
Roman-Punch Flavor	3 teaspoonfuls
FRENCH CREAM (*see*)	3 lb
Milk Chocolate	5 lb
Chopped Roasted Almonds	10 lb
Silver Sheet Gelatin (soaked in warm water)	½ oz

The sugar, corn syrup, salt, and cream are placed in a copper pan on the stove and stirred until dissolved. The sides are washed down and the batch is stirred and cooked to 242°F. It is then moved near a marble slab. The soaked gelatin is stirred in, the cream is added, together with the flavor, and the batch is stirred thoroughly. It is then turned out on a greased marble slab, kneaded, rolled into long thin rolls, and allowed to cool on the marble. For the coating, milk chocolate is cooled as for regular chocolate coating and the rolls are coated and rolled in the chopped almonds. They are set aside until cool and are then cut in small rollies and packed as usual.

This formula should yield approximately 32 pounds of candy.

ROMAN PUNCH SHERBET

Fresh Milk	20 qt
Sugar	20 lb
100 Grade Citrus Pectin	4 oz
Pink Color	¼ teaspoonful
Roman-Punch Flavor	2½ oz
Assorted Diced Fruits	5 lb

The pectin is mixed dry with 5 pounds of the sugar and placed in a copper pan with 4 quarts of the milk. The pan is placed on the stove over low heat, and the mixture is stirred and heated to the boiling point. It is then poured into the mixing can. The balance of the milk and sugar, the color, and flavor are added and stirred until all the sugar is dissolved. The mix is cooled rapidly and strained into the freezer. Freezing is started and, when the sherbet is about halfway frozen, the diced fruits are added and freezing is completed as usual. The storing and ripening of the sherbet are completed as under VANILLA ICE CREAM NO. 1 (*see*).

The batch will yield approximately 36 quarts.

ROMAN PUNCH TOFFEE

Crystal "A" Sugar	8 lb
Brown Sugar	5 lb
Corn Syrup	12 lb
Fresh 18% Cream	3 qt
Evaporated Milk	2 lb
Butter	1 lb
Coconut Butter	1 lb
Salt	2 tablespoonfuls
Chopped Pecans	2 lb
Roman-Punch Flavor	3 teaspoonfuls

The sugar, corn syrup, coconut butter, cream, and salt are placed in a copper pan on the stove and stirred until dissolved. When the batch begins to boil, the evaporated milk is stirred in, and stirring continued while adding the butter. The batch is cooked until it forms a medium ball when tested in cold water (*See* COLD-WATER TEST). The chopped pecans and the Roman-punch flavor are mixed in and the batch is poured out on a greased, cold slab between iron bars. The edges are folded together and the batch is cooled rapidly. When hard enough, the batch is cut and wrapped by hand or machine.

This batch will yield approximately 30 pounds of candy.

ROSE CHIPS

Sugar	20 lb
Water	3½ qt
Cream of Tartar	2 teaspoonfuls
Oil of Rose	2 teaspoonfuls
Pink Color	

The sugar, water, and cream of tartar are placed in a copper pan on the stove and stirred until dissolved. When the batch begins to boil, it is covered and allowed to steam for a few minutes. The sides are washed down and a thermometer is placed in the batch, which is cooked to 338°F. The batch is poured on a greased slab and colored a light pink. The edges are folded together while stirring in the oil of rose. Folding is continued until the batch is cool enough. It is then pulled over the hook partly and twisted until bright and glossy. It is taken over to the spinning table in front of the batch warmer, where it is formed flat, spun, and passed through a chip-cutting machine. When cool enough, the candy is packed in tins.

This batch will yield approximately 19 pounds.

ROSE FILBERT FILLED BUTTER CUPS: *See* FILBERT FILLED BUTTER CUPS, ROSE.

ROSTOPLAST, THE HANSELLA SUPER

This machine does the die forming and cutting automatically. A three-way cooling unit is attached. It is the best for making Vienna filled hard candies. (Courtesy of the Package Machinery Co., Springfield, Massachusetts.)

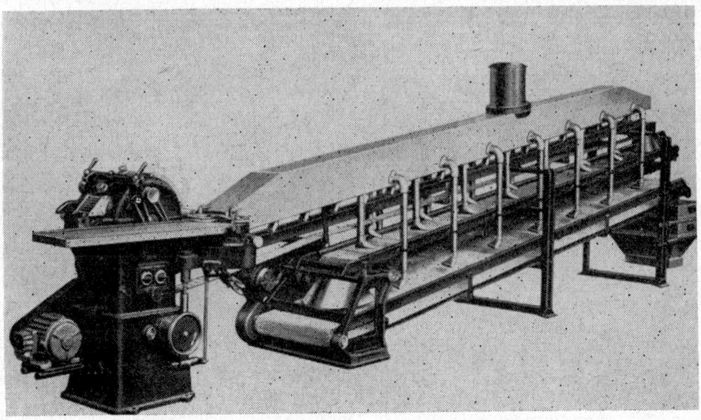

RUM BUTTER TOFFEE

Sugar	8 lb
Corn Syrup	12 lb
Coconut Butter	8 oz
Salt	1 teaspoonful

Fresh 18% Cream	2 qt
Butter	1 lb
Evaporated Milk	2 lb
Rum Flavor	3 teaspoonfuls

The sugar, corn syrup, coconut butter, salt, and fresh cream are placed in a copper pan on the stove and stirred until dissolved. When the batch begins to boil, the evaporated milk is added gradually and the batch is stirred constantly. When it begins to cook thick, the butter is added. Stirring and cooking are continued until the batch forms a medium ball when tested in cold water (*see* COLD-WATER TEST). The batch is poured on a greased slab between iron bars and the flavor is stirred in while folding the edges together. The batch is cooled rapidly. When it is cool enough to handle, the candy is cut and wrapped by hand or machine.

This batch will yield approximately 25 pounds.

RUM COFFEE TOFFEE

Sugar	8 lb
Corn Syrup	12 lb
Coconut Butter	8 oz
Salt	1 tablespoonful
Fresh Cream	2 qt
Evaporated Milk	2 lb
Butter	1 lb
Rum Flavor	1 teaspoonful
Coffee Extract	2 teaspoonfuls

The sugar, corn syrup, salt, coconut butter, and cream are placed in a copper pan on the stove and stirred until dissolved. When the batch begins to boil, the evaporated milk is stirred in and stirring is continued while the butter is added. The batch is cooked until it forms a medium ball when tested in cold water. It is then poured out on a cold, greased slab between iron bars and the edges are folded together while stirring in both flavors. The batch is then cooled rapidly and, when hard enough, it is cut and wrapped.

This batch will yield approximately 25 pounds.

RUM RAISIN CREAMS, HAND-ROLLED

The formula for VANILLA NUT CREAM CENTERS, HAND-ROLLED (*see*) is followed up to the cooling point. The marshmallow whip and invertase are then added. The batch is flavored with 5 teaspoonfuls of Roman-punch flavor. The beater is set in motion and 4 pounds of chopped seedless raisins are added while the batch is in the creaming process. It is then completed in the usual manner.

Note: The average ratio for using chocolate coating for hand-rolled centers is approximately 20 percent for enrober coating and 30 percent for hand coating.

RUM RAISIN ICE CREAM

Fresh 20% Cream	4 qt
Fresh 40% Cream	2 qt
Condensed Milk	2 qt
Sugar	8 lb
Grade 100 Citrus Pectin	3 oz
Preserved Seedless Raisins	4 qt
Roman-Punch Flavor	2 oz
Yellow Color	1 teaspoonful

The pectin is mixed dry with 2 pounds of sugar and placed in a double boiler on the stove together with 1 quart of the light cream. This mixture is stirred until dissolved and heated to the boiling point. It is then poured into the mixing can and the balance of the sugar and light cream, the condensed milk, and the heavy cream are added. The mix is stirred with a long paddle until the sugar is dissolved. The color and flavor are stirred in, and the mix is strained into the freezer. When it freezes about halfway, the preserved raisins are added. Otherwise the same process is used as for VANILLA ICE CREAM NO. 1 (*see*).

The batch will yield approximately 25 quarts.

RUM RAISIN SQUARES

Sugar	8 lb
Fresh 18% Cream	3 qt
Corn Syrup	4 lb

Salt	1 teaspoonful
Roman-Punch Flavor	4 teaspoonfuls
FRENCH CREAM (*see*)	4 lb
Small Seedless Raisins	5 lb

The sugar, corn syrup, salt, and cream are placed in a copper pan on the stove and stirred until dissolved. The sides are washed down and the batch is stirred and cooked to 244°F. It is then moved near the slab and the French cream and flavor are stirred in. The raisins are added and mixed in thoroughly. The batch is then poured out on a board lined with heavy wax paper and spread out evenly to approximately ½-inch thickness. The candy is allowed to set overnight. The following day, the board is turned upside down on the table, the wax paper is pulled off, and the candy is cut into small squares. When dry, the squares are packed in boxes.

This formula should yield approximately 25 pounds of candy.

RUM RAISIN TOFFEE

Sugar	12 lb
Corn Syrup	8 lb
Coconut Butter	8 oz
Salt	1 teaspoonful
Fresh 18% Cream	2 qt
Evaporated Milk	2 lb
Butter	1 lb
Chopped Soaked Raisins	2 lb
Jamaica-Rum Flavor	3 teaspoonfuls

The sugar, corn syrup, coconut butter, cream, and salt are placed in a copper pan on the stove and stirred until dissolved. When the batch begins to boil, the evaporated milk is added and stirring and cooking are continued while the butter is added. The batch is cooked until it forms a medium ball when tested in cold water (*see* COLD-WATER TEST) and the chopped raisins and Jamaica-rum flavor are mixed in. The batch is poured out on a greased, cold slab between iron bars and the edges are folded together. The batch is cooled rapidly and, when hard enough, it is cut to the desired size and wrapped.

This batch will yield approximately 25 pounds of candy.

RUNNING CREAM

Fondant for Cast Cream Centers

Sugar	60 lb
Water	12 qt
Cream of Tartar	4 teaspoonfuls
MARSHMALLOW WHIP (*see*)	5 lb
Invertase	4 teaspoonfuls

The sugar, water, and cream of tartar are placed in a large copper pan on the stove and stirred until dissolved. When the batch begins to boil, it is covered and allowed to steam for a few minutes. The sides are then washed down and a thermometer is placed in the batch, which is cooked to 244°F. The batch is poured out on a cooled, dampened cream beater and sprinkled with a few drops of cold water wherever a grain appears. The batch is cooled to 100°F, and the marshmallow whip and invertase are added. The machine is set in motion and allowed to run until the batch creams. The cream is then placed in a clean metal container and kept covered until it is ready for use in the making of cast cream centers for assorted chocolates.

This batch should yield about 72 pounds of cream.

RUSSIAN APPLE PASTE: *See* APPLE PASTE, RUSSIAN-TYPE.

RUSSIAN FRUIT CARAMELS: *See* FRUIT CARAMELS, RUSSIAN-TYPE.

RUSSIAN FRUIT MARMALADE: *See* FRUIT MARMALADE, RUSSIAN-TYPE.

RUSSIAN TOFFEE

Sugar	6 lb
Corn Syrup	4 lb
Fresh 18% Cream	3 qt
Salt	1 teaspoonful
Butter	1 lb
Coconut Butter	8 oz
Chopped Bitter Chocolate	18 oz
Vanilla Flavor	2 teaspoonfuls

The sugar, corn syrup, coconut butter, salt, and cream are placed in a copper pan on the stove and stirred until dissolved. The sides are washed down and stirring is continued as the butter and chocolate are added. The heat under the pan is reduced and the batch is stirred continuously and cooked until it forms a hard ball when tested in cold water (*see* COLD-WATER TEST). The batch is then removed from the stove and the vanilla flavor is stirred in. It is then poured out on a greased slab lined with iron bars and spread out to approximately ½-inch thickness. After it has cooled, it is sliced and cut into small oblongs. The toffee is then wrapped in moistureproof cellophane.

This formula should yield approximately 15 pounds of toffee.

S

SALTED ALMONDS NO. 1

5 Crown Valencia Almonds	10 lb
Soft Coconut Butter	15 lb
Salt	2 tablespoonfuls

The almonds are blanched rapidly, following the process that was used for SALTED PEANUTS *(see)*. The same process is followed also for cooking the almonds in the coconut oil. The almonds are roasted to a medium brown (not too dark). They are then drained, spread out on the marble lined with paper, and salted. They are then cooled and packed as usual.

The batch will yield approximately 10 pounds of salted almonds.

SALTED ALMONDS NO. 2

5 Crown Valencia Almonds	10 lb
GUM ARABIC SOLUTION *(see)*	2 tablespoonfuls
Salt	12 lb

The almonds are blanched and dried in the usual manner. The salt is placed in a large copper pan on the stove over low heat and stirred until it is hot. The almonds are placed in the salt pan and stirred constantly until they have a nice brown color. The almonds with the salt are then turned out into a brass wire sieve. The salt is brushed off, and the almonds are placed in the large copper pan, which now rests on a tripod. The gum arabic solution is sprinkled on top of the almonds and mixed through well. The almonds are then salted, allowed to cool, and packed in the usual manner.

Note: In contrast to the preceding method, this process is suitable for those who like a dry natural roast without grease. The salt may be used again for the same purpose.

The batch will yield approximately 10 pounds of salted almonds.

SALTED BRAZIL NUTS

Large Blanched Brazil Nuts	10 lb
GUM ARABIC SOLUTION (*see*)	2 tablespoonfuls
Salt	2 tablespoonfuls

The blanched Brazil nuts are covered with the gum arabic solution. They are then salted and toasted dry, using the method that was used for the SALTED WALNUTS (*see*). Great care must be taken to avoid overtoasting.

The batch will yield approximately 10 pounds of salted Brazil nuts.

SALTED CASHEW NUTS

Bombay All-Cashew Nuts	10 lb
Soft Coconut Butter	15 lb
Salt	2 tablespoonfuls

The coconut butter is heated in the cooking pan over low heat. The nuts are placed in the wire basket and the basket is placed in the cooking pan, which contains the hot oil. The cashews are cooked in the hot oil over very low heat until they are light brown. They are then drained, placed on a paper-lined marble slab, and salted. The nuts are then cooled and packed in the usual manner.

The batch will yield aproximately 10 pounds of salted cashews.

SALTED FILBERTS

Large Blanched Levant Filberts	10 lb
Soft Coconut Butter	15 lb
Salt	2 tablespoonfuls

The filberts must be blanched by a special blanching machine. Since they cannot be blanched manually, they are purchased already blanched from a supply house. The filberts are placed in a wire basket, which is placed into the hot oil under low heat. They are cooked to a light brown color. They are then drained, placed on a paper-lined marble, and salted. When cool, the filberts are packed as usual.

The batch will yield approximately 10 pounds of salted filberts.

SALTED NUTS

It does not take much experience and practice to make salted nuts, but is does take careful accurate work to develop a repeat business.

Three kinds of oil are used for roasting nuts. They are soft coconut butter, peanut oil, and pure olive oil. From a practical standpoint soft coconut butter is the best of these. It is a fairly good product and more effective than the other oils in preventing the nuts from becoming rancid and soggy. About 75 percent of all salted nuts is roasted in coconut oil. Peanut oil rates next to coconut oil for durability. However, it can only be used to roast peanuts because it does not blend well with other nuts. Pure olive oil is a luxury ingredient used mostly by quality retail firms. Its taste is excellent, but its shelf life is short. It turns rancid in ten days to two weeks and, therefore, can only be used in small quantities.

The loss of oil due to absorption is equivalent to about 4 percent of the weight of the cooked nuts. If the confectioner cooks 40 pounds of nuts, 1.6 pounds of oil will be lost through absorption. In order to allow for this loss of oil, the amount of nuts placed in the basket should be progressively decreased after each roast. Only if this is done will there be enough oil to cover the nuts. Each time the oil is heated it also loses part of its body. If it is heated above the cooking point, it decomposes and is unfit for roasting nuts. To avoid this, the oil should be heated over a small flame with the lowest heat.

Although it is more durable than the other oils, the life of coconut oil is nevertheless short. After it is used to roast nuts eight to ten times, the soft coconut butter should be discarded and the roasting pan well-cleaned and refilled with fresh oil. Peanut oil should also be discarded after it has been used eight to ten times. Olive oil should be discarded after it has been used about six times. Several suggestions have been made for prolonging the life of cooking oil. They involve filtering, purifying, and other processes. These are a false economy and will eventually hurt business. *To repeat:* Coconut oil should be used eight to ten times and then discarded. Olive oil should be used five to six times and then discarded.

Nuts for roasting come in white and dark colors. The white nuts are cashews, blanched peanuts, and blanched almonds. The dark nuts are unblanched peanuts and almonds. Although pecans do not come blanched, they are classed as blanched nuts because they do not blend well with oil that has been used for more than four roasts. It is important to roast blanched nuts in oil that has not been used more than four times because after that, the oil darkens and will give the blanched nuts a cloudy appearance. The confectioner should, therefore, roast blanched nuts first before the oil becomes discolored. He can then roast unblanched nuts in the discolored oil until it is ready to be discarded.

Three kinds of nuts do not blend well with oil and are roasted and salted without it. They are walnuts, blanched Brazil nuts, and pistachio nuts. These nuts are usually dampened with gum arabic solution so that the salt will adhere to them. They are toasted in a hot oven. The confectioner should guard against overtoasting, which will give the nuts a burned taste.

Medium-grained salt packed in bags is used for salting. This salt is better than the boxed salt, which has a fine grain and, therefore, does not stick to the surface of the nuts.

Salted nuts do not have a long shelf life and taste well only when fresh. When they grow stale they become soggy and inedible. Only small quantities should be carried in stock. Stale soggy nuts cannot be recooked.

For roasting the nuts in oil, a black, iron cooking pan is used which contains a wire basket. The roasting nuts are placed in the wire-screen basket, and the oil is heated in the cooking pan. The basket prevents the nuts from sticking to the bottom of the pan and makes for more uniform roasting. When the oil is heated, the basket containing the nuts is placed in the pan. The cooking oil penetrates the wire screen and cooks and roasts the nuts. When the cooking is done, the basket is picked up by the handle and hung up to drain on a hook attached to the cooking pan. The excessive oil drips back into the cooking pan and the nuts are removed for salting.

Most of the smaller and medium firms use this type of equipment. The larger producers use cooking pans and baskets which are ten to twelve times larger and square shaped instead of round. The square cooking pan is attached to the floor, the gas for heating is underneath the cooking pan, and the inside wire basket is moved mechanically—first for draining and then to the salting table.

Supply houses offer to the trade peanuts and almonds already blanched. These nuts, which are blanched mechanically, are offered at relatively low cost as a labor-saving proposition. However, if the peanuts and almonds are not roasted and salted shortly after they are blanched, they will absorb water and become stale. Roasting will not restore their freshness. Therefore, it is advisable to blanch these nuts on the premises so that they can be roasted and salted on the same day.

The confectioner who produces salted peanuts and almonds in large quantities will find the nut-blanching machine a good investment. This machine has two rubber rollers. After the peanuts and almonds are heated in boiling water, the water is drained off and the nuts are deposited in the hopper of the blanching machine. They are then passed through the rubber rollers, which remove their skins, and are released at the bottom of the machine, where they are spread out on clean boards and allowed to dry. The nuts are then ready for roasting and salting.

Since filberts and Brazil nuts are hard to blanch, it is advisable to buy them already blanched.

For the various salted nut formulas, *see* the *Individual Names,* such as SALTED PEANUTS, SALTED FILBERTS, etc.

SALTED PEANUTS

All-Jumbo Shelled Peanuts	10 lb
Water	4 qt
Soft Coconut Butter	15 lb
Salt	2 tablespoonfuls

The water is placed in a copper pan on the stove. The peanuts are placed in the hot water, which is allowed to boil. The pan is moved near a clean marble and a small quantity is taken out of the hot water with a skimmer and placed on the marble slab for blanching. The skins are removed and the blanched peanuts are set aside on the marble. This process is continued until all of the peanuts are blanched. They are then spread out on clean boards, which are placed in a warm spot. The peanuts are allowed to dry for about 1 hour and are then ready for roasting and salting.

For roasting, a cooking pan and wire basket, similar to that used for frying oysters, is used. The coconut butter is placed in the cooking pan over low heat and is heated until it melts and starts to simmer. The blanched peanuts are placed in the wire basket and the basket is placed in the cooking pan, which contains the hot oil. The peanuts are allowed to cook over low heat. While they are cooking, they are stirred with a small paddle so that they roast uniformly. They are cooked until they are medium brown. The heat is then turned off and the basket containing the peanuts is removed from the oil. The peanuts are allowed to drain for 2 minutes and are spread out on the marble, which has been lined with brown packing paper. The salt is sprinkled on top of the peanuts and mixed thoroughly. After they have cooled on the marble, the peanuts are packed in tins or jars.

The batch will yield approximately 10 pounds of salted peanuts.

SALTED PECANS

Jumbo Pecans	5 lb
Pure Olive Oil	10 lb
Salt	1½ tablespoonfuls

The oil is heated in the cooking pan almost to the boiling point. The pecans are placed in the wire basket and the wire basket is placed in the cooking pan, which contains the hot oil. The pecans are allowed to boil for 4 or 5 minutes and are then removed immediately. They are then drained, placed on a paper-lined marble slab, and salted. When they are cool, they are packed in the usual manner.

The batch will yield approximately 5 pounds of salted pecans.

SALTED PIGNOLIA NUTS

Italian Pignolia Nuts	5 lb
Soft Coconut Butter	10 lb
Salt	1½ tablespoonfuls

The nuts are very small and a basket with extremely small holes is needed so that the nuts do not fall through. The oil is heated almost to the boiling point and the nuts are placed in the basket, which is placed in the hot oil. The pignolias are cooked until they are a light brown color. They are then drained and spread out on a paper-lined marble. After they have been salted and cooled, they are packed as usual.

The batch will yield approximately 5 pounds of salted pignolia nuts.

SALTED PISTACHIO NUTS

Fancy Shelled Pistachio Nuts No. 1	5 lb
GUM ARABIC SOLUTION (*see*)	1½ tablespoonfuls
Salt	1½ tablespoonfuls

The pistachio nuts are placed in a small, clean copper pan and the gum arabic solution is sprinkled on top of them and mixed through. They are then salted and placed in a large flat pan. The pan is placed in a moderately hot oven and left there until the nuts are well dried and toasted. They are then packed as usual.

The batch will yield approximately 5 pounds of salted pistachio nuts.

SALTED WALNUTS

Chabert Walnut Halves	5 lb
Gum Arabic Solution	1½ tablespoonfuls
Salt	1½ tablespoonfuls

The walnuts are placed in a clean copper pan and the gum arabic solution is sprinkled on top of them and mixed thoroughly. They are then salted and placed in a large, flat pan. The pan is placed in the hot room or in a moderately hot oven and left there until the walnuts are well dried and toasted. They are then packed as usual.

The batch will yield approximately 5 pounds of salted walnuts.

SALT-WATER TAFFY, ANISE

The SALT-WATER TAFFY, VANILLA (*see*) formula is followed, but a black stripe is prepared and the batch is flavored with oil of anise. The stripe is fixed in the same manner as for the mint batch. It is finished as usual.

This batch should yield approximately 25 pounds.

SALT-WATER TAFFY, CHOCOLATE NO. 1

The SALT-WATER TAFFY, VANIILA (*see*) formula is followed, but 1½ pounds of finely cut bitter chocolate are stirred into the batch before it is poured. The marshmallow whip is increased to 2 pounds.

This batch should yield approximately 25 pounds of taffy.

SALT-WATER TAFFY NO. 2: *See* SAVING WASTE.

SALT-WATER TAFFY, GRAPE

The formula for SALT-WATER TAFFY, VANILLA (*see*) is followed, but the batch is colored violet and flavored grape. It is finished in the usual manner.

The batch should yield approximately 25 pounds.

SALT-WATER TAFFY, LEMON

The formula for SALT-WATER TAFFY, VANILLA (*see*) is followed, but the batch is colored yellow and flavored with oil of lemon.

It should yield approximately 25 pounds of taffy.

SALT-WATER TAFFY, LIME

The formula for SALT-WATER TAFFY, VANILLA (*see*) is used, but the batch is colored green and flavored with oil of lime. It is finished in the usual manner.

This batch should yield approximately 25 pounds.

SALT-WATER TAFFY, MINT

The formula for SALT-WATER TAFFY, VANILLA (*see*) is followed, but a red stripe is prepared and the batch is flavored with oil of peppermint. It is decorated all around with thin, red stripes.

This batch should yield approximately 25 pounds.

SALT-WATER TAFFY, ORANGE

The formula for SALT-WATER TAFFY, VANILLA (*see*) is followed, but the batch is colored orange and flavored with oil of orange. It is finished in the usual manner.

This batch should yield approximately 25 pounds.

SALT-WATER TAFFY, STRAWBERRY

The formula for SALT-WATER TAFFY, VANILLA (*see*) is followed, but the batch is colored pink and flavored with strawberry. It is finished in the usual manner.

The batch should yield approximately 25 pounds.

SALT-WATER TAFFY, VANILLA

Sugar	8 lb
Corn Syrup	12 lb
Water	1 qt
Salt	1 teaspoonful
Coconut Butter	8 oz
Evaporated Milk	1½ lb
Vanilla Flavor	3 teaspoonfuls
MARSHMALLOW WHIP (*see*)	1½ lb

The sugar, corn syrup, coconut butter, water, and salt are placed in a copper pan on the stove and stirred until dissolved. When the batch begins to boil, it is covered and allowed to steam for a few minutes. The sides are washed down and a thermometer is placed in the batch, which is cooked to 258°F. The heat is shut off and the batch is reduced with the evaporated milk to 255°F. The marshmallow whip is then stirred in and the batch is poured out on a cold, greased slab between iron bars. The edges are folded together while the flavor is stirred in. The batch is cooled rapidly until it is hard enough, when it is taken over to the pulling machine and pulled for about 6 minutes. It is then transferred to the spinning table, where it is formed round and put through a wrapping machine.

This formula should yield approximately 25 pounds of candy.

SALT-WATER TAFFY, WINTERGREEN

The formula for SALT-WATER TAFFY, VANILLA (*see*) is followed, but before the batch is cooled on the slab, a small portion is colored red and saved for a stripe. The main batch is flavored with oil of wintergreen. Before the batch is shaped, it is striped with two wide red stripes, one on each side. It is finished in the usual manner.

This batch should yield approximately 25 pounds of taffy.

SAUCES: *See* FOUNTAIN SYRUPS and *Individual Sauces,* e.g., VANILLA CARAMEL SAUCE, VANILLA FUDGE SAUCE.

SAVING WASTE

Many candy factories fail because they tolerate excessive scrap and waste. The departments in some of these factories may operate at full speed and even work overtime. Every corner in the factory may be busy. The superintendent or foreman rushes to get the orders out. But eventually the rush stops. Christmas or Easter arrives. After taking stock, the owner is disappointed. He cannot locate the leak; yet the leak is everywhere apparent. Every empty barrel in the place is filled with scrap. Some of the leak is in the sink, where the spoiled, melted batches were poured. The confectioner's profit literally went down the drain.

Due to carelessness and mismanagement, tons of good sugar and other costly ingredients are wasted in candy factories every year. The author has worked in plants where men were hired as spinners on filled goods and spoiled batch after batch day after day for months. Because they were friendly with the foreman they got away with it. The author saw one superintendent send several barrelfuls of scrapped but still useful chocolate straws down to the engine room to burn as fuel because he did not know what to do with them.

Saving waste means saving profits. Or, to put it more succinctly, saving the very life of the plant. Wasteful and extravagant production causes more grief than any other single problem. Every pound of sugar, corn syrup, nuts, and other essential material can be turned into candy instead of going into the refuse barrels. The suggestions and formulas that follow are meant to show the confectioner that he can cut scrap to a minimum and still make quality candy.

To begin: Machinery, cooking kettles, tools, utensils, pans, slabs, and marbles should be arranged systematically. Every item should be spotlessly clean and in its proper place. Every department head should know his job thoroughly. He should be able to make the most of the personnel and equipment he supervises. Logical, systematic arrangement of personnel and equipment and intelligent supervision will help keep scrap down.

For example, most candy factories do not have air-conditioned hard-candy departments. In these factories, the hard-candy department should be situated on a middle floor (neither the ground nor the top floor) because a middle floor is not likely to be damp and this location helps keep the finished hard candy fairly dry before it is packed in cans. The spinning table should always be situated away from drafts, because although the batch is always kept in front of a batch warmer or on a batch roller, excessive draft will cool the surface of the batch too rapidly and sometimes cause scrap.

The chocolate department is best situated on the ground floor of the plant or, if this floor is used for shipping and office space, on the second floor (i.e., it should always be on the lowest production floor). This is suggested because the lowest floor is usually the coolest and, even with air conditioning, the chocolate department should be on the floor that is naturally coolest. Also almost every department produces some centers for chocolate coating and it is more convenient to send these centers down for finishing. After the chocolate department gives all chocolate goods the finishing touch, they go to the shipping and stock room. Placing the chocolate department on the lowest production floor directly over the shipping and stock room saves time and labor.

In most plants, the soft candy departments (e.g., caramels, nougats, jellies, pastes, cream centers) are situated close together and the crystal department is included in the setup. The production of most of these candies involves the same or similar machinery. The same depositing machine and starch buck are used for all starch-cast goods. They may also share the same hot room. In addition, many soft goods are crystallized and, therefore, share the crystal room.

Sometimes employees need to be arranged as carefully as machines. If they do not work harmoniously, production is affected. This is particularly true in the chocolate department. For example, it takes a group of ten to twelve girls to run the chocolate enrober. All work closely together. Without complete harmony, the enrober will not operate properly. If the girl who regulates the flow of chocolate from the tank does not work properly, the chocolates will either be too thin or too heavy. If the stringers do not use all their skill in forming the strings, the appearance of the chocolates will suffer. It is up to the forelady to arrange the group with an eye to individual temperament as well as ability. Where chocolates are dipped by hand, it is even more important that the forelady knows the personalities of her chocolate dippers so that she can distribute assignments diplomatically and the group can work with a minimum of friction.

Every kind of scrap should be kept separately in clean, well-covered cans. The cans should be properly labelled and placed in a cool convenient spot so that the scrap can be promptly located and used at the first opportunity. The different kinds of scrap can be arranged as follows: hard candy scrap, creams, gums and jellies, chewing candy scrap, chocolate scrap, mint scrap, fruit and nut scrap, etc. In order to keep sanitation under control, the confectioner should guard against excessive accumulations of scrap. The cans should never overflow.

The various departments within the plant should cooperate in making the best possible use of scrap. If scrap from the chocolate department can be used to make centers for filled goods, the hard-candy man should welcome it instead of shrugging his shoulders. "I did not make this scrap" is a wasteful attitude. Some hard-candy scrap can be used in making molasses candy. Chocolate scrap can be used to make chocolate caramels or fudge. Mint scrap can be used in making all sorts of mint candies. Scrap which contains fruits and nuts should be strained after it is remelted. The remelted scrap can best be used in making fudge for chocolate coating. When chocolate-covered fruits and nuts are scrapped, the fruits and nuts are soggy, but when the scrap is remelted and then strained, the chocolate taste comes back. Because the chocolate taste is restored, the liquid is used, but the soggy fruits and nuts are discarded. After it has been used several times for crystallizing, crystal syrup can be used again in making nougat or cream fondant. Whenever scrap can be used without cheapening the quality of the product, its use is justified.

It is well known that hard-candy production creates more scrap than the manufacturer of any other candy. Most solid hard candy is made with stripes in all colors. The finished candy is acceptable only if it is striped. However, in producing this candy, the beginning and the end of each batch do not come perfectly striped and must be scrapped. In making filled goods, each piece must be filled with the center. Again the beginning and end of the filled-goods batch do not have filling. Though experts can minimize the amount of scrap, it cannot be completely eliminated.

If making hard candy entails a large amount of scrap, it is also true that considerable scrap can go into it. For example, in making filled goods, where a 16-pound center is desired and the formula calls for 5 pounds of sugar, 5 pounds of corn syrup, 1 pint of water, and 6 pounds of nuts, the confectioner can use 6 pounds of hard candy scrap, 4 pounds of corn syrup, 1 quart of water, and 6 pounds of nuts and get the same results.

Some hard-candy scrap cannot be recooked. When fruit drops are made and citric or tartaric acid is used as flavor, the scrap that results from the sieving is usually kneaded into the next batch and can only be used in this way. However, the scrap that remains in the shop overnight cannot be used again, for if it is kneaded into the batch the following morning, the batch will grain.

Hard candy scrap that contains citric or tartaric acid cannot be recooked and must be discarded, unless it is kneaded into another batch the same day. All other scrap that is kept clean and uncontaminated has some potential use.

If the scrap is kept in clean, well-covered containers in a cool, dry room, as suggested, it will keep for a reasonable time. However, if it is carelessly dumped into empty barrels or drums, poorly covered, and kept in warm rooms, the scrap will become contaminated, wormy, and unusable during the summer months.

Several plants have organized scrapless contests with some success. The foreman who runs his department with the least amount of scrap for the season is awarded first prize. Second and third prizes are also awarded to department heads. The foreman explains the contest to his candymakers and popularizes it. The candy maker who works with the least amount of scrap could be placed first on the promotion list or other incentives (cash prizes) could be offered.

An intelligent hiring policy helps create a scrapfree plant. Only competent candy makers should be hired and the right man should be placed in the right job. Expert cream men sometimes make poor hard-candy men. A man may be good at making caramels, but will not do on nougats. That man is a good cooker but a poor spinner. A small retail manufacturer who employs only one candy maker should select an expert all-around man capable of making all kinds of candy. If every man understands his job thoroughly and gets intelligent supervision, the biggest leak is plugged. The element of human fallibility is reduced to a minimum.

Scrap is listed as an ingredient in the following 13 formulas. Although they contain scrap, the quality of these products will not be affected if the formulas are followed. These formulas are by no means the only outlet for scrap, but will give the alert confectioner some idea of its potential uses. They are arranged alphabetically.

BURNT SUGAR

Clean Hard-Candy Scrap	25 lb
Water	5 qt

The scrap and water are placed in a large clean copper pan on the stove and stirred continuously until all the scrap is dissolved. When the batch begins to boil, a thermometer is placed into it and the batch is cooked to 344°F. The batch is then removed to a stand near open windows, and 4 quarts of water are poured into it in a slow, steady stream. While adding the water the batch should be stirred constantly. It is then cooled rapidly and the burnt sugar is poured in a clean can and covered. It is then ready for use.

The batch will yield approximately 35 pounds of burnt sugar.

CHOCOLATE COATING CARAMELS

Clean Chocolate Scrap	25 lb
Corn Syrup	15 lb
Fresh Light 18% Cream	4 qt

Evaporated Milk	12 lb
Salt	1 tablespoonful
Vanilla Flavor	4 teaspoonfuls

The chocolate scrap and the cream are placed in a copper pan on the stove over low heat and stirred until all the chocolate scrap is dissolved. The scrap is then strained into another large copper pan, which is placed on the stove over medium heat. The batch is stirred continuously while the corn syrup and the evaporated milk are added. The batch is cooked until it forms a medium ball when tested in cold water (*see* COLD-WATER TEST). It is then removed from the stove and the salt and flavor are stirred in. The batch is poured out on a greased slab between iron bars and spread out evenly to approximately ¾-inch thickness. It is allowed to cool overnight. The caramel is then sliced and cut into small squares, which are coated with milk chocolate.

The batch will yield approximately 70 pounds of chocolate-covered caramels.

CHOCOLATE FUDGE FOR COATING

Clean Chocolate Scrap	25 lb
Corn Syrup	5 lb
Fresh 18% Light Cream	4 qt
Evaporated Milk	10 lb
Salt	1 tablespoonful
RUNNING CREAM (*see*)	10 lb
Vanilla Flavor	4 teaspoonfuls

The chocolate scrap and the cream are placed in a copper pan on the stove over low heat and stirred until all the chocolate scrap is melted. The melted chocolate scrap is then strained into another large copper pan, which is placed on the stove over medium heat. The corn syrup is stirred in rapidly and the batch is stirred continuously while the evaporated milk is added. Stirring is continued, and the batch is cooked until it forms a soft ball when tested in cold water (*see* COLD-WATER TEST). It is then moved to a stand near the slab, and the salt, flavor, and running cream are stirred in slowly. When the cream is entirely melted, the batch is poured out on a greased slab lined with iron bars and spread out evenly to approximately ½-inch thickness. It is allowed to set and is then marked into squares with a roller knife. When cool enough, the squares are broken apart and the fudge is coated with milk chocolate.

The batch will yield approximately 70 pounds of chocolate fudge.

CHOCOLATE SALT-WATER TAFFY NO. 2

Clean Hard-Candy Scrap	25 lb
Corn Syrup	15 lb
Coconut Butter	2 lb
Water	4 qt
Salt	1 tablespoonful
Chopped Bitter Chocolate	3 lb
Vanilla Flavor	2 teaspoonfuls
MARSHMALLOW WHIP (*see*)	2 lb

The scrap, corn syrup, coconut butter, water, and salt are placed in a large copper pan on the stove and stirred until all the scrap is melted. When the batch begins to boil, it is covered and allowed to steam for a few minutes. The sides are washed down, and a thermometer is placed in the batch, which is cooked to 250°F. The batch is then moved to a stand and the chocolate and flavor are stirred in. When the chocolate is melted, the marshmallow whip is stirred in. The batch is then poured out on a cooled, greased slab and the edges are folded together. The batch is cooled rapidly and placed on the pulling machine. It is pulled for about 5 minutes and is then transferred to the wrapping machine. It is then cut, wrapped, and packed in boxes.

The batch will yield approximately 50 pounds of salt-water taffy.

CHOP-SUEY JELLIES

The formula for LICORICE JELLIES (*see*) is followed with a few changes, but after the jelly scrap is stirred into the larger cooked jelly batch, this batch is colored a bright pink. It is then flavored with 6 teaspoonfuls of sherry flavor. It is finished as the licorice jellies. The pink color is substituted for the licorice paste and the sherry flavor is substituted for oil of anise.

The batch will yield approximately 75 pounds of jellies.

FRUIT NOUGAT

Crystal-Syrup Scrap (from the crystallizing department)	25 lb
Sugar	10 lb
Corn Syrup	15 lb
Water	2 qt
Coconut Butter	2½ lb
Liquid Red Color	2 teaspoonfuls
Jamaica-Rum Flavor	6 teaspoonfuls
Clean Gumdrop and Jelly Scrap	15 lb
MARSHMALLOW WHIP (*see*)	10 lb

The crystal-syrup scrap, sugar, water, corn syrup, and coconut butter are placed in a large copper pan on the stove and stirred until dissolved. When the batch begins to boil, it is covered and allowed to steam for a few minutes. The sides are washed down, and a thermometer is placed in the batch, which is cooked to 260°F. The batch is then removed to a stand near the slab and the marshmallow whip is added. Then the color and flavor are added. After cooling for about 10 minutes, the batch is stirred, beaten, and thoroughly mixed. The gum and jelly scrap is then stirred in. The batch is poured out on a greased slab lined with iron bars and allowed to set overnight. It is then sliced, cut into small squares or oblongs, and wrapped in moistureproof cellophane. The nougat may also be coated in a dark sweet chocolate.

The batch will yield approximately 120 pounds of chocolate-covered fruit nougat.

HOREHOUND DROPS NO. 2

Clean Hard-Candy Scrap	15 lb
Corn Syrup	7 lb
Sugar	8 lb
Water	4 qt
Burnt Sugar	1 oz
Extract of Horehound	1½ oz

The sugar, corn syrup, water, and scrap are placed in a large, clean copper pan on the stove and stirred until completely dissolved. When the batch begins to boil, it is covered and allowed to steam for a few minutes. The sides are washed down and a thermometer is placed in the batch, which is cooked to 315°F.

The batch is then poured out on a cooled greased slab lined with iron bars and colored with burnt sugar until it takes on a dark color. The edges are folded together while stirring in the extract of horehound. Folding is continued until the batch is cool enough. It is then transferred to the spinning table in front of the batch warmer and passed through a drop-roller machine. The drops are allowed to cool on the table. They are then broken apart and mixed with a half-and-half mixture of magnesia powder and powdered sugar. The drops are then put through a sieve and packed in tins.

This batch will yield approximately 30 pounds of candy.

LICORICE CARAMELS

Clean Cream and Chewing Candy Scrap	18 lb
Corn Syrup	12 lb
Fresh 18% Light Cream	4 qt
Evaporated Milk	15 lb
Salt	1 tablespoonful
Coconut Butter	3 lb
Licorice Paste	2 oz
Oil of Anise	4 teaspoonfuls

The scrap, corn syrup, coconut butter, salt, and fresh cream are placed in a large copper pan on the stove over medium heat. The batch is stirred continuously until all the scrap is dissolved. When the batch begins to boil, the evaporated milk is stirred in gradually. Stirring is continued and the batch is cooked until it forms a medium ball when tested in cold water (*see* COLD-WATER TEST). The licorice paste is stirred in and the batch is removed from the stove. The oil of anise is then stirred in and mixed thoroughly. The batch is poured on a greased, cooled slab between iron bars and spread out to 1-inch thickness. It is allowed to cool overnight. The following day, the batch is sized and cut into caramels, which are wrapped as usual.

The batch will yield approximately 50 pounds of candy.

LICORICE JELLIES

Agar-Agar	8 oz
Water	12 qt
Sugar	20 lb
Corn Syrup	10 lb

MARSHMALLOW WHIP (*see*)	3 lb
Licorice Paste	1½ oz
Oil of Anise	4 teaspoonfuls
Clean Jelly Scrap	20 lb

The agar-agar and water are placed in a large copper pan on the stove and stirred until dissolved. When the batch begins to boil, the sugar is stirred in gradually. When the batch again reaches the boiling point, the corn syrup is stirred in rapidly and the batch is cooked to the jelly test (about 224°F). About 6 pounds of the batch are set aside in a small pan and the marshmallow whip is stirred into this small batch and mixed thoroughly. The licorice paste, oil of anise, and the jelly scrap are added to the large batch on the stove and mixed thoroughly. Two clean boards are then lined with wrapping paper. About one half of the white jelly (small batch) is spread out as a bottom layer on the two boards. This white layer is allowed to set. The licorice jelly (larger batch) is then divided and poured out on both boards, spread out evenly, and allowed to set. The remainder of the white jelly is then spread out evenly on top of the licorice jelly and allowed to set. Fine granulated sugar is sprinkled on top of the jelly, the boards are turned upside down on the table and the paper is removed. The jelly is again sprinkled on top with sugar. It is then sliced and cut into small squares, which are rolled in granulated sugar.

The cut jellies are spread out on clean boards. When dry, they are placed in crystal pans, which are only half filled and they are crystallized in an overnight crystal (*see* CRYSTALLIZING). The following morning, they are drained and allowed to dry. They are then packed in boxes as usual. The batch will yield approximately 75 pounds of jellies.

LICORICE SALT-WATER TAFFY

Clean Hard-Candy and Cream Scrap	25 lb
Corn Syrup	15 lb
Coconut Butter	2 lb
Water	4 qt
Salt	1 tablespoonful
Licorice Paste	2 oz
Oil of Anise	4 teaspoonfuls
MARSHMALLOW WHIP (*see*)	2 lb

The scrap, corn syrup, coconut butter, water, and salt are placed in a large copper pan on the stove and stirred until all the scrap is dissolved. When the batch begins to boil, it is covered up and allowed to steam for a few minutes. The sides are washed down, and a thermometer is placed in the batch, which is heated to 252°F. The licorice paste is stirred in. The batch is removed from the stove and the marshmallow whip and oil of anise flavor are stirred in. It is then poured out on a greased, cooled slab and the edges are folded together. The batch is cooled rapidly and a small piece (about 1 pound) is saved for a stripe. The main batch is pulled for 5 minutes and is then returned to the slab. This batch is formed round and striped with the clear candy. Fine stripes are run throughout the batch. It is then transferred to the wrapping machine and cut and wrapped as usual.

The batch will yield approximately 50 pounds of salt-water taffy.

MOLASSES PEPPERMINT SALT-WATER TAFFY

Clean Hard-Candy and Cream Scrap	25 lb
Corn Syrup	20 lb
Coconut Butter	2 lb
Water	4 qt
Salt	1 tablespoonful
Molasses	4 qt
MARSHMALLOW WHIP (*see*)	2½ lb
Oil of Peppermint	5 teaspoonfuls

The scrap, corn syrup, molasses, coconut butter, water, and salt are placed in a large copper pan on the stove and stirred continuously until all the scrap is dissolved. When the batch begins to boil, it is covered and allowed to steam for a few minutes. The sides are washed down and a thermometer is placed in the batch, which is heated to 252°F. The batch is removed from the stove and the marshmallow whip and flavor are stirred in. The batch is then poured out on a greased slab lined with iron bars and the edges are folded together. The batch is cooled rapidly and a small piece of the clear batch is saved for a stripe. When the batch is cool enough to handle, it is transferred to the pulling machine and pulled for 5 minutes. The batch is then taken back to the slab and striped in fine stripes with the clear candy. The batch is then spun out, and cut into small kisses or oblong salt-water taffy. The taffies are then wrapped and packed in boxes.

The batch will yield approximately 55 pounds of salt-water taffy.

MOLASSES PLANTATIONS

Water	4 qt
Clean Hard-Candy and Cream Scrap	20 lb
Corn Syrup	20 lb
Coconut Butter	2½ lb
Molasses	4 qt
Salt	1 tablespoonful
Vanilla Flavor	6 teaspoonfuls
MARSHMALLOW WHIP (*see*)	8 lb

The water, scrap, corn syrup, coconut butter, molasses, and salt are placed in a large copper pan on the stove and stirred until all the scrap is melted. When the batch begins to boil, it is covered and allowed to steam for a few minutes. The sides are washed down and a thermometer is placed in the batch, which is heated to 250°F. The batch is then moved to a stand and the flavor and marshmallow whip are stirred in and mixed thoroughly. The batch is tested in cold water to see if it forms a soft ball (*see* COLD-WATER TEST). If the cook is too high, it is reduced carefully with evaporated milk. The batch is then poured out on a cooled, greased slab lined with iron bars and spread out evenly to approximately ¾-inch thickness. It is allowed to cool until hard. It is then sliced and cut into small oblongs, which are coated with a dark sweet or milk chocolate.

The batch will yield approximately 75 pounds of chocolate-covered plantations.

OLD-FASHIONED COCONUT STRIPS NO. 2

Powdered Sugar	5 lb
150 Grade Citrus Pectin	4 oz
Clean Hard-Candy and Cream Scrap	20 lb
Corn Syrup	20 lb
Molasses	4 qt

Salt	1 tablespoonful
Evaporated Milk	8 lb
Vanilla Flavor	4 teaspoonfuls
Macaroon Coconut	40 lb
Coconut Butter	2½ lb
Water	4 qt

The pectin is mixed dry with the powdered sugar and placed in a large copper pan. The molasses is added, and the pan is then placed on the stove and stirred until the ingredients are dissolved. The scrap, water, corn syrup, salt, milk, and coconut butter are added, and stirring is continued until all the scrap is dissolved. The batch is still stirred continuously and cooked until it forms a medium ball when tested in cold water (*see* COLD-WATER TEST). It is then removed from the stove and poured into an open mixing machine. The machine is started at low speed and the flavor is added. The coconut is stirred in gradually. Sufficient coconut is used to form a stiff paste and absorb all the liquid. The batch is then turned out on a greased slab lined with high iron bars, spread out evenly, and rolled out to approximately 2-inch thickness. It is allowed to cool overnight and is then cut in lengths of 4 inches. It is then sliced and cut into strips and rolled in fine granulated sugar. When the coconut strips are dry, they are packed in boxes.

The batch will yield approximately 100 pounds of coconut strips.

SCRAP: *See* SAVING WASTE.

SEA-FOAM FUDGE, CHERRY

The formula for SEA-FOAM FUDGE, VANILLA (*see*) is followed, but the batch is colored a deep red and flavored cherry. Following the fondant and marshmallow whip, 4 pounds of cherry pieces are stirred into the batch, which is then poured out on boards lined with wax paper and sprinkled on top with more cherry pieces. It is finished in the usual manner and should yield 65 pounds of candy.

SEA-FOAM FUDGE, CHOCOLATE

White Sugar	16 lb
Brown Sugar	4 lb
Corn Syrup	5 lb
Coconut Butter	4 oz
Fresh Light Cream	4 qt
Evaporated Milk	3 lb
Salt	1 tablespoonful
Chopped Bitter Chocolate	5 lb
Cream Fondant	12 lb
Vanilla Flavor	3 teaspoonfuls
MARSHMALLOW WHIP (*see*)	4 lb

Both sugars, corn syrup, coconut butter, cream, and salt are placed in a large copper pan on the stove and stirred until dissolved. When the batch begins to boil, the sides are washed down. The evaporated milk is stirred in, and stirring is continued. The batch is cooked until it forms a soft ball when tested in cold water (*see* COLD-WATER TEST). The chocolate is then stirred in rapidly and the batch is moved to a stand near the slab. The fondant is added and stirred until melted. Then the marshmallow whip and flavor are added and mixed through. The batch is poured out on boards lined with wax paper and allowed to set overnight. It is then marked in squares and cut as desired.

The batch should yield 50 pounds of candy.

SEA-FOAM FUDGE, CHOCOLATE NUT

The formula for SEA-FOAM FUDGE, CHOCOLATE is followed, but 4 pounds of pecan pieces are added while stirring in the fondant and the marshmallow whip. After mixing thoroughly, the batch is poured out on boards lined with wax paper and sprinkled with pecans on top. The batch is finished as usual and should yield 54 pounds of candy.

SEA-FOAM FUDGE, MAPLE WALNUT

White Sugar	15 lb
Maple Sugar	5 lb

Corn Syrup	5 lb
Coconut Butter	8 oz
Light Fresh Cream	4 qt
Evaporated Milk	4 lb
Salt	1 tablespoonful
Maple Flavor	4 teaspoonfuls
Cream Fondant	12 lb
MARSHMALLOW WHIP (*see*)	4 lb
Walnut Pieces	4 lb

Both sugars, corn syrup, coconut butter, fresh cream, and salt are placed in a large copper pan on the stove and stirred until dissolved. When the batch begins to boil, the sides are washed down. The evaporated milk is stirred in gradually. Stirring is continued and the batch is cooked until it forms a medium firm ball when tested in cold water (*see* COLD-WATER TEST). The batch is then moved to a stand near the slab and the fondant is stirred in rapidly. The flavor is added while stirring in the marshmallow whip. The walnuts are added and mixed well. The batch is then poured out on boards lined with wax paper and sprinkled on top with more walnuts. The fudge is allowed to set overnight and then marked in squares and cut as desired.

This batch should yield 50 pounds of candy.

SEA-FOAM FUDGE, PINEAPPLE

The formula for SEA-FOAM FUDGE, VANILLA (*see*) is followed, but the batch is colored a delicate straw yellow and flavored with pineapple flavor. While stirring in the fondant, 4 pounds of chopped, glacéed pineapple are added to the batch. The marshmallow whip is stirred in and mixed through. The batch is then poured out on boards lined with wax paper, sprinkled on top with chopped pineapple pieces, and finished in the same manner as the original.

The batch should yield 65 pounds of candy.

SEA-FOAM FUDGE, PISTACHIO

The formula for SEA-FOAM FUDGE, VANILLA (*see*) is followed, but the batch is colored green and flavored pistachio while the fondant is stirred in. While stirring in the marshmallow whip, 2 pounds of finely chopped pistachio nuts are added. The batch is then poured out on boards lined with heavy wax paper and more chopped pistachio nuts are sprinkled on top. The batch is finished in the usual manner, and should yield 61 pounds of candy.

SEA-FOAM FUDGE, VANILLA

Sugar	22 lb
Corn Syrup	8 lb
Coconut Butter	8 oz
Fresh 18% Light Cream	6 qt
Evaporated Milk	4 lb
Salt	1 tablespoonful
Vanilla Flavor	4 teaspoonfuls
Cream Fondant	12 lb
MARSHMALLOW WHIP (*see*)	5 lb

The sugar, corn syrup, coconut butter, fresh cream, and salt are placed in a large copper pan on the stove and stirred until dissolved. When the batch begins to boil, the sides are washed down, and stirring is continued while the evaporated milk is added gradually. The batch is cooked until it forms a medium firm ball when tested in cold water (*see* COLD-WATER TEST). It is then moved to a stand near the slab, the fondant is stirred in rapidly, and the marshmallow whip and flavor are added and stirred until completely mixed. The batch is then turned out on boards lined with heavy wax paper and allowed to set overnight. It is then marked in squares and cut as desired.

The batch should yield 55 pounds.

SEA-FOAM FUDGE, VANILLA NUT

The formula for SEA-FOAM FUDGE, VANILLA (*see*) is followed, but 4 pounds of black walnuts are added to the batch before pouring. When the batch is poured on the wax paper-lined boards, some black walnuts are sprinkled on the top and spread all over. The batch is finished in the usual manner and should yield 59 pounds.

SEA-FOAM KISSES, CHERRY

The formula for SEA-FOAM KISSES, COCONUT (*see*) is followed to the flavoring point. The batch is colored red and flavored with 2 teaspoonfuls of cherry flavor; 1 pound of chopped cherries is also added. The batch is finished as usual and should yield approximately 15 pounds of candy.

SEA-FOAM KISSES, CHOCOLATE NUT

The formula for SEA-FOAM KISSES, COCONUT (*see*) is followed, but the batch is flavored with 8 ounces of melted bitter chocolate and 12 ounces of Brazil nuts are added. The batch is finished as usual and should yield approximately 15 pounds of candy.

SEA-FOAM KISSES, COCONUT

Sugar	6½ lb
Corn Syrup	1½ lb
Water	1½ qt
Slow-Set Pectin	1 oz
Macaroon Coconut	18 oz
Vanilla Flavor	1 teaspoonful
Glycerin	½ teaspoonful
MARSHMALLOW WHIP (*see*)	½ lb

The pectin is mixed dry with half of the sugar, and it is placed on the stove in a copper pan containing half of the water. This mixture is stirred until dissolved and the balance of the sugar and then water and corn syrup are added. Stirring is continued until the batch boils. The sides are then washed down and the batch is cooked to 242°F. It is then removed from the heat and cooled for about 10 minutes. The marshmallow whip is then stirred in gradually, and the coconut and flavor are mixed in. The glycerin is added and mixed thoroughly. The batch is then spooned out into small kisses and deposited on dipping trays. these are allowed to cool until dry, when they are packed in boxes as usual.

This formula should yield approximately 14 pounds of candy.

SEA-FOAM KISSES, MAPLE WALNUT

The formula for SEA-FOAM KISSES, COCONUT (*see*) is followed all the way, but when the marshmallow is stirred in, the batch is flavored with 2 teaspoonfuls of maple flavor and 12 ounces of chopped walnuts are added. The batch is finished as usual and should yield approximately 15 pounds of candy.

SEA-FOAM KISSES, PINEAPPLE

The formula for SEA-FOAM KISSES, COCONUT (*see*) is followed to the flavoring point. The batch is then colored a light yellow and flavored with 2 teaspoonfuls of pineapple flavor; 1 pound of chopped, glacéed pineapple is also added. The batch is finished in the usual manner and should yield approximately 15 pounds of candy.

SEA-FOAM KISSES, VANILLA PECAN

The formula for SEA-FOAM KISSES, COCONUT (*see*) is followed all the way through with the following exception: 12 ounces of chopped pecans are mixed into the batch instead of the coconut. The batch is then finished in the usual manner and should yield approximately 15 pounds.

SEQUOIA ROLLIES

The formula for ICELAND ROLLIES (*see*) is followed, but batch No. 1 is colored orange and flavored with orange emulsion. The batch is then formed in long thin rolls as usual. Batch No. 2 is also made according to the Iceland Rollies formula. The small sauce batch is colored green and flavored pistachio. The rolls are dipped in the green sauce and rolled in chopped walnuts spread out on the marble. They are then finished in the usual manner and should yield approximately 38 pounds of rollies.

SESAME SQUARES

Sugar	8 lb
Corn Syrup	8 lb

Water	1 qt
Salt	1 tablespoonful
Butter	1 lb
Sesame Seeds	5 lb

The sugar, water, corn syrup, and salt are placed in a clean copper pan on the stove and stirred until dissolved. When the batch begins to boil, it is covered and allowed to steam for a few minutes. The sides are washed down and a thermometer is placed in the batch, which is cooked to 260°F. The thermometer is removed, the butter is stirred in, and some of the sesame seeds are stirred in and cooked gradually. This process is continued until all the seeds are in, roasted well, and the candy breaks crisp when tested in cold water (*see* COLD-WATER TEST). The batch is poured out on a greased slab, the edges are folded together, and the batch is kept soft in center of the slab. One half of the batch is rolled out thin, cut with an adjustable cutter in 2 inch squares, and moved to the edge of the slab. The same process is continued with the other half of the batch. When cool enough, the candy is packed in square tins lined with waxed paper between layers.

This batch will yield approximately 20 pounds of candy.

SHERBETS: *See* FROZEN CONFECTIONS and *Individual Sherbet Formulas,* e.g., LEMON SHERBET, PINEAPPLE SHERBET.

SHORT-DAY CRYSTAL: *See* CRYSTALLIZING.

SILVER CHOCOLATE STRINGS

Center

Milk Chocolate	4 lb
Vanillin	1 g
6X Powdered Sieved Sugar	2½ lb
Ground Roasted Almonds	1 lb

The chocolate is melted in a clean steam kettle or in a double boiler. The powdered sugar, vanillin, and the ground almonds are all mixed in to make a stiff paste. The center is kept warm until ready to use.

Jacket

Sugar	10 lb
Water	1½ qt
Cream of Tartar	1 teaspoonful
Vanillin	1 g

The sugar, water, and cream of tartar are placed in a copper pan on the stove and stirred until dissolved. When the batch begins to boil, it is covered and allowed to steam for a few minutes. The sides are washed down and a thermometer is placed in the batch, which is cooked to 330°F. The batch is poured out on a greased slab. The edges are folded together while stirring in the vanillin. Folding and kneading are continued until the batch is cool enough. Then it is partly pulled over the hook and twisted to a glossy silver color. Then the batch is taken over to the spinning table in front of the batch warmer, where it is spread out and formed flat. The center is placed in it, the jacket is folded over all around and the ends are all closed. It is then spun and put through a cutting machine, setting the size at about 2 inches. The candy is then allowed to cool on the table and packed in tins.

The batch will yield approximately 16 pounds.

SILVER DRAGÉES

Center

Sugar	25 lb
Corn Syrup	5 lb
Water	3 qt
Vanilla Flavor	3 teaspoonfuls

The sugar, corn syrup, and water are placed in a copper pan on the stove and stirred until dissolved. When the batch begins to boil, it is covered and allowed to steam for a few minutes. The sides are washed down and a thermometer is placed in the batch, which is cooked to 300°F. The batch is then poured out on a greased, cold slab lined with iron bars. The edges are folded together while the flavor is stirred in. Folding is continued until the batch is cool enough. It is then taken over to the pulling machine, pulled for 7 minutes, and then transferred to the spinning table in front of the batch warmer, where it is passed through a very small, round drop-roller machine (about 300 pieces to the pound). The small drops are allowed to cool on the table and are then broken apart and put through a sieve. They are then ready for the panning process.

Panning

For this operation 1½ pounds of powdered gum arabic are dissolved in 1½ quarts of water, heated slowly, and put through a fine sieve. Then 20 pounds of sugar, 3 quarts of water, and 1 teaspoonful of cream of tartar are placed in a pan on the stove, stirred, and heated to 240°F. One teaspoonful of oil of lemon, together with the gum solution, is stirred into the syrup, which is then divided into two portions. One part remains white and the other is colored a deep bright silver. The little drops are placed in the revolving pan, which is set in motion. The gummed white syrup is then used for coating the dragées. About 1 quart is applied slowly to the candies, stirred through by hand, and then covered with 4 pounds of powdered sugar. They are allowed to run in the pan until dry, when another charge of the syrup is applied. This process is continued until all the white syrup is used up and the coated candies run dry. Then the same process is followed with the silver-colored syrup, allowing the coated dragées to dry between each charge. After the final charge, they are removed and allowed to set about 1 hour.

Polishing

The silver dragées are placed in the polishing pan and polished as usual (*see* POLISHING PANNED CANDIES).

SIMPLE SYRUP

Sugar	25 lb
Water	12 qt

The sugar and 8 quarts of the water are placed in a copper pan on the stove and stirred until dissolved. The sides are washed down and the batch is heated to the boiling point. The balance of water is added and the batch is removed from the stove and cooled off. The syrup is then ready for use.

This batch will yield approximately 20 quarts of simple syrup.

SLAB-GELATIN MARSHMALLOWS: *See* MARSHMALLOWS.

SNOWDRIFT MARSHMALLOWS

Fine Powdered Gelatin (soaked in 8 qt of water)	3 lb
6x Powdered Sugar	40 lb
Corn Syrup	40 lb
Granulated Sugar	20 lb
Water	3 qt
Pearl Cooking Starch (dissolved in 3 qt of water)	4 lb
Vanilla Flavor	2 oz

The soaked gelatin and powdered sugar are heated very slowly until dissolved and then poured into the beater. The corn syrup is then added cold and the beater is started at low speed.

Then the granulated sugar and water are placed in a large copper pan on the stove and stirred until dissolved. After heating to 240°F, the soaked pearl starch is added to this mixture and allowed to boil again for about 2 minutes. This is then poured into the original beating batch in a fine stream. The beater is then shifted to high speed, the vanilla is added, and beating is continued until the batch is white and light. It is then placed into the depositor and cast in small, round impressions molded in dry, warm starch. The marshmallows are now placed in the hot room until they are dry and firm; then they are removed, sifted, and packed.

This batch should yield approximately 125 pounds of candy.

SNOWDROP MARSHMALLOWS

Sugar	40 lb
Corn Syrup	40 lb
Water	8 qt
Powdered Gelatin (dissolved in 6 qt of water)	2 lb
Pure Vanilla Flavor	1½ oz

The sugar, corn syrup, and water are placed in the marshmallow beater and stirred at low speed for about 5 minutes. The soaked gelatin is added and the speed is shifted to high. Beating is continued until the batch is white and light; then the flavor is mixed in. The marshmallow is then poured into the depositor and cast in impressions of round snow drops molded in clean dry starch. These are left in a hot room of 100°F overnight, or until they are firm and dry. They are then removed, brushed free of starch, and finally packed. These marshmallows are excellent for coating with chocolate or for toasting in coconut.

This batch should yield approximately 85 pounds of marshmallows.

SNOWFLAKE MARSHMALLOWS

Powdered Gelatin (soaked in 3 qt of water)	2 lb
Sugar	60 lb
Corn Syrup	40 lb
Water	8 qt
Instant Egg Albumen (soaked in 1 qt of water)	8 oz
Vanilla Flavor	2 oz

The soaked gelatin is placed into the beater. The sugar, corn syrup, and water are placed in a large copper pan on the stove and stirred until dissolved. When the batch begins to boil, it is covered and allowed to steam for a few minutes. The sides are washed down and a thermometer is placed in the batch, which is heated to 240°F. The batch is then poured on top of the gelatin in the beater, which is operated at a low speed for about 5 minutes. The albumen and 1 quart of the water are now added to the beaten batch and the beater is shifted to high speed. Beating of the batch continues until it is white and light; then the vanilla flavor is mixed in. The marshmallow is now poured into the warm depositor and cast in molded, clean, dry starch. The filled starch boards are placed in a hot room (not over 100°F). The marshmallows are allowed to set overnight. They are then removed, sifted, and packed.

This batch should yield approximately 110 pounds of candy.

SODA-FOUNTAIN SYRUPS: *See* FOUNTAIN SYRUPS.

SOFT CANDY DEPARTMENT: *See* SAVING WASTE.

SOFT GOODS

Jellies and gum drops, fast-selling summer items, are classed as soft goods. This classification may be accurate for jellies, but it is not always accurate when applied to gum drops. Although some quality sweet shops sell gum drops that are soft and delicious, huge quantities of cheap gum drops are made. For the most part they contain corn syrup, cooking starch, and imitation flavors. They never were a credit to the industry. There is no profit in their production and, even at the lowest price, the consumer does not get his money's worth. The wise confectioner will always refuse to make these "soft" goods.

Quality gum drops are made with a moderate amount of sugar. The best cooking starch, bright colors, and fine flavors will give a quality product that the confectioner will be proud to sell and the consumer will remember with pleasure. Some old-fashioned sweet shops still make the old-time "gum arabic" gum drops, which sell for 1 dollar a pound and more. Once the customer buys them, he will go out of his way to find them again.

Jellies are made in three ways. First, there is the starch jelly in which cooking starch gives the jellied body. These jellies are cast in starch and crystallized or sugar-sanded for shelf-life protection. Although these goods are still important, the current prejudice against high-caloried foods and the wide-spread interest in reducing have affected their popularity somewhat. It is likely that the demand for starch jellies will decrease.

Jellies made with citrus pectin have a distinct advantage because they are a quality product that is low in calories. If real fruit is used in all fruit flavors, a substantial demand can be developed for these goods.

Finally, there are fruit jellies made with agar-agar. They are somewhat difficult to make, but a larger amount of fruit pulp can be used in the formula as well as a larger amount of citric acid to give a distinct tart flavor. There is always a certain class of trade that prefers these jellies.

Most pastes are made with cooking starch as the body builder. Some pastes are sold in bulk, but most are coated in chocolate. In view of the suspicious attitude toward starchy foods and high-caloried sweets already noted, it would seem that modern confectioners would try to develop new paste formulas with less starch or perhaps eliminate starch altogether as an ingredient. By using pectin in the same way as in the jelly formulas that follow, by cooking the batch about two to four degrees higher, and by incorporating a small amount of gum arabic solution, the author has obtained fairly good results along these lines. These formulas are not included in the book because they are still in the experimental stage.

To maintain quality in making pastes, jellies, and gum drops means to follow formulas accurately, use the best ingredients, and keep essential equipment in good working condition. Equipment and machinery that is worn out or obsolete must be overhauled or replaced. The cooking equipment is important in making soft goods. Whereas pectin or agar jellies may be cooked by the open fire or steam methods, jellies, gum drops, and pastes which are made with cooking starch as the jellying agent must be cooked in steam kettles to get good results.

When casting soft goods in starch, most retail manufacturers use the manual method, i.e., they use the funnel for casting and level the starch boards by hand. The medium wholesale manufacturer, however, usually handles this process with three units: the depositing machine, the starch buck, and the starch printer. The depositor casts the jellies, pouring out one row at a time. The starch buck automatically sieves and cleans the starch off one whole board of candy and, in the same operation, it refills the board with starch and levels it so that it is ready for the hand printer. The hand printer makes the impressions in the starch, printing one whole board at a time. The large manufacturer uses the "Mogul." This giant machine does the work of the just-described three units in one rapid operation. The latest models operate electronically.

The smaller and medium manufacturer who is interested in maintaining quality will find the happy medium in the three-unit combination.

The life of most automatic machinery is considered to be ten years. After that period, the machinery is bound to weaken and affect production. For instance, the pump on the depositor may not function properly and will no longer deposit the correct amount of liquid for the number of impressions in one row of the starch board. This will result in impressions which are overrun or half filled. In either case, the result is scrap. Another example: In the hopper part of the depositor a thermometer registers the temperature of the tank that holds the liquid for casting. If this thermometer is out of order and the heat is not properly regulated, the batches will be either too cold or too hot. In either case, the result is again scrap. In the starch board, several mechanisms handle the soft candies until they are completely cleaned of starch. When any of these delicate mechanisms wears out or breaks down, the result is again waste. That is why it is important to see that machinery is always in perfect condition. It is wiser to replace a worn-out piece of equipment with a new machine than with a second-hand unit, which is usually guaranteed for only one year.

Crystallizing soft goods is important since appearance and shelf life are involved. *See* CRYSTALLIZING and study it closely.

For the various soft-goods formulas, *see* the *Individual Names,* such as ORANGE JELLY, PINEAPPLE FRAPPÉ, TURKISH FIG PASTE, etc.

SOUR BALLS, CHERRY

The formula for SOUR BALLS, LEMON (*see*) is followed, but the white stripe is prepared while the main batch is colored red and flavored with wild cherry flavor. It is finished as usual.

The batch will yield approximately 23½ pounds.

SOUR BALLS, LEMON

Sugar	20 lb
Corn Syrup	5 lb
Water	2½ qt
Yellow Color	1 teaspoonful
Lemon Flavor	3 teaspoonfuls
Powdered Citric Acid	2 oz

The sugar, water, and corn syrup are placed in a copper pan on the stove and stirred until dissolved. When the batch begins to boil, it is covered and allowed to steam for a few minutes. The sides are washed down and a thermometer is placed in the batch, which is cooked to 310°F. The batch is then poured out on a greased slab and the edges are folded together while stirring in the citric acid and oil of lemon. A small piece (approximately 1 pound) is taken off the main batch and pulled white. This is kept warm in front of the batch warmer. The main batch is kneaded until cool enough, when it is taken to the spinning table in front of the batch warmer, where it is shaped and kept square while it is striped with fine thin, white stripes all around. Then the batch is shaped round, spun out in sticks, cut to size, and put through a ball machine. The candy is allowed to cool on the other end of the machine and then packed in tins.

The batch will yield approximately 23½ pounds.

SOUR BALLS, LIME

The formula for SOUR BALLS, LEMON (*see*) is followed, but a small white piece is set aside and pulled white for the stripe and the main batch is colored green and flavored lime. It is finished in the usual manner.

This batch will yield approximately 23½ pounds.

SOUR BALLS, ORANGE

The formula for SOUR BALLS, LEMON (*see*) is followed, but while the white stripe is prepared, the main batch is colored and flavored orange. It is completed as usual.

The batch will yield approximately 23½ pounds.

PEPPERMINT BALLS

The formula for SOUR BALLS, LEMON (*see*) is followed, omitting the citric acid. For the stripe a small part of the batch is taken off, colored a deep red and kept clear and warm. The main batch remains white. Oil of peppermint is stirred in while folding the edges together. When the batch is cool enough, it is pulled on the hook to a satin white and then transferred to the spinning table in front of the batch warmer, where it is kneaded and shaped square. It is then striped with fine, thin, red stripes all around, cut, and finished as usual. The balls are packed in tins.

This batch will yield approximately 23½ pounds.

SPANISH APRICOT NOUGAT: *See* CHOCOLATE-COVERED SPANISH APRICOT NOUGAT.

SPANISH PEANUT BARS: *See* PEANUT BARS, SPANISH-TYPE.

SPEARMINT GUMDROPS: *See* GUMDROPS, SPEARMINT.

SPEARMINT LEAVES

Sugar	20 lb
Corn Syrup	10 lb
Water	3½ qt
Green Color	1 teaspoonful
Oil of Spearmint	4 teaspoonfuls

The water, sugar, and corn syrup are placed in a copper pan on the stove and stirred until dissolved. When the batch begins to boil, it is covered and allowed to steam for a few minutes. The sides are washed down and a thermometer is placed into the batch, which is cooked to 290°F. The batch is poured on a greased slab and colored a deep green. The edges are folded together while stirring in the oil of spearmint. The folding is continued until the batch is cool enough, then it is taken to the spinning table in front of the batch warmer. It is then passed through the peach-stone roller machine and allowed to cool on the table. The candies are broken apart, put through a sieve, and then placed on a tin-lined table. The candies are dampened and stirred through with ½ pint simple syrup and then covered with about 15 pounds of fine granulated sugar. They are allowed to dry for 1 hour. Then they are sieved again and placed in boxes. The batch will yield approximately 33 pounds of candy.

Note: This batch will lose 2 pounds after cooking, but will gain 5 pounds after being sugared.

SPEARMINT MELLOWS

Sugar	25 lb
Cream of Tartar	2 teaspoonfuls
Water	3½ qt
Green Color	1 teaspoonful
Oil of Spearmint	4 teaspoonfuls

The sugar, water, and cream of tartar are placed in a copper pan on the stove and stirred until dissolved. When the batch beings to boil, it is covered and allowed to steam for a few minutes. The sides are washed down and a thermometer is placed in the batch, which is heated to 268°F. The batch is then poured out on a cold, greased slab and the edges are folded together while the color and flavor are added and mixed through. The batch is cooled rapidly and then taken over to the pulling machine and pulled for 7 minutes. It is then taken over to the spinning table, dusted with powdered sugar, and formed flat. Now it is passed through the mint-cutting machine and then placed on boards lined with wax paper and dusted with powdered sugar. The boards are placed in a warm room overnight. When creamed, the mints are broken apart and packed in airtight tins.

This batch should yield approximately 25 pounds of candy.

SPICED STRING JELLIES, ANISE

Sugar	35 lb
Corn Syrup	35 lb
Water	12 qt
Tartaric Acid (dissolved in ¾ oz of water)	¼ oz
Cooking Starch (dissolved in 12 qt of water)	10 lb
Anise Flavor	1½ oz

The sugar, corn syrup, and water are placed in a large, steam cooking copper kettle and stirred until dissolved. When the batch begins to boil, the tartaric acid is added and the dissolved cooking starch is stirred in slowly. The batch is cooked until heavy strings break off the palette knife when it is dipped into the jelly, or to 224°F. The batch is flavored and poured out on a warm depositing machine. It is cast in string impressions in warm starch, placed in the hot room (about 150°F), and allowed to set for about 2 days. The jellies are then removed, sifted free of starch, dampened, and sugared. When thoroughly dry, they are crystallized as usual (*see* CRYSTALLIZING).

This batch will yield about 100 pounds of candy.

SPICED STRING JELLIES, CINNAMON

The formula for SPICED STRING JELLIES, ANISE (*see*) is followed, but the batch is colored a deep red and flavored with oil of cassia. It is then finished in the usual manner and should yield approximately 100 pounds of candy.

SPICED STRING JELLIES, SPEARMINT

The formula for SPICED STRING JELLIES, ANISE (*see*) is followed, but the batch is colored a bright green and flavored with oil of spearmint. It is finished in the usual manner and should yield approximately 100 pounds of candy.

SPICED STRING JELLIES, VIOLET

The formula for SPICED STRING JELLIES, ANISE (*see*) is followed, but the batch is colored a bright violet and flavored with violet flavor. It is finished in the usual manner and should yield approximately 100 pounds of candy.

STARCH BUCK, G-2 SIMPLEX

This machine automatically cleans all centers from starch. It processes one full tray at a time. (Courtesy of the National Equipment Corporation, New York, New York, and Springfield, Massachusetts.)

STARCH CAST MARSHMALLOWS: *See* MARSHMALLOWS.

STICK CANDY, CHERRY

The formula for STICK CANDY, LEMON is followed, but red color, citric acid, and wild-cherry flavor are stirred into the batch while the edges are folded together.

The batch will yield approximately 23½ pounds.

STICK CANDY, LEMON

Sugar	25 lb
Water	4 qt
Cream of Tartar	2½ teaspoonfuls
Yellow Color	1 teaspoonful
Lemon Flavor	3 teaspoonfuls
Citric Acid	1 oz

The sugar, water, and cream of tartar are placed in a copper pan on the stove and stirred until dissolved. When the batch begins to boil, it is covered and allowed to steam for a few minutes. The sides are washed down and a thermometer is placed in the batch, which is heated to 330°F. The batch is then poured out on a greased slab, and the edges are folded together while stirring in the citric acid and the oil of lemon. A small piece (about 1 pound) is taken off the batch, pulled white, and kept warm for a stripe. Then the main batch is kneaded and folded until it is cool enough and transferred to the spinning table in front of the batch warmer, where it is kneaded and formed square. The white stripe is drawn out in fine, thin lines and placed all around the batch. Then the batch is formed round and spun out in sticks about ¼ inch in diameter to the entire length of the table. The sticks are rolled round, allowed to cool on the table, and are then cut with shears to desired size. The candy is packed in tins or glass jars.

The batch will give about 23½ pounds.

STICK CANDY, LICORICE

The formula for STICK CANDY, LEMON (*see*) is followed, omitting the citric acid. Black color and oil of anise are stirred into the batch while the edges are folded together. The batch is completed according to the original formula and will yield approximately 23½ pounds.

STICK CANDY, LIME

The formula for STICK CANDY, LEMON *(see)* is followed, but green color, citric acid, and oil of limes are stirred into the batch while the edges are folded together. The batch is completed in the usual manner.

It will yield approximately 23½ pounds.

STICK CANDY MACHINE

This machine sizes, twists, and cuts stick candy automatically. (Courtesy of the Vacuum Candy Machinery Co. and the Racine Confectioners' Machinery Co., New York, New York.)

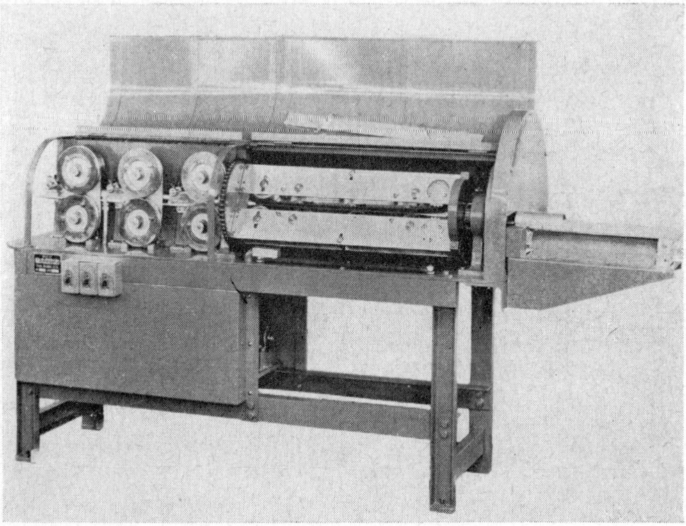

STICK CANDY, PEPPERMINT

The formula for STICK CANDY, LEMON *(see)* is used, but a small portion of the batch (about 1½ pounds) is colored red for the stripe and oil of peppermint is stirred in while the edges are folded together. The main batch is pulled white and striped with thin red stripes all around. It is finished in the usual manner.

The batch will yield approximately 23½ pounds.

STICK CANDY, WINTERGREEN

The formula for STICK CANDY, LEMON (*see*) is followed, omitting the citric acid. A small portion of the batch (about 1½ pounds) is colored a deep red and kept warm for a stripe. The main batch is flavored with oil of wintergreen and then pulled on the hook and twisted to a glossy white. It is then transferred to the spinning table in front of the batch warmer, kneaded and formed square. The red stripe is drawn into one 4 inches wide and three fine stripes. These are placed on the batch on two sides. The batch is formed round, spun, and finished as usual. It will yield approximately 23½ pounds.

STRAWBERRY CHOCOLATE STRAWS: *See* CHOCOLATE STRAWS, STRAWBERRY.

STRAWBERRY COCONUT KISSES: *See* COCONUT KISSES, STRAWBERRY.

STRAWBERRY COCONUT PATTIES: *See* COCONUT PATTIES, STRAWBERRY.

STRAWBERRY CREAM COCONUT SQUARES: *See* CREAM COCONUT SQUARES, STRAWBERRY.

STRAWBERRY CREAM WAFERS

The BANQUET MINT WAFERS (*see*) formula is followed, but the batch is heated to 248°F. While cooling on the cream beater, the batch is colored a delicate pink and flavored with 3 teaspoonfuls of strawberry flavor and 3 tablespoonfuls of strawberry jam. It is handled and finished the same way as the mint wafers.

The batch will yield approximately 45 pounds of wafers.

STRAWBERRY CREAMS, HAND-ROLLED

The formula for ORANGE CREAMS, HAND-ROLLED is followed up to the cooling. Then the marshmallow whip, invertase, and citric acid are added and the beater is set in motion. While in creaming, the batch is flavored with 3 teaspoonfuls of strawberry flavor and 3 pounds of pure strawberry pulp. It is then finished in the usual way.

Seventeen pounds of centers will yield about 25 pounds of chocolates after dipping.

STRAWBERRY EGGS, HAND-ROLLED

The formula for PINEAPPLE EGGS, HAND-ROLLED (*see*) is followed, but the batch is colored pink when it is on the cream beater and flavored with 3 pounds of pure strawberry pulp, 2 teaspoonfuls of strawberry flavor, and 2 teaspoonfuls of citric acid solution. The beater is then set in motion and creaming is continued until the batch forms a lump. It is then transferred to the work table and kneaded until soft. The batch is then put through the cut-roll machine and is cut and rolled to the desired size. The eggs are allowed to set on the boards until cool and they are then coated in a high-quality, dark, semisweet chocolate and finished with the mark "S" on top. The finished eggs are allowed to set on the dipping trays and are then packed in boxes.

The batch will yield about 70 pounds.

STRAWBERRY FRENCH CHEWING TAFFY: *See* FRENCH CHEWING TAFFY, STRAWBERRY.

STRAWBERRY FRENCH CRISP: *See* FRENCH CRISP, STRAWBERRY.

STRAWBERRY FRUIT AND NUT PILLOWS: *See* FRUIT AND NUT PILLOWS, STRAWBERRY.

STRAWBERRY ICE CREAM

Fresh 20% Cream	4 qt
Fresh 40% Cream	2 qt
Condensed Milk	2 qt
Sugar	8 lb

Pure Strawberry Pulp	6 lb
100-Grade Citrus Pectin	2 oz

The citrus pectin is mixed dry with 2 pounds of the sugar and placed in a double boiler on the stove together with 1 quart of the light cream. This mixture is stirred until it dissolves and is almost boiling. It is then poured into the mixing can. The balance of the light cream and sugar, the condensed milk, and heavy cream are added into the mixing can and the mix is stirred with a long paddle until the sugar is dissolved. The mix is then strained into the freezer, and the process described under VANILLA ICE CREAM NO. 1 is followed. When the cream is halfway frozen, the strawberry pulp is added.

The batch will yield approximately 30 quarts.

STRAWBERRY JELLY

Canned Strawberries	10 lb
Sugar	10 lb
Sugar	12 lb
Corn Syrup	4 lb
Agar-Agar	4 oz
Water	8 qt
Pink Color	1 teaspoonful
Strawberry Flavor	3 teaspoonfuls

The strawberry fruit and the 10 pounds of sugar are placed in a copper pan on the stove, stirred, and cooked to a firm jelly or about 222°F. The fruit is then set aside.

Now the agar-agar and water are placed in a large copper pan on the stove and stirred constantly. When boiling for a few minutes, the sugar is stirred in gradually, the corn syrup is added rapidly, and the batch is heated to 223°F. It is then moved to a stand and the strawberry fruit, color, and flavor are stirred in thoroughly. The jelly is then cast in starch, run into varishaped impressions which are printed on starch boards, and allowed to set overnight. The jellies are then taken out and sifted free from starch, dampened with a little syrup, rolled in fine granulated sugar, and allowed to dry. Then they are placed in crystal pans and crystallized as usual (*see* CRYSTALLIZING).

This batch should yield approximately 50 pounds of candy.

STRAWBERRY JELLY FRUIT CUTS: *See* JELLY FRUIT CUTS, STRAWBERRY.

STRAWBERRY MARSHMALLOW SQUARES: *See* MARSHMALLOW SQUARES, STRAWBERRY.

STRAWBERRY MARSHMALLOWS, GELATIN-TYPE

The formula for VANILLA MARSHMALLOWS, GELATIN-TYPE (*see*) is followed, but the batch is colored a bright pink and flavored with strawberry while still in motion. It is then finished in the usual manner and should yield approximately 125 pounds.

STRAWBERRY SALT-WATER TAFFY: *See* SALT-WATER TAFFY, STRAWBERRY.

STRAWBERRY SHERBET

Fresh Milk	20 qt
Sugar	20 lb
100-Grade Citrus Pectin	4 oz
Lemon Juice	1 qt
Red Color	1/4 teaspoonful
Pure Strawberry Flavor	1/2 oz
Preserved Strawberries	6 qt

The pectin is mixed dry with 5 pounds of the sugar and placed in a copper pan with 4 quarts of the milk. The pan is placed on the stove over low heat and the mixture is stirred and heated to the boiling point. It is then poured into the mixing pan. The balance of the milk and sugar, the juice, color, and flavor are all added and stirred until all the sugar is dissolved. The mix is then cooled rapidly and strained into the freezer. The freezing process is started, and when the sherbet is about halfway frozen, the preserved strawberries are added, and the freezing is completed as usual. The storing and ripening of the sherbet is completed as under VANILLA ICE CREAM NO. 1 (*see*).

The batch will yield approximately 36 quarts.

STRAWBERRY SYRUP

Fresh or Frozen Strawberries	1 qt
Sugar	2 lb
SIMPLE SYRUP (*see*)	4 qt
FRUIT ACID SOLUTION (*see*)	2 oz
Liquid Pink Color	½ teaspoonful

The berries are sprinkled with the sugar, placed in a wooden mixing bowl, and allowed to remain there overnight. The following morning, the berries are mashed into a pulp. The simple syrup is placed in a stone crock and the pulp is stirred into the syrup. The acid solution, flavor, and color are then added and mixed throroughly. The syrup is covered well and is then ready for use.

The batch will yield approximately 6 quarts of strawberry syrup.

STRAWBERRY WALNUT PILLOWS: *See* WALNUT PILLOWS, STRAWBERRY.

SUGAR-COATED PEANUTS

Sugar	15 lb
Water	2 qt
Roasted Virginia Peanuts	10 lb
Vanilla Flavor	½ oz

The sugar and water are placed in a standard revolving pan (under which a small gas burner has been placed) and stirred until dissolved. The pan is allowed to run in slow motion and the sugar batch is cooked to a soft ball. The peanuts are then added and allowed to cook until they are well roasted with the sugar. The gas is then turned off, the flavor is added, and the pan is allowed to revolve until the peanuts dry. The gas is turned on again (low heat) and the peanuts are stirred with a paddle while the pan is in motion until the sugar is melted down from the sides of the pan and the peanuts look glossy. The heat is then turned off and the peanuts are rapidly removed from the pan, placed on a marble, and separated quickly. They are allowed to cool on the marble and packed in boxes.

This batch should yield approximately 35 pounds of candy.

SUCKER (LOLLIPOP) MACHINE

This machine automatically feeds the sticks into the suckers. (Courtesy of the Vacuum Candy Machinery Co. and the Racine Confectioners' Machinery Co., New York, New York.)

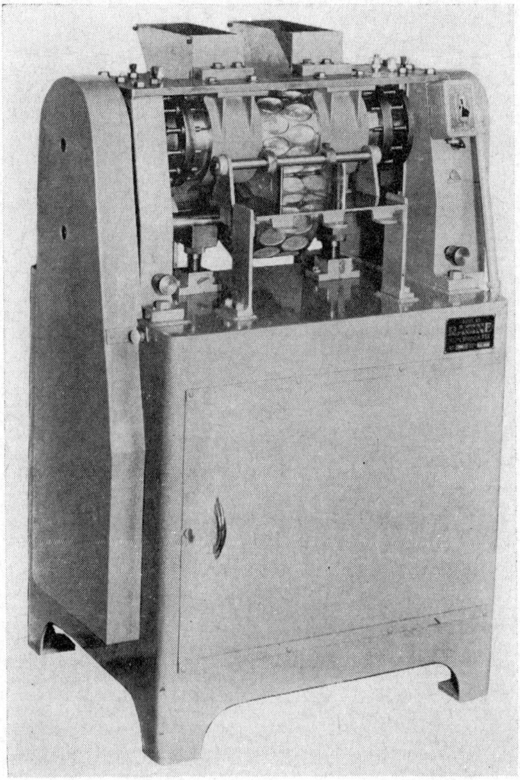

SUNDAES: *See* FOUNTAIN SYRUPS and *Individual Nut Toppings,* e.g., CHOCOLATE ALMOND CRUNCH, MAPLE WALNUT SUNDAE TOPPING.

SWEATING: *See* HARD CANDY.

SWEET POTATOES: *See* MARZIPAN SWEET POTATOES.

SWISS FONDANT CHOCOLATES, CHOCOLATE

The formula for SWISS FONDANT CHOCOLATES, VANILLA is followed all the way through but the fondant is flavored with 3 pounds of melted bitter chocolate in addition to the vanilla flavor. The batch is then processed in the usual manner.

The batch will yield approximately 65 pounds of chocolates.

SWISS FONDANT CHOCOLATES, COFFEE

The SWISS FONDANT CHOCOLATES, VANILLA (*see*) formula is followed down to the creaming process. Instead of vanilla flavor, 1½ ounces of concentrated coffee flavor are added. The fondant is then processed and cut as usual. The oblongs are coated in a dark vanilla chocolate.

The batch will yield approximately 65 pounds of chocolates.

SWISS FONDANT CHOCOLATES, VANILLA

Sugar	30 lb
Invert Sugar	2 lb
Melted Corn Syrup	3 lb
Salt	2 teaspoonfuls
Fresh 20% Light Cream	8 qt
Silver Sheet Gelatin (soaked in 2 cups of warm water)	1 oz
Vanilla Flavor	3 teaspoonfuls
Invertase	1 teaspoonful

The sugar, invert sugar, salt, and one half of the cream are placed in a large copper pan on the stove and strired until dissolved. When the batch begins to boil, the sides are washed down. The balance of the cream is added gradually and the batch is stirred continuously and heated to 244°F. The melted corn syrup is stirred in rapidly and the batch is poured out on a clean, cooled, dampened cream beater and cooled rapidly to approximately 100°F. The flavor and invertase are added and the machine is set in motion. The soaked gelatin is added to the batch while it is in the creaming process. After the batch has creamed to a lump, it is allowed to sweat back for 20 minutes. It is then kneaded soft, put through a cut-roll machine and shaped into small oblongs. After cooling on the boards, these oblongs are dipped in a high-quality fondant-processed dark chocolate. Finely ground roasted almonds are added and mixed with the coating. After the oblongs are dipped and set, they are packed in boxes.

The batch will yield approximately 65 pounds of chocolates.

T

TAFFIES: *See* CHEWING CANDIES and *Individual Taffy Formulas,* e.g., FRENCH CHEWING TAFFY, SALT-WATER TAFFY.

TAFFY CRISP: *See Individual Taffy Crisp Formulas,* e.g., FIG TAFFY CRISP, FRUIT AND NUT TAFFY CRISP.

TAFFY-WRAPPING MACHINE, TYPE K

This machine cuts and wraps taffies, toffee, and kisses. (Courtesy of the Package Machinery Co., Springfield, Massachusetts.)

THROAT PASTILLES: *See* GLYCERIN THROAT PASTILLES.

THROAT TABLETS: *See* JAGUAR THROAT TABLETS.

TOASTED COCONUT LOLLIPOPS: *See* LOLLIPOPS, TOASTED COCONUT.

TOFFEE CHIPS, CHOCOLATE-COVERED: *See* CHOCOLATE-COVERED TOFFEE CHIPS.

TOFFEES: *See* CHEWING CANDIES and *Individual Toffee Formulas,* e.g., COGNAC TOFFEE, DUTCH HOPJES TOFFEE.

TURKISH FIG PASTE

Best Brand Cooking Starch	5 lb
Water	12 qt
Sugar	20 lb
Corn Syrup	15 lb
Cream of Tartar	3 teaspoonfuls
Ground Turkish Pressed Figs	5 lb
Oil of Lemon	1 teaspoonful

The starch is dissolved in the water, placed in a large, clean, copper steam kettle and heated to the boiling point. The sugar, corn syrup, and cream of tartar are then added, cooked and stirred constantly. The batch is cooked to a firm jelly string, or to about 224°F. The ground figs are then stirred in rapidly and the batch is drawn out into a clean copper pan and allowed to cool while the flavor is stirred in. The batch is poured out into heavily dusted starch boards and allowed to set overnight. The paste is then sliced and cut into small oblongs, rolled in a mixture of powdered sugar and starch, and allowed to dry. It is then packed in boxes as usual.

The batch will yield approximately 65 pounds of paste.

TURKISH FRUIT NOUGAT

Sugar	10 lb
Corn Syrup	10 lb
Water	1½ qt

Coconut Butter	12 oz
Honey	2 lb
MARSHMALLOW WHIP (*see*)	5 lb
Glacéed Cherries	2 lb
Glacéed Pineapple	2 lb
Assorted Diced Fruits	2 lb
Roman Punch Flavor	3 teaspoonfuls

The sugar, corn syrup, and water are placed in a large copper pan on the stove and stirred until dissolved. When the batch begins to boil, it is covered and allowed to steam for a few minutes. The sides are washed down and a thermometer is placed in the batch, which is cooked to 260°F. The coconut butter and honey are then stirred in and the marshmallow whip is added gradually, while beating the batch with a large spatula. Beating is continued until the batch is thick and the glacéed fruits and flavor are added and mixed thoroughly. The batch is then turned out on trays, which are lined and covered with wafer paper, and allowed to set overnight. It is then cut and wrapped as usual.

This batch should yield approximately 35 pounds of candy.

TURKISH NOUGAT, SEMI-CHEWY TYPE

Sugar	12 lb
Corn Syrup	12 lb
Coconut Butter	1 lb
Water	2 qt
Instant Egg Albumen	4 oz
Water	½ qt
Chopped Almonds	4 lb
Vanilla Flavor	2 teaspoonfuls
Honey Flavor	1 teaspoonful

The instant egg albumen, 4 pounds of the corn syrup and 1 pint of the water are placed in a medium-size beater and beating is started at medium speed. In the meantime, the balance of the corn syrup, the sugar, coconut butter, and water are placed in a copper pan on the stove and stirred until dissolved. When the batch begins to boil, it is covered and allowed to steam for a few minutes. The sides are washed down and a thermometer is placed in the batch, which is cooked to 260°F. The batch is then poured in a slow stream on the beater and beating is continued until the batch is thick. The chopped almonds and the flavors are then added and mixed through and the batch is turned out on trays about 2½ inches deep, lined, and covered with wafer paper. After setting overnight, it is cut and wrapped as usual.

This batch should yield approximately 35 pounds of candy.

TURKISH RAHAT LACOUM PASTE, LEMON

Sugar	35 lb
Corn Syrup	15 lb
Water	16 qt
Cooking Starch	5 lb
(best grade, dissolved in 5 qt of water)	
Cream of Tartar	2 teaspoonfuls
Tartaric Acid (dissolved in water)	1 oz
Oil of Lemon	5 teaspoonfuls

The sugar, corn syrup, and water are placed in a clean, large steam-cooking kettle and stirred until dissolved. When the batch begins to boil, the soaked cooking starch is stirred in gradually. Stirring is continued while the cream of tartar and the tartaric acid are added and the batch is cooked to a good firm jelly (or 226°F) which, when tested between two fingers, can be pulled to a long firm string and will leave one finger easily and snap back to the other. The batch is then drawn out of the steam kettle into a clean copper pan and allowed to cool for a while. The flavor is then stirred in and the batch is poured out into heavily dusted starch boards and allowed to set overnight. The paste is then sliced, cut into small squares, and rolled in a mixture of three quarters of powdered sugar and one quarter of starch. It is then allowed to dry and packed in boxes.

The batch will yield approximately 67 pounds.

TURKISH RAHAT LACOUM PASTE, MINT

The formula for TURKISH RAHAT LACOUM PASTE, LEMON is followed, but the batch is colored a bright green and flavored with 4 teaspoonfuls of triple-distilled oil of peppermint. It is completed in the usual manner, and should yield approximately 67 pounds of candy.

TURKISH RAHAT LACOUM PASTE, ORANGE

The formula for TURKISH RAHAT LACOUM PASTE, LEMON is followed, but the batch is colored a deep orange and flavored with oil of orange. It is then completed in the usual manner, and should yield approximately 67 pounds of candy.

TURKISH RAHAT LACOUM PASTE, RASPBERRY

The formula for TURKISH RAHAT LACOUM PASTE, LEMON is followed, but the batch is colored a bright deep red and flavored with 3 pounds of raspberry pulp and 3 teaspoonfuls of raspberry flavor. The batch is then completed as usual, and should yield approximately 67 pounds of candy.

TUTTI-FRUTTI CREAM CHOCOLATES

Sugar	35 lb
Invert Sugar	2 lb
Melted Corn Syrup	3 lb
Water	7 qt
Glacéed Cherries and Pineapple	2 lb
Assorted Diced Fruits	4 lb
MARSHMALLOW WHIP (*see*)	4 lb
Jamaica Rum Flavor	2 teaspoonfuls
Invertase	2 teaspoonfuls

The sugar, invert sugar, and water are placed in a copper pan on the stove and stirred until dissolved. When the batch begins to boil, it is covered and allowed to steam for a few minutes. The sides are washed down, and a thermometer is placed in the batch, which is cooked to 248°F. The melted corn syrup is stirred in rapidly and the batch is poured out on a cooled, dampened cream beater. The top is sprinkled with cold water and the batch is cooled down rapidly to 110°F. The marshmallow whip and invertase are added and the beater is set in motion. The fruit and flavor are added while creaming is continued. The batch is creamed to a lump and allowed to sweat back for 15 minutes. It is then kneaded until soft, passed through a cut-roll machine, and formed into small balls. They are spread out on trays and, when dry, they are coated with high-quality milk chocolate. The creams are packed as usual.

The batch will yield approximately 85 pounds of chocolates.

TUTTI-FRUTTI ICE CREAM

Chopped Walnuts	8 oz
Chopped Pecans	8 oz
Chopped Brazil Nuts	8 oz
Diced Citron	2 lb
Preserved Pineapple	1 lb
Chopped Maraschino Cherries	1 qt
Sherry Emulsion	1 oz
Jamaica Rum Flavor	½ oz
Fresh 20% Cream	10 qt
Sugar	8 lb

All of the fruits and nuts are mixed in a mixing bowl and set aside. The cream is poured into a mixing can and the sugar is added. The mixture is then stirred with a long paddle until the sugar is dissolved. The sherry emulsion and rum flavor are stirred in and the mix is strained into the freezer. When the cream freezes about halfway, the mixed fruits and nuts are added. Otherwise the same process is used as under VANILLA ICE CREAM NO. 1 (*see*).

The batch will yield approximately 20 quarts.

V

VACUUM COOKER, D-3 SIMPLEX OPEN-KETTLE

This machine is used to cook quality hard candy. (Courtesy of the Vacuum Machinery Co. and the Racine Confectioners' Co., New York, New York.)

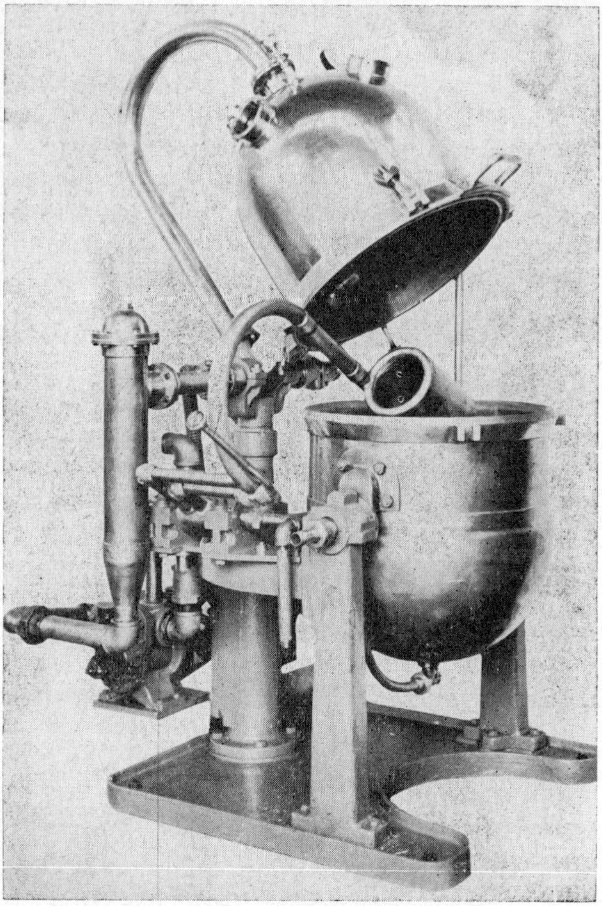

VACUUM UNIT, WITH A PRECOOKING KETTLE AND OPEN-TYPE CREAM BEATER, E-3 SIMPLEX

This is a continuous system for cooking, cooling, and creaming fondant cream. (Courtesy of the Vacuum Machinery Co. and the Racine Confectioners' Machinery Co., New York, New York.)

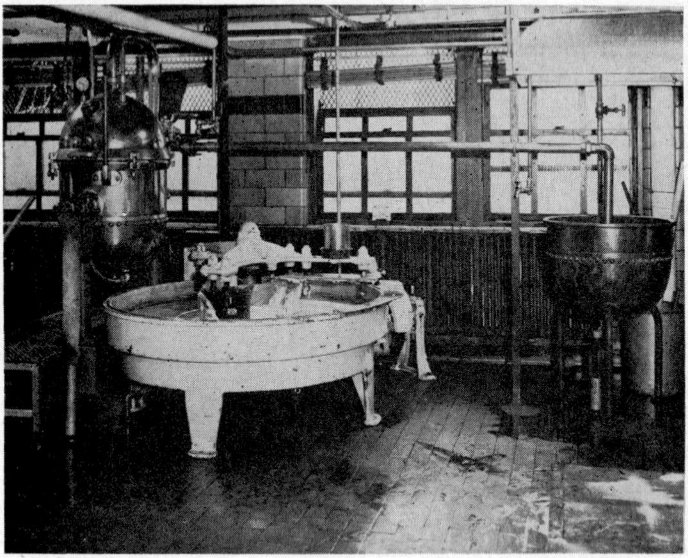

VALENCIA ALMONDS: *See* CHOCOLATE-COVERED ALMONDS.

VANILLA BOARDWALK FUDGE: *See* BOARDWALK FUDGE, VANILLA.

VANILLA BUTTER CREAMS: *See* BUTTER CREAMS, VANILLA.

VANILLA BUTTERSCOTCH DAINTIES: *See* BUTTERSCOTCH DAINTIES, VANILLA.

VANILLA BUTTERSCOTCH WAFERS: *See* BUTTERSCOTCH WAFERS, VANILLA.

VANILLA CANDY CURLS: *See* CANDY CURLS, VANILLA.

VANILLA CARAMELS, CHOCOLATE-COVERED: *See* CHOCOLATE-COVERED VANILLA CARAMELS.

VANILLA CARAMEL SAUCE

Lecithin	½ oz
Sugar	8 lb
Corn Syrup	5 lb
Salt	1 teaspoonful
Fresh 18% Cream	3 qt
150-Grade Citrus Pectin	2 oz
Vanilla Flavor	3 teaspoonfuls
Butter	2 lb

The pectin is mixed dry with 3 pounds of the sugar. A clean copper pan containing 1 quart of the cream is placed on the stove. The pectin sugar is added and stirred until dissolved. The salt, lecithin, corn syrup, and the balance of sugar and cream are added to the batch. The batch is stirred over medium heat and the butter is added gradually. After it is heated to 220°F, the batch is removed to a cool spot and the flavor is stirred in. The batch is cooled rapidly, poured into a clean stone crock, and covered. It is then ready for use.

The batch will yield approximately 10 quarts of vanilla caramel sauce.

VANILLA CARO WHIRLS: *See* CARO WHIRLS, VANILLA.

VANILLA CHOCOLATE STRAWS: *See* CHOCOLATE STRAWS, VANILLA.

VANILLA COCONUT PATTIES: *See* COCONUT PATTIES, VANILLA.

VANILLA COCONUT STARS: *See* COCONUT STARS, VANILLA.

VANILLA COLONIAL NOUGAT: *See* COLONIAL NOUGAT, VANILLA.

VANILLA COMMERCIAL FUDGE: *See* COMMERCIAL FUDGE, VANILLA.

VANILLA CORDIALS

Sugar	10 lb
Water	2 qt
Cream of Tartar	1 teaspoonful
Slow-Set Citrus Pectin	1 oz

Invertase	1 teaspoonful
Vanilla Flavor	1 teaspoonful
Roman Punch Flavor	½ teaspoonful

The pectin is mixed dry with 3 pounds of the sugar and then placed in a copper pan on the stove with 1 quart of water and the cream of tartar. It is stirred until dissolved. The balance of the sugar and water are added and stirring is continued until the sugar dissolves. When the batch begins to boil, it is chilled with a cupful of cold water. The scum is removed, the sides are washed down, and the batch is heated to 234°F. The batch is then removed from the stove and the flavors and invertase are stirred in rapidly. The batch is cooled for about 5 minutes and then cast in bottle-shaped impressions molded in clean, warm, dry starch. These are placed in a warm room and allowed to set overnight. They are then removed from starch and brushed off carefully. They are coated with a dark vanilla sweet chocolate. Extra-heavy coating is used and the cordials are handled very carefully. After the cordials have set, they are wrapped in colored aluminum foil and then packed in boxes.

The batch will yield approximately 20 pounds of cordials.

VANILLA CREAM COCONUT SQUARES: *See* CREAM COCONUT SQUARES, VANILLA.

VANILLA CREAMS, CAST

Approximately 30 pounds of running cream are placed in a steam-bath copper pan over medium heat and stirred constantly. Stirring is continued while 4 teaspoonfuls of pure vanilla extract are added. The cream is remelted to the right fluidity until the batch becomes lukewarm or about 120°F. It is then run into round impressions by funnel or depositing machine and molded in clean, dry starch. The creams are allowed to set overnight and are then removed, brushed free of starch, and placed on clean boards lined with wax paper. They are coated in dark semisweet or high-quality milk chocolate.

Thirty pounds of cream centers will yield about 40 pounds of chocolates when chocolate coated by enrobing machine, or 45 pounds when hand dipped.

VANILLA CREMO FUDGE: *See* CREMO FUDGE, VANILLA.

VANILLA DE LUXE CARAMELS: *See* DE LUXE CARAMELS, VANILLA.

VANILLA FILLED BUTTER CUPS: *See* FILLED BUTTER CUPS, VANILLA.

VANILLA FRENCH CHEWING TAFFY: *See* FRENCH CHEWING TAFFY, VANILLA.

VANILLA FRENCH CRISP: *See* FRENCH CRISP, VANILLA.

VANILLA FUDGE SAUCE

Lecithin	1/4 oz
Corn Syrup	1 lb
Sugar	6 lb
150-Grade Citrus Pectin	1 1/2 oz
Salt	1/2 teaspoonful
Fresh 18% Cream	2 1/2 quarts
Vanilla Flavor	1 teaspoonful
Glycerin	2 teaspoonfuls
Butter	1/2 lb

The pectin is mixed dry with 2 pounds of the sugar, and 1 quart of the cream is placed in a copper pan on the stove over medium heat. The pectin-sugar mix is added and stirred until dissolved. The lecithin, remaining sugar and cream, and the salt are added to the batch on the stove, which is stirred continuously while the butter is added, and it is cooked to 218°F. It is then removed to a cool spot. The glycerin and flavor are then stirred in and mixed thoroughly. The sauce is cooled rapidly, poured into a clean stone crock, and covered. It is then ready to use.

The batch will yield approximately 6 quarts of vanilla fudge sauce.

VANILLA ICE CREAM NO. 1

Fresh 20% Cream	16 qt
Fresh 40% Cream	4 qt
Sugar	12 lb
Pure Vanilla Flavor	1 1/2 oz
100-Grade Citrus Pectin	4 oz

The pectin is mixed dry with the sugar. The sugar and the creams are placed in a large steam-bath copper pan over low heat and stirred until dissolved. When the mixture begins to boil, it is removed from the stove, and the vanilla flavor is stirred in. The mixture is then strained, covered, and allowed to cool for about 1 hour. It is then placed in the freezing machine, which is run at low speed for about 6 minutes. The machine is then shifted to medium speed and run for 7 minutes. Mixing is finished at high speed for 7 minutes. The freezing time for a mixture of this size is 20 minutes. The cream is then removed from the freezer and placed into ice-cream cans. The filled cans are placed in the storage cabinet overnight and the cream is then ready to serve. This formula will produce a good, smooth ice cream.

It will yield approximately 40 quarts.

VANILLA ICE CREAM NO. 2

Fresh 20% Cream	8 qt
Fresh 40% Cream	2 qt
Condensed Milk	2 qt
100-Grade Citrus Pectin	2 oz
Sugar	8 lb
Pure Vanilla Extract	1 oz

The pectin is mixed dry with 3 pounds of the sugar and placed in a double boiler with 2 quarts of the light cream. This is set on the stove and stirred until dissolved. The mixture is then poured into a large ice-cream can. The balance of the light cream, the condensed milk, the heavy cream, and the balance of the sugar are all added into the mixing can and stirred with a large paddle until the sugar is dissolved. The mix is then strained and placed in the freezing machine. The same freezing process is followed as for VANILLA ICE CREAM NO. 1 (*see*). The ice cream is then finished in the usual way. This formula will produce a light-texture ice cream, which will increase about 50 percent in bulk.

It will yield approximately 25 quarts.

VANILLA ITALIAN CARAMELS: *See* ITALIAN CARAMELS, VANILLA.

VANILLA LADY CARAMELS: *See* LADY CARAMELS, VANILLA.

VANILLA LONDON CARAMELS: *See* LONDON CARAMELS, VANILLA.

VANILLA MARSHMALLOW PAN CARAMELS: *See* PAN CARAMELS, VANILLA MARSHMALLOW.

VANILLA MARSHMALLOW SQUARES: *See* MARSHMALLOW SQUARES, VANILLA.

VANILLA MARSHMALLOWS, GELATIN TYPE

Batch 1

Sugar	40 lb
Corn Syrup	40 lb
Water	6 qt
Powdered Gelatin (soaked in 5 qt of water)	2½ lb

Batch 2

Sugar	20 lb
Water	4 qt
Pure Vanilla	1½ oz

The sugar, corn syrup, and water are placed in a large clean copper pan on the stove and stirred until dissolved. When the batch begins to boil, it is covered and allowed to steam for a few minutes. The sides are washed down and a thermometer is placed in the batch, which is heated to 242°F. The soaked gelatin is then placed in the beater and the batch is poured out on top of the gelatin. The beater is run at a low speed for about 5 minutes and then shifted to high speed and beating is continued.

The 20 pounds of sugar and the 4 quarts of water are then placed in a copper pan on the stove, stirred until dissolved, and heated to 240°F. This cooked syrup is poured in a fine stream into the beater and beating is continued for another 3 minutes while the vanilla is mixed in. The power is then shut off and the marshmallow is poured out into a warmed depositing machine (120°F), cast in square impressions which are molded in clean, dry starch, and placed overnight in the hot room. The marshmallows are then brushed free of starch and are ready for packing.

This batch should yield approximately 125 pounds.

VANILLA NUT BOARDWALK FUDGE: *See* BOARDWALK FUDGE, VANILLA NUT.

VANILLA NUT BUTTERSCOTCH WAFERS: *See* BUTTERSCOTCH WAFERS, VANILLA NUT.

VANILLA NUT CREAMS, HAND-ROLLED

Sugar	50 lb
Water	10 qt
Cream of Tartar	3 teaspoonfuls
Melted Corn Syrup	6 lb
MARSHMALLOW WHIP (*see*)	4 lb
Vanilla Flavor	6 teaspoonfuls
Invertase	2 teaspoonfuls
Chopped Hand-Picked Black Walnuts	5 lb

The sugar, water, and cream of tartar are placed in a large copper pan on the stove and stirred until dissolved. When the batch begins to boil, it is covered and allowed to steam for a few minutes. The sides are then washed down and a thermometer is placed in the batch, which is cooked to 250°F. The melted corn syrup is stirred in gradually and the batch is poured out on a cooled, dampened cream beater. The batch is cooled rapidly for about 7 minutes or to about 120°F. The marshmallow whip, flavor, invertase, and nuts are added and the beater is set in motion and allowed to run until the batch creams to a lump. The batch is allowed to sweat back for about 15 minutes and it is then transferred to the working table, where it is kneaded until it is soft and smooth. It is then passed through a cut-roll machine or cut by hand. The creams are placed on trays lined with wax paper, and, when cool, they can be dipped in dark semisweet vanilla chocolate. This formula should yield about 72 pounds of centers.

Note: The average ratio for using chocolate coating for hand-rolled centers is approximately 20 percent for enrober coating and 30 percent for hand coating.

VANILLA NUT DE LUXE CARAMELS: *See* DE LUXE CARAMELS, VANILLA NUT.

VANILLA NUT LONDON CARAMELS: *See* LONDON CARAMELS, VANILLA NUT.

VANILLA NUT SEA-FOAM FUDGE: *See* SEA-FOAM FUDGE, VANILLA NUT.

VANILLA PAN CARAMELS: *See* PAN CARAMELS, VANILLA.

VANILLA PARFAIT

Fresh 20% Cream	16 qt
Instant Powdered Egg Yolks	1 lb
Sugar	15 lb
Pure Vanilla Flavor	2 oz

The egg yolks are mixed dry with 5 pounds of the sugar. This mixture is placed in the mixing can with half of the cream and stirred until the sugar is dissolved. The balance of the sugar and cream are added and the mix is again stirred until all the sugar is dissolved. The flavor is then stirred in and the mix is strained into the freezer as usual. The same freezing process is followed as for VANILLA ICE CREAM NO. 1 (*see*).

The batch will yield approximately 35 quarts.

VANILLA PECAN SEA-FOAM KISSES: *See* SEA-FOAM KISSES, VANILLA PECAN.

VANILLA PECAN SUNDAE TOPPING

Sugar	9 lb
Corn Syrup	1 lb
Water	2½ qt
150-Grade Citrus Pectin	1 oz
Chopped Pecans	5 lb
Vanilla Flavor	4 teaspoonfuls
Glycerin	4 teaspoonfuls

The pectin is mixed dry with 3 pounds of the sugar and 1 quart of the water is placed in a small copper pan on the stove. Then the pectin-sugar mixture is added and stirred until dissolved. The remaining sugar and water are added to the batch and again stirred until the sugar is completely dissolved. The corn syrup is stirred in rapidly and the batch is heated to 218°F.

It is then moved to a cool spot and the chopped pecans are stirred in rapidly. The flavor and glycerin are added and mixed thoroughly. The pecan sundae topping is then poured into clean glass jars and closed with airtight lids. It is then ready for use.

The batch will yield approximately 10 quarts of vanilla pecan sundae topping.

VANILLA SALT-WATER TAFFY: *See* SALT-WATER TAFFY, VANILLA.

VANILLA SEA-FOAM FUDGE: *See* SEA-FOAM FUDGE, VANILLA.

VANILLA VARIETY CARAMELS: *See* VARIETY CARAMELS, VANILLA.

VANILLA VARIETY FUDGE: *See* VARIETY FUDGE, VANILLA.

VANILLA WALNUT PILLOWS: *See* WALNUT PILLOWS, VANILLA.

VANILLA YELLOW JACK: *See* YELLOW JACK, VANILLA.

VARIETY CARAMELS, COCONUT

The formula for VARIETY CARAMELS, VANILLA (*see*) is followed, but 2½ pounds of macaroon coconut are stirred into the batch before pouring. It is finished in the same way as the vanilla batch and should yield approximately 43 pounds.

VARIETY CARAMELS, DATE

The formula for VARIETY CARAMELS, VANILLA (*see*) is followed, with the addition of 4 pounds of chopped dates to the batch before pouring. The batch is finished in the usual manner and should yield approximately 44 pounds of candy.

VARIETY CARAMELS, FIG

The formula for VARIETY CARAMELS, VANILLA (*see*) is followed, adding 4 pounds of chopped figs when the batch has almost finished cooking. The cooking and remaining processes are completed in the usual manner.

The batch should yield approximately 44 pounds of candy.

VARIETY CARAMELS, RAISIN

The formula for VARIETY CARAMELS, VANILLA (*see*) is followed, stirring 3½ pounds of raisins into the batch before pouring. The batch is completed as usual and should yield approximately 43 pounds of candy.

VARIETY CARAMELS, VANILLA

Sugar	10 lb
Corn Syrup	12 lb
Coconut Butter	1 lb
Salt	1 tablespoonful
Fresh 18% Light Cream	3 qt
Evaporated Milk	9 lb
Vanilla Flavor	4 teaspoonfuls
CARAMEL PASTE (*see*)	4 lb

The sugar, corn syrup, coconut butter, salt, and cream are placed in a large copper pan on the stove and stirred until dissolved. When the batch begins to boil, the evaporated milk is stirred in gradually. Stirring is continued and the caramel paste is added. The batch is cooked to a medium firm ball when tested in cold water (*see* COLD-WATER TEST). The vanilla flavor is then stirred in and the batch is poured out on a greased slab between iron bars. It is spread out evenly with a palette knife and allowed to cool until hard enough to handle. It is then passed through a sizing machine, cut into square caramels, and wrapped in moistureproof cellophane.

This batch should yield approximately 40 pounds of caramels.

VARIETY FUDGE, CHOCOLATE

The formula for VARIETY FUDGE, VANILLA (*see*) is followed, but the batch is reduced to 234°F. with evaporated milk. Two pounds of finely cut bitter chocolate are stirred into the batch before the fondant is added. The batch is finished as usual and should yield 32 pounds of candy.

VARIETY FUDGE, VANILLA

Sugar	12 lb
Corn Syrup	8 lb
Water	1½ qt
Evaporated Milk	7 lb
Salt	1 tablespoonful
Cream Fondant	6 lb
Vanilla Flavor	3 teaspoonfuls

The sugar, water, salt, and 6 pounds of the corn syrup are placed in a copper pan on the stove and stirred until dissolved. When the batch begins to boil, it is covered and allowed to steam for a few minutes. The sides are washed down and a thermometer is placed in the batch, which is heated to 260°F. The batch is then reduced with the evaporated milk to 236°F and the fondant is then stirred in and mixed thoroughly. The flavor and the balance of the corn syrup are added and mixed through again. The batch is then poured out on a greased slab with low bars, spread out evenly with a palette knife to about ¼-inch thickness, and marked into squares with an adjustable cutter. It is allowed to cool and packed in boxes.

This batch should yield 30 pounds of candy.

VERMONT MAPLE CRYSTALS

Batch 1

Maple Sugar	15 lb
Crystal "A" Sugar	25 lb
Corn Syrup	10 lb
Water	8 qt

The sugars, corn syrup, and water are placed in a copper pan on the stove and stirred until dissolved. When the batch begins to boil, it is covered and allowed to steam for a few minutes. The sides are washed down and a thermometer is placed in the batch, which is heated to 244°F. The batch is then poured out on a cooled, dampened open-type cream beater and the top is sprinkled with a few drops of cold water. The batch is allowed to cool to about 115°F. The machine is set in motion and the batch is creamed to a white lump. It is then ready for the remelting process.

Batch 2

Sugar	20 lb
Water	5 qt
Caramel Color	1 tablespoonful
Maple Flavor	3 tablespoonfuls

The sugar and water are placed in a large copper pan on the stove and stirred until dissolved. When the batch begins to boil, it is covered and allowed to steam for a few minutes. The sides are washed down and a thermometer is placed in the batch, which is cooked to 240°F. The fondant cream (batch 1) is then placed in the syrup (batch 2). The caramel color and maple flavor are added and mixed thoroughly. The batch is heated to about 170°F and is then cast into dry, clean starch, using fancy molds for the impressions. The cast creams are allowed to set overnight. They are then sieved, brushed free of starch, and placed in half-filled crystal pans. They are crystallized in a cold day crystal (*see* CRYSTALLIZING).

The batch will yield approximately 90 pounds.

VIENNA FILLED HARD CANDIES, APPLE

APPLE PASTE (*see*)	10 lb
Sugar	17 lb
Corn Syrup	3 lb
Water	3 qt
Pure Apple Extract	3 teaspoonfuls
Powdered Citric Acid	2 teaspoonfuls

The apple paste is placed in a small pan on the stove over low heat and stirred constantly until moderately hot. The batch is then removed and kept hot near the stove. This is to be used for the center.

The sugar, corn syrup, and water are placed in a copper pan on the stove and stirred until dissolved. When the batch begins to boil, it is covered and allowed to steam for a few minutes. The sides are washed down and a thermometer is placed in the batch, which is heated to 330°F. The batch is then poured out on a cooled, greased slab lined with iron bars. The flavor and citric acid are stirred in while the edges are folded together. The folding process is continued until the batch is firm enough to handle. The batch is then transferred to the spinning table in front of the batch warmer. One third of the batch is taken off, kneaded until smooth, and then flattened out large enough to hold the center. The warm apple paste is placed in the middle of the flat batch and the batch is wrapped over and closed all around so that the center remains sealed in the batch. The remainder of the batch is then kneaded and formed flat. The small center batch is placed in the middle of this flat batch. The flat batch is wrapped all around the center batch so that it forms one batch. The batch is formed into the shape of a large bottle and kept in front of the batch warmer. It is spun into medium-sized sticks with gloved hands. It is then passed through a plastic hard-candy machine, with a set of dies which are suitable for forming apple candies. The candies are allowed to cool on the table and are then packed in tins.

The batch will yield approximately 28 pounds of candy.

VIENNA FILLED HARD CANDIES, APRICOT

The formula for VIENNA FILLED HARD CANDIES, APPLE (*see*) is followed, but 10 pounds of apricot pulp are heated for the center as usual; the jacket batch is colored orange and flavored with apricot extract. The batch is then handled and finished as usual.

It will yield approximately 28 pounds of candy.

VIENNA FILLED HARD CANDIES, CHOCOLATE

Center

Peanut Butter	6 lb
Melted Sweet Chocolate	3 lb
Dutch Cocoa Powder	1 lb

Jacket

Sugar	17 lb
Corn Syrup	3 lb
Cream of Tartar	1 teaspoonful
Water	3 qt
Melted Bitter Chocolate	1½ lb
Vanillin Powder	¼ teaspoonful

The peanut butter, sweet chocolate, and cocoa are heated together carefully until warm enough for the center. The sugar, corn syrup, cream of tartar, and water for the hard candy are cooked and processed as usual. When the batch is poured on the slab, the bitter chocolate and vanillin are stirred in. The batch is then finished in the usual manner. A peanut-shaped die is used.

The batch will yield approximately 30 pounds of candy.

VIENNA FILLED HARD CANDIES, LEMON

The formula for VIENNA FILLED HARD CANDIES, APPLE (*see*) is followed with the exception of the coloring and flavoring. The batch is colored a delicate yellow and flavored with pure oil of lemon.

The batch will yield approximately 28 pounds of candy.

VIENNA FILLED HARD CANDIES, LIME

The formula for VIENNA FILLED HARD CANDIES, APPLE (*see*) is followed with the exception of the color and flavor. The batch is colored a delicate green and flavored with pure oil of limes.

It will yield approximately 28 pounds of candy.

VIENNA FILLED HARD CANDIES, PEACH

The formula for VIENNA FILLED HARD CANDIES, APPLE (*see*) is followed, but 10 pounds of peach jam is used instead of the apple paste. The jam is heated for the center as usual, the jacket batch is colored yellow and flavored with peach extract.

The batch will yield approximately 28 pounds of candy.

VIENNA FILLED HARD CANDIES, PINEAPPLE

The formula for VIENNA FILLED HARD CANDIES, APPLE (*see*) is followed, but 10 pounds of pineapple jam is used instead of the apple paste. The jam is heated for the center as usual, the hard candy jacket is colored yellow and flavored with pineapple true-fruit flavor. The batch is processed, handled, and finished as usual.

It will yield 28 pounds of candy.

VIENNA FILLED HARD CANDIES, PLUM

The formula for VIENNA FILLED HARD CANDIES, APPLE (*see*) is followed but 10 pounds of plum jam is used instead of the apple paste. The jam is heated for the center as usual, the jacket batch is colored purple and flavored with extract of plum. Otherwise the batch is processed, handled, and finished as usual.

It will yield approximately 28 pounds of candy.

VIENNA FILLED HARD CANDIES, RASPBERRY

The formula for VIENNA FILLED HARD CANDIES, APPLE (*see*) is followed, but 10 pounds of raspberry jam is used in place of the apple paste. The jam is heated for the center as usual. The hard candy jacket is colored red and flavored with true-fruit raspberry flavor.

This batch will yield approximately 28 pounds of candy.

VIENNA FILLED HARD CANDIES, RUM RAISIN

Center

Raisins Processed in Syrup	7 lb
Sugar	5 lb
Water	½ qt

The raisins, sugar and water are placed in a copper pan on the stove over medium heat, stirred constantly, and heated to 222°F. This center is kept warm to be ready when wanted. For ingredients and directions for the jacket batch, *see* VIENNA FILLED HARD CANDIES, APPLE. The hard candy is colored pink and flavored with Roman punch.

The batch will yield approximately 28 pounds of candy.

VIENNA FILLED HARD CANDIES, STRAWBERRY

The formula for VIENNA FILLED HARD CANDIES, APPLE is followed, but 10 pounds of strawberry jam is used instead of the apple paste. The jam is heated for the center as usual. The hard candy jacket is colored pink and flavored with pure strawberry extract.

This batch will yield approximately 28 pounds of candy.

VIOLET BONBONS: *See* BONBONS, VIOLET.

VIOLET CANDY CURLS: *See* CANDY CURLS, VIOLET.

VIOLET CHOCOLATE STRAWS: *See* CHOCOLATE STRAWS, VIOLET.

VIOLET FILLED BUTTER CUPS: *See* FILLED BUTTER CUPS, VIOLET.

VIOLET GUMDROPS: *See* GUMDROPS, VIOLET.

VIOLET WALNUT PILLOWS: *See* WALNUT PILLOWS, VIOLET.

VIRGINIA PEANUT CRISP: *See* PEANUT CRISP, VIRGINIA-TYPE.

W

WALNUT BUTTER TOFFEE

Sugar	17 lb
Corn Syrup	8 lb
Water	3 qt
Butter	12 oz
Chopped English Walnuts	4 lb
Salt	1 tablespoonful

The sugar, water, and corn syrup are placed in a clean copper pan on the stove and stirred until dissolved. When the batch begins to boil, it is covered and allowed to steam for a few minutes. The sides are washed down and a thermometer is placed into the batch, which is heated to 310°F. The butter, salt, and chopped walnuts are then stirred in. The batch is poured on a greased slab and spread out with a palette knife very thin. It is then loosened underneath with a large knife and cut with a roller knife to 1½-inch squares. When cool enough, the candy is packed in tins.

The batch will yield approximately 27 pounds.

Note: This is a fine center for chocolate coating. A dark sweet or milk chocolate coating is used.

WALNUT GOODIES: *See* MARZIPAN WALNUT GOODIES.

WALNUT NUT SQUARES: *See* NUT SQUARES, WALNUT.

WALNUT PILLOWS, LIME

The WALNUT PILLOWS, VANILLA (*see*) formula is followed, but the batch is colored a light green and flavored with oil of limes while the edges are folded together. It is finished in the usual manner.

The batch will yield approximately 31 pounds.

WALNUT PILLOWS, STRAWBERRY

The WALNUT PILLOWS, VANILLA (*see*) formula is used, but the batch is colored pink and flavored strawberry. It is finished in the usual manner.

The batch will yield approximately 31 pounds.

WALNUT PILLOWS, VANILLA

Center

Sugar	10 lb
Corn Syrup	15 lb
Molasses	1½ qt
Water	1½ qt
Salt	2 tablespoonfuls
Hand-Picked Black Walnuts	25 lb

The sugar, water, molasses, corn syrup, and salt are placed in a copper pan on the stove and stirred until dissolved. When the batch begins to boil, it is covered and allowed to steam for a few minutes. The sides are washed down and a thermometer is placed in the batch, which is heated to 240°F. The heat is shut off and the black walnuts are stirred in to make a stiff paste. This is kept warm until ready for use. It is sufficient for four batches.

Jacket

Sugar	20 lb
Water	3½ qt
Cream of Tartar	2 teaspoonfuls
Vanilla Flavor	2 teaspoonfuls

The sugar, water, and cream of tartar are placed in a copper pan on the stove and stirred until dissolved. When the batch begins to boil it is covered and allowed to steam for a few minutes. The sides are washed down and a thermometer is placed in the batch, which is cooked to 338°F. The batch is poured out on a greased slab, and the vanilla flavor is stirred in while the edges are folded together. Folding and kneading are continued until the batch is cool enough. Then it is pulled partly and twisted on the hook until it becomes glossy and silvery. The batch is then transferred to the spinning table in front of the batch warmer, where it is kneaded and formed into a flat batch large enough to hold the center. Twelve pounds of the center are then placed in the middle of the jacket, which is wrapped over, ends closed, and formed into a round batch. Then it is spun and cut on a buttercup cutting machine. After cooling on the table, it is packed in tins.

The batch will yield about 31 pounds.

WALNUT PILLOWS, VIOLET

The WALNUT PILLOWS, VANILLA (*see*) formula is used, but the batch is colored a bright purple and flavored with violet flavor. It is finished in the usual manner.

The batch will yield approximately 31 pounds.

WALNUT-PASTE BONBON CENTERS

Sugar	25 lb
Water	5 qt
Cream of Tartar	2 teaspoonfuls
Melted Corn Syrup	3 lb
Ground Walnuts	5 lb
Vanilla Flavor	3 teaspoonfuls
Invertase	1 teaspoonful
Glycerin	1 teaspoonful to every 5 lb of bonbon cream

The sugar, water, and cream of tartar are placed in a copper pan on the stove and stirred until dissolved. When the batch begins to boil, it is covered and allowed to steam for a few minutes. The sides are then washed down and a thermometer is placed in the batch, which is heated to 248°F. The melted corn syrup is stirred in rapidly, and the batch is poured out on a cooled, dampened cream-beating machine and allowed to cool for about 7 minutes. The nuts, flavor, and invertase are added, and the beater is set in motion and allowed to run until the batch creams to a lump. It is then allowed to sweat back for about 15 minutes; then it is transferred to the rolling table, which is dusted with flour. The batch is then kneaded, rolled out, and cut into small balls. These are then ready to be used as bonbon centers for white and yellow colors.

This batch will yield about 40 pounds of centers.

WALNUT-PASTE BONBONS: *See* BONBONS, WALNUT-PASTE.

WALNUTS, CHOCOLATE-COVERED: *See* CHOCOLATE-COVERED WALNUTS.

WALNUTS, SALTED: *See* SALTED WALNUTS.

WATERMELON ICE

Water	20 qt
Sugar	20 lb
Lemon Juice	1½ qt
Pink Color	½ teaspoonful
Watermelon Pulp	6 qt

The watermelon pulp is obtained from ripe watermelons. The seeds are picked out carefully, and the pulp is chopped up fine in a large mixing bowl and set aside. The water, sugar, lemon juice, and color are placed in the mixing can and stirred until all the sugar is dissolved. The mix is strained into the freezer and, when the ice is about halfway frozen, the watermelon pulp is added. The ice is then completed according to directions under VANILLA ICE CREAM NO. 1 (*see*).

The batch will yield approximately 32 quarts.

WILD-CHERRY TABLETS

The formula for LEMON TABLETS (*see*) is followed but wild-cherry flavor is used for flavoring and good deep red for color. For the form, some prefer a round cherry roller.

The batch will yield approximately 20½ pounds of candy.

WINTERGREEN BARLEY SUGAR CUTS: *See* BARLEY SUGAR CUTS, WINTERGREEN.

WINTERGREEN SALT-WATER TAFFY: *See* SALT-WATER TAFFY, WINTERGREEN.

WINTERGREEN STICK CANDY: *See* STICK CANDY, WINTERGREEN.

WINTERGREEN YELLOW JACK: *See* YELLOW JACK, WINTERGREEN.

Y

YELLOW JACK, PEPPERMINT

The YELLOW JACK, VANILLA (*see*) formula is used, but the portion taken off for the stripes is colored a deep red and the main batch is flavored with oil of peppermint.

The batch will yield approximately 15 pounds of candy.

YELLOW JACK, VANILLA

Sugar	6 lb
Corn Syrup	10 lb
Molasses	4 qt
Coconut Butter	8 oz
Water	½ qt
Salt	1 tablespoonful
Vanilla	3 teaspoonfuls

All of the ingredients, excluding the vanilla flavor, are placed in a large copper pan on the stove and stirred until dissolved. When the batch begins to boil, the sides are washed down, stirring is continued, and the batch is cooked until it forms a hard ball when tested in cold water (*see* COLD-WATER TEST). The batch is then poured out on a greased slab and the edges are folded together while stirring in the extract of vanilla. Folding is continued until the batch is cool enough. Then it is taken to the pulling machine and pulled for approximately four minutes. It is then transferred to the spinning table, formed into a flat batch, stretched out to three-inch width, and cut in bars, about 4 inches in length. It is wrapped as usual.

This will yield about 15 pounds.

YELLOW JACK, WINTERGREEN

The YELLOW JACK, VANILLA (*see*) formula is followed, but a small clear piece is taken from the batch before pulling and saved for a stripe. The main batch is flavored wintergreen before pulling the bars. The batch is striped with four stripes in the center. It is finished as usual.

The batch will yield approximately 15 pounds of candy.

Food Tech. Office
Shannon Hall